WORLD CANALS

D0936444

BOOKS BY CHARLES HADFIELD

British Canals. An illustrated history (7th ed)
The Canal Age (2nd ed)
The Canals of the East Midlands (2nd ed)
The Canals of North West England (with Gordon Biddle)
The Canals of South and South East England
The Canals of South Wales and the Border (2nd ed)
The Canals of South West England (2nd ed)
The Canals of the West Midlands (3rd ed)
The Canals of Yorkshire and North East England
William Jessop, Engineer (with A. W. Skempton)

WORLD CANALS

Inland Navigation Past and Present

CHARLES HADFIELD

Facts On File Publications
New York, New York • Oxford, England

Copyright © 1986 by Charles Hadfield

All rights reserved. No part of this book may
be reproduced or utilized in any form or by any
means, electronic or mechanical, including
photocopying, recording or by any information
storage and retrieval systems, without permission
in writing from the Publisher.

Library of Congress Cataloging-in-Publication Data

Hadfield, Charles, 1909–
World Canals.

Bibliography: p.
Includes index.
1. Canals—History. 2. Inland navigation—History.
I. Title.
HE526.H33 1986 387.4′09 85-29272
ISBN 0-8160-1376-4

Printed in Great Britain

10 9 8 7 6 5 4 3 2 1

387.409 Hadfield, Charles
HAD World canals :
1986 inland navigation
past and present

CONTENTS

LIST OF ILLUSTRATIONS

Maps

PREFACE

Before I had written anything about British canals, I had in 1938 bought my first books about those abroad: Vernon-Harcourt's engineering classic *Rivers and Canals* in its 1896 edition and Roberts's *The Middlesex Canal*, then newly published by Harvard University Press. Thereafter I began to collect books and ephemera, and to file material, against the time when I could see, maybe one day travel on, some of the world's great waterways.

I saw the Nile in 1950, travelled briefly on the Congo in 1960, and in 1961 looked for the first time from the deck of a Köln-Düsseldorfer cruise boat at the traffic passing through the Rhine gorge before the Binger Loch had been removed. That did it. My wife and I set out to spend waterway holidays overseas: the Netherlands, Germany, Sweden, Norway, Portugal, France, the Rhine again, the Danube, the United States and Canada. So it went on until Ron and Joan Oakley of the Inland Waterways Association began to organise regular waterway explorations abroad: then we went with them.

Books begin as ideas. Some remain so, but others move on to become shapes, which with luck develop into patterns. I cannot now date when the idea of writing an account of world inland navigations first became a pattern, but my wife noted in her diary that I began to write it on 22 November 1971. One drafts and redrafts synopses, clarifies and edits pieces of the pattern, till slowly a whole emerges. But because one only learns about the pattern by working at it, so one rewrites over and over again—until one day it is done.

World Canals tries to give the reader—perhaps already a canal enthusiast—an impression of the pattern that waterway transport has imposed upon time and space. To do more would be beyond my ability to write, the limits of words, maps and illustrations laid down for me by my generous publisher, and, probably, readers' pockets. The result, I fear, is only too easy to criticise: 'What a pity Mr Hadfield has not ...'; 'It is difficult to credit that Mr Hadfield has not even mentioned ...', 'If only Mr Hadfield had taken the trouble to study ...' All are painfully true. I can only plead that to my knowledge no one has previously attempted what is here offered, at any rate in

English. L. F. Vernon-Harcourt in *Rivers and Canals* (1st ed, 1882, 2nd ed, 1896), J. S. Jeans in *Waterways and Water Transport* (1890) and Robert Payne in *The Canal Builders* (1959) showed the way.

I have faced many practical difficulties: changes of political boundaries, of place names and orthography, of units of weight and currency, and have tried to solve them sensibly if not always logically. In one respect I have failed to find a solution; and so decided to use the metric system throughout the Old World chapters, but imperial measurements for the New, though I know both Canada and Latin America are metricated. In the Old World the abbreviation 'm' stands for 'metres', in the New for 'miles', but 'M' for 'million' throughout.

Given my limit of pages available, and because I was writing for the general reader, I sadly chose not to give references, except when quoting verbatim, or provide a bibliography.

Elsewhere I have tried to thank some of the very many who over the years have helped me. But two debts should be paid here. In 1938, I bought those first two overseas waterway books from the red-headed girl whom I afterwards married. Since then she has been taken half round the world to look at canals and rivers, barges and ships, locks, inclines and lifts, and to meet people who have only one thing in common—talk about inland waterways. She encouraged me to start, work at, and indeed finish this book. Greater love has no woman than that she marries a canal man.

My other debt is to my subject. Could any writer wish for better?

Charles Hadfield

PART I

THE OLD WORLD

I
BEGINNINGS TO 1519

Artificial canals to carry water for irrigation or town supply are as old as civilisation; indeed, to some extent civilisation depended upon them. So people learned the art of building channels that held water, could be graded over long distances, carried over valleys on aqueducts, and given sluices to control water flow.

Before the Romans

Such canals date back at least to 3000 BC in Egypt, and some may have been used also for navigation. During the Sixth Dynasty (c2300–2180 BC), however, Pepi I sent expeditions up the Nile, and in doing so, wished to by-pass the first cataract near Aswan. He therefore ordered navigation canals to be built by Uni, governor of Upper Egypt, who left an inscription:

> His Majesty sent me to dig five canals in the South and to make three cargo boats and four tow-boats of acacia wood. Then the dark-skinned chieftains... drew timber for them, and I did the whole in a single year. (1)

An alternative was found later when about 1700 BC a slipway some 3 km long was built to enable boats to pass the second cataract near Wadi Halfa.

Later more irrigation canals were built in Egypt, some clearly used also for navigation: indeed, in *The Book of the Dead*, 'to sail for ever in a boat along those intricate canals where the reeds are continually bending in the heavenly wind'(2) is one of the pleasure of Paradise. Let us hope that it still is.

Navigable irrigation canals, and perhaps others used only by boats, probably date back to Assyrian, maybe earlier, times in the land between Tigris and Euphrates, in what is now Iraq. Some of those that served Babylon on the Euphrates were navigable; the ships of Alexander the Great used them in 331 BC.

The most famous canal of antiquity linked the Nile near Cairo to the northern part of the Bitter Lakes, with access thence to the Red Sea. Rameses II is said to have built it in the twelfth century BC.

Periods of disuse and reconstruction followed; indeed, the canal was rebuilt four times between c600 BC and the second century AD, then rehabilitated once again by the Arab 'Amr ibn-al-'As in 641–2. Surviving remains suggest that it was about 97km long, 46m wide and 5m deep, though dimensions probably varied at different periods.

The canal's considerable depth was due to varying flood- or low-water conditions in the Nile itself. These explain the letter written in 710 by the Arab governor of Egypt to the administrator of Aphrodito up the Nile, urging him to send needed supplies down the canal:

> If you fail to send any of the said materials and provisions and the water has subsided, you will have to carry them by road as far as Suez, paying the expense of porterage out of your private substance.(3)

At the Red Sea end Diodorus Siculus wrote of Ptolemy Philadelphus' rebuilding (early third century BC):

> ... and in the most suitable spot constructed an ingenious kind of lock. This he opened, whenever he wished to pass through, and quickly closed again, a contrivance who usage proved to be highly successful. (4)

This must have been a type of flash-lock*. The canal was not particularly successful. At the Cairo end, silt carried down the Nile repeatedly blocked it and had to be cleared. At the Suez end the Red Sea, liberally supplied with reefs and shoals, was also difficult to navigate upwards against the prevailing northerly winds. Many ships therefore called at ports on Egypt's east coast, whence cargoes were carried overland to the Nile. Another canal with varying water-levels from the Canopic branch of the Nile to the port of Alexandria is often ascribed to Cleopatra, but may be dated anywhere between the late fourth and the first century BC.

In about 600 BC Periander, who had also considered making a canal, began the *diolkos* or ship-railway across the isthmus of Corinth, which remained in use for many centuries—maybe until the ninth century AD. Small warships or empty cargo craft were carried on wheeled cradles running in grooves cut in a paved causeway. Goods from laden vessels had to be carried overland between Kenchreai and Lechaion. The *diolkos* was 3.5m to 5m wide overall, with a track gauge of 1.5m. A portion was excavated in 1956 and can be seen.

Demetrius Poliorcetes (fourth century BC), Julius Caeser, Caligula, Nero and Herodes Atticus (second century AD) also considered

*From earliest days dams were built across rivers and water channels to increase water depth above them. Flash-lock is a convenient term to describe gated openings made in such dams to pass boats. If the gate rises vertically, it is usually called a staunch. For more information see p. 30ff.

making a canal here. Of these, Nero made a serious start and might have completed one had he survived. Demetrius and Caligula were deterred, as Nero nearly was, by engineers' reports that the sea at the proposed western end was higher than at the eastern, and that a current would therefore be created which might wash away the town of Aegina. The engineers were right: the water to the west can be 51 cm higher, and there is a permanent west–east current through the present Corinth Canal. Nero's men, in the three or four months they were at work, shifted half a million cu. m out of the necessary $13\frac{1}{2}$ million. His works were visible till the modern canal was built, with cuttings up to 30 m deep and 50 m wide for 2 km at the western end and 1.5 km at the eastern.

Meanwhile, an earlier isthmian canal had been built. Xerxes of Persia, intending to attack Greece, with his army crossed the Hellespont by a bridge of boats, and then, to avoid the dangerous passage round the 48 km-long promontory of Mount Athos, in 480 BC cut a 4000 m canal to take his ships across the isthmus joining it to the mainland, giving it breakwaters at each end to prevent silting. Traces can still be seen of what was perhaps the first sea-to-sea canal.

Herodotus, writing of this canal, gives the earliest account I know of canal navvies at work:

> When the trench reached a certain depth, the labourers at the bottom carried on with the digging and passed the soil up to others above them, who stood on ladders and passed it on to another lot, still higher up, until it reached the men at the top, who carried it away and dumped it. Most of the people engaged in the work made the cutting the same width at the top as it was intended to be at the bottom, with the inevitable result that the sides kept falling in, and so doubled their labour. Indeed they all made this mistake except the Phoenicians, who in this—as in all other practical matters—gave a signal example of their skill. They, in the section allotted to them, took out a trench double the width prescribed for the actual finished canal, and by digging at a slope gradually contracted it as they got further down, until at the bottom their section was the same width as the rest.(5)

Roman Times

Much of the Roman world was linked together by sea—mainly Mediterranean—routes. As extensions of sea routes the Romans navigated rivers for both military and commercial purposes, in some cases using the same craft: Cologne, for instance, traded with London. Many reliefs and similar carvings show river boats from the Rhine and its tributaries, drifting, sailing or being towed, poled or hauled by

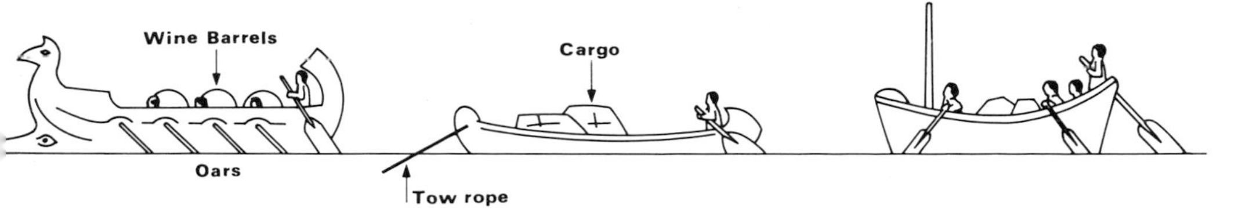

Fig 1 Rhine vessels of the Roman period: (left) the Neumagen boat (3rd century); (centre) from the Iglau memorial column (3rd century); and (right) Blussus' boat (1st century). The Neumagen boat, which resembles a Mediterranean type, may represent the kind of craft that worked between the Rhine and Britain, probably carrying grain.

men*, while actual craft from about 15m to 34m in length and 2.9m to 5.3m beam have been found at sites from Holland to Lake Neuchâtel. Generally they were flat-bottomed, shallow, straight-sided and with sloping ends. Rivers were especially used for the carriage of heavy bulk goods such as grain, timber, building stone, wine and pottery, for land haulage by mules, donkeys or oxen over very poor roads was inadequate and expensive.

On smaller rivers, the Romans certainly built flash-locks. For instance, aerial surveys in 1978–9 produced considerable evidence of river navigation systems in northern England, which included dams and associated canals, seemingly near Roman forts.

Busy ports usually grew up at transhipment points: Alexandria for the Nile, Marseille for the Rhône, Ostia for the Tiber, a special case. Although this is quite a small river, about 100m wide and a metre deep at Rome, and liable to floods, the 22.5km between the city and the sea were much needed to bring goods from Ostia to Rome itself. Thanks to bow-hauling by men (in the first century the voyage took up to three days), and the use of sails when possible, the river was made reasonably navigable for boats and some seagoing ships. The Tiber boat (there were about 300 at one time) was distinctive: a round hull with a curious upturned bow and stern; a deck; a forward mast carrying the towrope on upstream trips and lowered for drifting downstream, able also to carry a spritsail; and a capstan at the stern to wind in the towrope that came to it from the masthead, or even to warp the boat upstream if necessary. Monte Testaccio, beside the Tiber in Rome, emphasises the city's huge river traffic. It consists entirely of sherds of jars and *amphorae* that had once held wine, oil and similar liquids, and which had been broken in transit or in the warehouses: 869m round and 37m high, it probably contains about 1½ million tons of fragments.

*Towing of river barges by animals seems to have been rare in Roman times.

Before we leave the Tiber, let us mention the floating corn-mill. When Belisarius was besieged in Rome by the Ostrogoths in 537, and waterpower was cut off, he mounted corn-mills and waterwheels on moored barges. In the Middle Ages mills were often placed under the arches of bridges (eg the Grand Pont of Paris). Floating mills could still be seen on the Danube well into our own century, an enduring link with Rome.

To supplement or link their rivers, the Romans built canals. In the first century BC one ran for 26km across the Pontine marshes near Rome, parallel to the Appian Way, and carried passengers when the road was damaged by floods: in one of Horace's *Satires* we have an account of travel on it in a passage boat. Others for combined drainage and navigation were built in north Italy, based on the Po, Adige and Brenta, and in England, where the best known is the Fossdyke from the Trent to the Witham at Lincoln.

There were in Roman times three principal outfalls of the Rhine. The southernmost, the Helinium, was much the same as today; the middlemost ran from what is now the Lek past Utrecht and Leiden; and the Ijssel branch to the present Ijsselmeer. The Romans cut a canal, the Fossa Corbulonis, inside the sand-dunes parallel with the coast, to link the Old Rhine near Leiden to the Helinium near Hook of Holland. This canal became the Vliet, and the northern portion still exists.

Drusus, probably to be able to move troops quickly, in 12 BC cut another, the Fossa Drusiana, for about 14.5km from the Rhine to the Ijssel, its likely course being from above Arnhem to near Doesburg. In southern Gaul Caius Marius about 101 BC built the Fossa Mariana, used at first for military but later for commercial purposes, from Arles on the lower Rhône to the Mediterranean, thus by-passing the difficult delta. The now-disused Canal d'Arles à Fos (the former Canal du Rhône à Marseille) follows much the same course, and Fos-sur-Mer was itself named after the Roman cut.

To the south-east, the Danube for long formed the imperial boundary. Before Trajan the Romans had in part of the Carpathian gorges made a towing path, in one stretch cut out of the cliff, on another supported on wooden beams, the ends of which were held in holes drilled into the rock. When Trajan was preparing to invade Dacia across the river, he strengthened the path, and also cut one or more canals near Sip at the Iron Gates, up which boats could be towed by ox-teams. Thereafter, the two Roman Danube fleets, the *Classis Pannonica* based on Belgrade, and the *Classis Moesica* near its mouth, could make contact with each other.

Fig 2 Traces of Trajan's road through the Carpathian gorge of the Danube had not disappeared in 1969. They now lie beneath river levels raised by the Iron Gates locks.

A proposed canal in Gaul to link the Moselle and the Saône (as the Canal de l'Est does today), and another in north-western Asia Minor from Lake Sabanca near Izmit to the sea, involved considerable changes of level, and so cause us to speculate upon whether Roman engineers knew of pound-locks*. However, no indisputable evidence has been found that they did; differences of level were probably overcome by forms of flash-lock or inclined plane. Nevertheless, we have only to study the 500km or so of artificial waterchannel that provided Rome with its water supply to understand what Roman hydraulic engineers could achieve.

Change in China

At the other end of the Old World, in China, large-scale building of navigation canals was taking place before, during and after Rome's heyday. It was favoured by the country's great size; the fact that its rivers ran, very broadly speaking, parallel to each other from the mountains down to the sea, so that transverse transport links were

*A pound-lock is a chamber enclosed at either end by movable gates, water being admitted or drained through sluices to raise or lower a boat. For more information, see p. 30ff.

needed; a strong central government that could maintain a consistent development policy; competent administrators and technically advanced engineers. Somewhat similar factors brought about rapid canal-building in Imperial Germany from the 1870s onwards.

Here we can only glance at a few examples, like the Pien (Bian) Canal in Henan, built in about the fourth century BC. It left the Yellow river a little west of Xinyang, then a grain-growing area, and ran almost level to the Huaihe and Hongze lake. In 219 BC came the 'Magic Canal', built in Guangxi. Linking the upper reaches of two rivers by way of an interconnecting saddle, it is the oldest contour transport canal known.

One of the world's most famous waterways, the Grand Canal, developed out of the Pien Canal. Extended by the Sui dynasty (581–617), it began near Hangzhou, then ran north across the Yangtse and Yellow rivers to end near Beijing (Peking). Part of its route was along rivers, part by lateral canals running beside and fed from them. But a short section in Shandong was cut through slightly rising land: a summit cutting was excavated to give a water surface elevation of 6 m, and fed by diverting the flow of two small rivers. This compensated for water loss through the single gates at either end. It is the earliest summit-level canal of which we know. The Grand Canal's primary purpose was to carry grain junks from central China to feed Beijing, though it was also much used by passengers.

By the last half of the first century BC the Chinese were using staunches on both canals and river navigations, in the latter case flashes* being provided as later in England. Such staunches were usually of the log type and probably derived from sluices on irrigation canals. A short wall was built on either side of a narrowed section of canal. Vertical grooves were then cut opposite each other, into which logs were slid, these being raised or lowered by ropes running over the pulleys of primitive cranes. Sometimes the logs were joined together to form a gate, raised or lowered with the help of counterweights.

Double slipways developed at least from AD348. These enabled a canal to change level without losing water. A dam was built across the waterway, with inclined ramps on each side. Boats were then hauled up and over the summit by ox-powered capstans, to be lowered again into the canal on the other side.

While helpless on the slipways boats were more open to theft or banditry, and it was because of this that a Chinese work records that in AD984 the engineer Chhaio Wei-Yo

*Releases of water from dams, sometimes in sequence, to help boats over shallows.

> ... therefore first ordered the construction of two gates at the third dam along the West River (near Huai-yin). The distance between the two gates was rather more than 50 paces (76.2 m), and the whole space was covered over with a great roof like a shed. The gates were 'hanging gates'; (when they were closed) the water accumulated like a tide until the required level was reached, and then when the time came it was allowed to flow out.(6)

By putting two staunches with vertically rising gates close together, he had built a pound-lock, the first of which there is clear evidence. Thereupon, according to a text of 1086, some double slipways on the Pien Canal section of the Grand Canal were replaced by pound-locks, which enabled much larger boats to navigate. A text of 1072 mentions a staircase pair of locks—two sharing a common centre gate or gates—and one of 1089 a pound-lock on a canalised river. From references of 1098 and 1120 we find batteries of hand-pumps to supply water to summit-levels, and it also seems likely that pumps mounted on pontoons were used to pump back water used at pound-locks.

From the thirteenth century, however, grain increasingly moved by sea and not on the Grand Canal. Its pound-locks, no longer needed for large craft, became disused, and this section of canal reverted to staunches and slipways. With it ended China's interest in developing canal technology. As to whether their technology could have become known to the West, or whether the pound-lock was separately developed there, we have no certain answer. A relation is possible, but unlikely.

Development in the Dark Ages

West of China, invading Swedes in the second half of the eighth century had established themselves south of Lake Ladoga at Novgorod*, near the headwaters of the Volga and Dnepr, and from here traded down these great rivers to the Caspian's further shore towards Baghdad, and to the Black Sea and Constantinople, establishing on their way a trading centre at Kiev. The Tartar invasion of the thirteenth century ended the Swedish-influenced principalities, though by the fourteenth century Novgorod was within the sphere of the Hanseatic merchants. One of Europe's main trade routes ran from the Baltic via Riga and Novgorod to the Volga and Nijni Novgorod, then further south to the Don for Azov and the Black Sea. But soon all this was swallowed up by the rise of the Muscovite

* Novgorod is on Lake Ilmen, south of Leningrad, not to be confused with Nijni Novgorod on the Volga, now Gorki.

power—trading gave way to colonisation, and expansion had to wait for Peter the Great.

In the West, in the Dark Ages after the Romans, large and also quite small rivers continued to be used for navigation by shallow-draught craft. Any boat that could float would probably then carry more goods further than could a pack-animal or a primitive cart.

Einhard in his *Annals* tells us that the emperor Charlemagne had been 'convinced by some persons who claimed to be experts that if a navigable moat were constructed between the Rivers Regnitz and Altmühl it would be possible to voyage in comfort from the Danube to the Rhine. So he went with all his followers to the place, and ordered a large number of labourers to be procured.' This was in 793. Though the cut was to be only 2000 paces long and 9m wide, the site was marshy, and canal-cutting techniques in bog unknown. And so after an autumn's work the effort ended. Given the navigation value of small streams in those days, we should surely think of this scheme as much more an attempt to link two useful local systems than two great rivers.

In the western world of that time, people were few and trade minimal, for the manor, upon which most people lived, was largely self-sufficient. Agricultural productivity was low, partly because the scratch-plough, which the Romans had brought from the Near East, was unsuited to Europe's heavy soils. Thus, some ten workers on the land were needed to support one off it: town-dwellers were few, and most craftsmen were part-time farmers. When the heavy wheeled plough was introduced from Slav lands to the Po valley in 643 and spread northwards, changed farm organisation greatly increased output; more food became available to feed townsmen and transport workers, and two-way trade grew. Another new tool helped the process. Roman scythes are almost unknown, but by Charlemagne's time hay and cereal crops could be efficiently reaped.

The Romans had been curiously inefficient in using horses as working animals, for lack of proper harness. By adapting an ox's yoke, unsuited to a horse's anatomy, they enabled a pair to pull only about 220 kilos, equivalent to a light cart with two men. In Charlemagne's time, the horsecollar attached to shafts or traces appeared, to be followed by horseshoes, necessary for wet lands but unknown in Europe outside Britain until about 900, and the whipple-tree by 1077, to equalise the pull of a load between two horses. With these innovations, horse-haulage of heavy waggons and horse-towing of boats began, the animals themselves being fed by oats grown on deep-ploughed land and raped by scythemen.

We know little of the craft then used. The commonest was probably the log-raft, floated down river and then broken up and sold for fuel and building. These were widespread then and later in Europe, but especially so on the broad, slow rivers of Lithuania and Poland. We may also imagine simple barge-like craft, some able to be rowed, some sailed, and both to be towed when possible. It is unlikely that many were at all large. Not long ago, however, a trading boat dating from Charlemagne's time was discovered in a former Rhine backwater at Krefeld-Gellep. Flat-bottomed, it was some 14m long and over 1.8m wide.

River Navigation in the early Middle Ages

Though it is difficult to judge the relative importance of river and road transport in medieval Western Europe, it seems clear that, at any rate in France, whereas the Seine and Loire were much used (the latter because its current was sluggish enough to let boats sail upstream), boats on the smaller or difficult rivers met with too many impediments. There could be overmuch water or not enough; fallen trees and overgrown banks; no towing path; charges for passage by riparian owners or bridge proprietors; and above all, dams made for fishponds or to impound water for driving mill-wheels. Water-mills for grinding corn were widespread in Europe by the eleventh century—England's *Domesday Book* notes nearly 6000. As medieval industry developed, waterpower began to be used also for fulling cloth, smelting, forging and working iron, sawmilling, and a host of other operations. More mills meant more obstacles. Finally, boats and men to work them were not so easily available as their land equivalents.

Freight-carrying, therefore, on most rivers was mainly of really heavy goods preferably downstream: timber, marble, and above all wine, which by water arrived at its destination in better condition and with much less leakage. People seem normally to have travelled by land, though one factor did sometimes influence them towards water: boats were safer, less liable to robbery.

Some rivers were, however, improved. The burghers of St Omer caused such effective work to be done on the Aa between their town and the sea at Gravelines that whereas about 1100 it was being used by 6-ton boats, sixty years later it could take 300-ton and larger seagoing ships. When the area belonged to England, Edward III (1327–77) and his son the Black Prince had the channel of the River Lot, a tributary of the Garonne in south-west France, dredged and a towpath built so that wine could be shipped downstream from

Fig 3 A loaded salt boat (below) descends the Salzach river from Laufen, loading wharf for salt from the Salzburg area, and then the Inn to Passau at its confluence with the Danube; (above) the boat is bow-hauled and poled back against the current.

Cahors. Meanwhile from about 1100 the north Italian system of irrigation and transport waterways built by the Romans was being revived, while in England Henry I (1100–35) seems to have re-cut the Fossdyke.

By the mid-fifteenth century traffic began to move to water, one reason being an increasing number of road tolls. Channels were improved on the smaller rivers and tolls lessened or abolished, while the diverse products of waterpowered mills not only provided traffic, but enabled flash- and, later, pound-locks to be built, their timber sawn and ironwork furnished, and then to be used by men trained in the millwrights' craft.

Larger rivers carried passengers, merchandise or rafts. Navigation had to cope as best it could with drought and flood, unimproved channels and, on rivers like the Elbe or the lower Rhine, shifting courses through undrained lands. Silt had by then blocked the mouth of the Old Rhine, flooding followed, and large lakes were formed in the peat, similar to England's Broads. Water transport in the Low Countries often therefore involved portages; the Dutch word 'overdrach' is probably the origin of place names ending in '-drecht'.

Human obstacles took the form of tolls, imposed by local lords who found river traffic attractively easy to tax. The Rhine had 19 toll points at the end of the twelfth century and over 60 in the fourteenth, the Elbe about 35 in the fifteenth. Yet passenger and light goods craft ran regularly in the fourteenth century between Koblenz and Andernach, Mainz and Oppenheim, Mainz and Frankfurt. In the following century Ruhr coal was moving on the Rhine. There was, however, little traffic on the Elbe between Magdeburg and Hamburg, and almost none on the Danube below Vienna.

As in Roman times, so in the Dark Ages and afterwards, the same close links existed between coastal and short-sea trading that still apply today. Some craft moved from sea to river or canal and back again, others transhipped their cargoes at ports for transport upstream. Those cities grew and flourished that were sited at interchange points, or beside navigable rivers. One is reminded of the foolish man:

In his journeys he noticed,
How well the providence of God had disposed,
Who put rivers in all the places
Where big cities happened to be.(7)

From the tenth century onwards the bigger rivers were important in enabling traders, led by the 'emperor's men', the Easterlings*, to reach the rising cities and international fairs of Europe. Out of this early Germanic trading activity, based on Hamburg and Lübeck, emerged the Hanseatic League (*hansa* = a defensive association) of cities whose merchants built up trade routes that extended from London and Bruges to Marseille, from Bergen to Venice and Danzig to Novgorod. The League reached its flowering time in the fourteenth century. Advances in marine technology like the substitution early in

*Our word 'sterling' remains to remind us of the Easterlings.

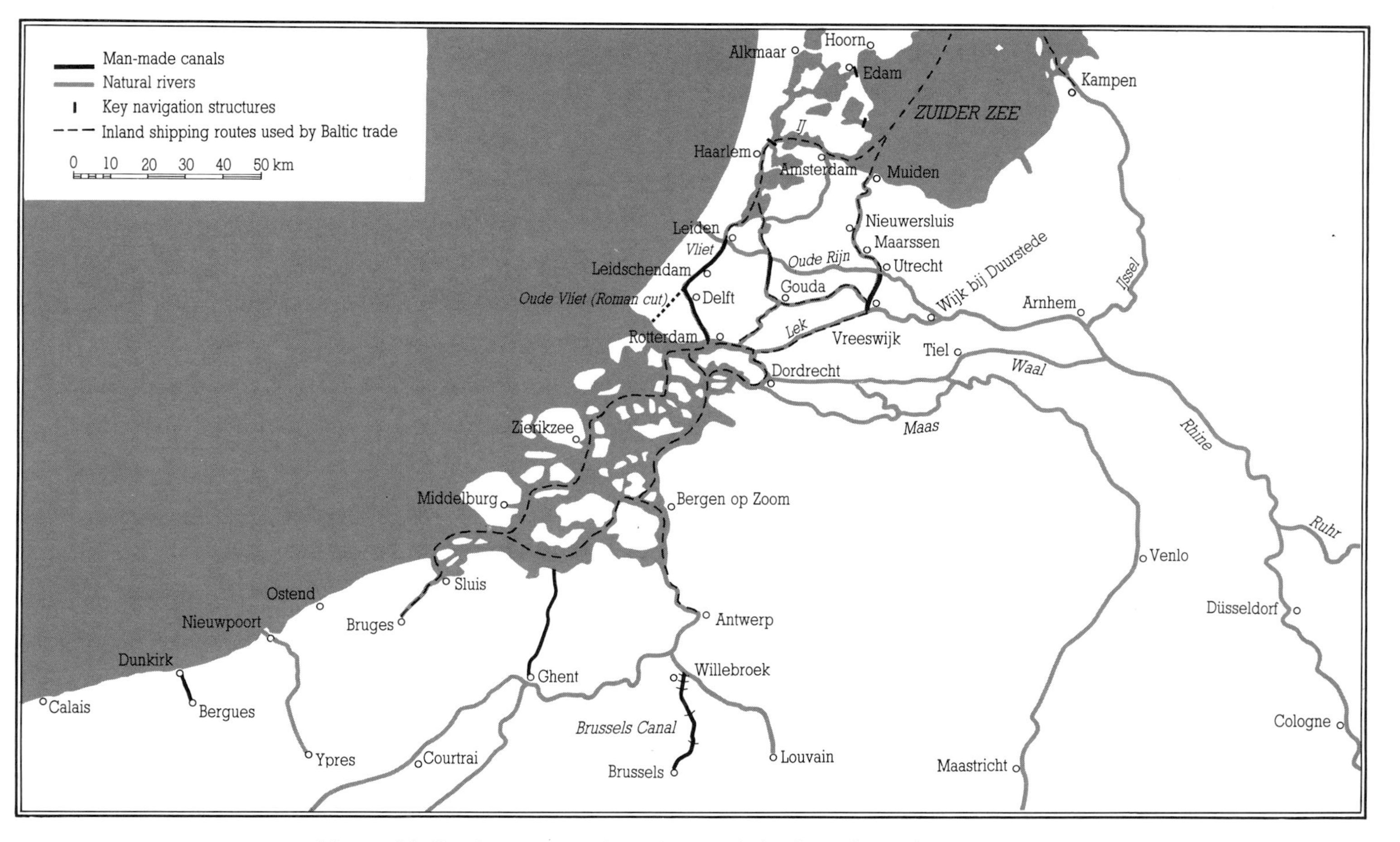

Map 1 Medieval waterways in and around the Low Countries

the thirteenth century of the rudder for the steering oar, and navigable rivers, Seine, Rhine, Elbe, Dvina, Volga and the rest, were the enablers of this growing and beneficial medieval trade, which in turn encouraged waterway improvements. Trade worked, for instance, up the Seine to Rouen and Paris, where there were Hansa corporations of merchants. That of Paris obtained from King Philip Augustus (1180–1223) the right to navigate the Seine, build a port at Bercy above Paris, and charge tolls.

The small and only moderately seaworthy vessels that carried on the Baltic trade—grain, timber, pitch and furs inwards, and exports of fish, cheese, salt and manufactured goods back—used three routes through the Netherlands inland waters from what later became the Zuiderzee. One was via Kampen—to develop into an important trading centre—and the Gelderse Ijssel to the upper Rhine; a second by the Utrechtse Vecht, and the third by the IJ.

The Utrechtse Vecht route began at Muiden, and ran up the Vecht to Maarssen (where transhipment to lighters took place), Utrecht* and the Lek, earlier via the Kromme Rijn and Dorestad (now Wijk bij Duurstede), later by a cut dug about 1148 to Gein. Thence by the Hollandse Ijssel or Vreeswijk, ships could reach the Merwede and Dordrecht, whence by winding routes through the islands they could reach Antwerp on the Schelde or Sluis and Bruges. A later alternative, the Hollandse *binnenvaart* (inland waterway), took craft via the IJ and the Spaarne to the Haarlemmermeer. Most then passed via the Oude Rijn and Gouwe to Gouda or Moordrecht, and so to join the route through the islands. Its popularity caused Utrecht to rebuild its section of waterway soon after 1300.

Antwerp's future was to brighten, that of Bruges to dim. At this time a sea-inlet, the Zwijn, cut into the land to make Sluis and Damme (north-east of Bruges) seaports, linked to Bruges by the River Roya. Ships from all over Europe traded to Damme, whence lighters worked up the Roya to the wharves of the Waterhalle, a port building spanning the channel. Then in 1404–5 a great storm caused the Zwijn to start silting up, and Damme to become unattainable. Antwerp was to inherit what Bruges lost.

Counts, bishops and rulers could make grants or give privileges, but the cities of the Low Countries provided most of the money and energy for waterway improvements, and collected most of the tolls, though embankment and drainage boards were also concerned. Waterway development was, however, hampered by intense inter-

*The main canal at Utrecht, made about 1150, still exists as the Oude Gracht.

urban rivalries, which could include the placing of obstructions and the destruction of sluices.

Staunches, flash-locks and inclines

Staunches with a single vertically-rising gate were an early means of controlling water-levels: they existed in Holland from 1065, in France in 1080–5, and in Italy in 1198, but whether also used for navigation is unknown. By the end of the thirteenth century they were widespread in Europe. In the form of flood or tidal gates, to enable craft to pass between still and tidal water, they are known at Damme in 1168 and Nieuwpoort in 1184.

The main obstacles to river navigation were in Holland dykes, and elsewhere the mill-dams each built to impound water needed to drive a waterwheel. Cargoes could be transferred past them by unloading and reloading, by using some form of crane, or by a primitive inclined plane, perhaps a slipway or a *pont à rouleaux*, using logs or rollers. A more efficient way to pass a dam, however, was by some form of gate. A common design had a balance beam that could be swung across the opening, behind which vertical bars (in England called rimers) engaged with slots in the sill below. Between these rimers paddles or needles (rectangular flat pieces of wood) each with a handle, could be raised or lowered, the handles of rimers and paddles being kept under control by a detachable horizontal guide-bar. The current was held back until a boat wished to pass. Then the paddles were lifted out, followed by the rimers, and the beam swung aside. Depending upon the height of the weir, a sizeable flow of water (called a flash) rushed through the gap. When the section of river above the weir had fallen sufficiently, the boat would either be winched* upwards through the gap, or shoot down past it. If going downwards, the flash would help it over shallows below a weir: if going upwards, it might have to wait almost until the river pound had refilled itself. Such flash-locks are mentioned in France as early as the eighth century. The earliest post-Roman reference we have in England is in a charter of Richard I, dated 1197, concerning the Thames. Types of flash-lock other than paddle and rimer were staunches with vertically rising gates and watergates or half-locks, using one or two swinging gates.

From staunches and flash-locks there evolved the pound-lock: that is, a chamber enclosed at either end by gates, and placed between two

*Winches for hauling boats upwards through such dams are mentioned from the thirteenth century on the Yonne in France.

pounds or stretches of water which do or can vary in their height relative to each other. Water can be admitted to the chamber from the upper pound, or drawn from it into the lower pound, through sluices opened or closed by paddles, to enable craft in the chamber to alter level as necessary. A pound-lock saves water, and is safer to work and quicker to use than a flash-lock.

A pound-lock may have been constructed, probably between 1285 and 1315, at Spaarndam near Haarlem when a 'spoije' was built through the dyke. We do not know what was meant, but clearly it embodied some new principle, for its building required an order, first from Count Floris V in 1285, then from Willem III in 1315. But what? M. Thélu* considers that it was not a pound-lock, but that instead and for the first time the opening had a swinging gate the full height of the dyke. Seagoing vessels could therefore pass it without lowering their masts or being compelled to tranship their cargoes. Such an innovation, he thinks, would have justified the counts' order and grant of special tolls.

When Utrecht rebuilt its waterway after 1300, the city authorities included a large 3-gate impounded area between it and the Lek at Vreeswijk. What is often called the first pound-lock in Europe, thus created in 1373, was not truly so. It was in fact a basin holding twenty or thirty craft, enclosed by a river flood gate on one side and a canal gate on the other, and was, we learn from regulations of 1378 and 1412, operated three times a week. Delfzijl in 1389 and Schiedam in 1395 had similar arrangements. Damme, however, had what amounted to an entrance lock, with a chamber 30.5 m × 10.4 m. Built in 1394–6, it may have been the first such to be worked whenever passage was needed.

It seems likely that at Gouda in Holland about 1413 two staunches were built close together to enclose a chamber, as soon afterwards near Amsterdam and, before 1435, at Utrecht. We do not, however, certainly know whether these were indeed pound-locks, or single gates brought together in pairs and used as such.

Flanders also had its equivalents of China's double slipways. These were the *overdragen*, known from the twelfth century—inclined planes or slipways of wood set into river dams, down which water constantly ran. Boats, tied fore and aft to a cable, could then be slid down them or hauled up by a man and a windlass, a horse-gin or a waterwheel. On the medieval canal from Ypres to Nieuwpoort a sketch shows four such, three counterbalanced and one single-track and powered

*To whose researches on early locks and inclined planes I am greatly indebted.

by two treadmills. Another existed at Watten on the Colme until it was destroyed at the siege of St Omer in 1638.

One such, seemingly at some time fitted with rollers, survived until early in the nineteenth century. In the fifteenth century Haarlem was only accessible from Amsterdam by a roundabout route, so a new canal was dug and provided with a lock. The people of Haarlem destroyed it and built a dam instead, whereupon the Amsterdammers constructed an *overtoom* for small barges, the canal itself becoming known as the *Overtoomse vaart*. The *vaart* survived until closed in 1921 to make a street, itself still called Overtoom.

It was therefore similar to the kind of incline Smeaton more than once saw in the Low Countries in the 1750s for getting boats over dyke walls:

> Near the watering Port I saw what they call a Windlass, for drawing Small Vessels over a Dyke, from one canal to another. The difference of Level of the 2 Canals was about 4 feet; the obliquity of the inclined plains being about 8 to one the purchase was by an Axis in peritrochio*, with a Rouet à hérisson** at each end; Diamr of the wheel about 10 feet, that of the Roller about 10 inches besides the rope. The Axis of the Windlass was placed about 10 feet perpendicularly above the highest part of the Dam, and the dam reached about 8 or 10 Inches perpendicular above the surface of the water in the higher canal, and equally sloped both ways; the Top of the dam was not an obtuse Angle made by the 2 inclined plains, but a little rounded; the plains were smooth planked and the vessels were drawn upon moveable rollers of about 5 Inches Diamr.(8)

Such early slides developed into the first true inclined plane, that which enabled boats to pass a wooden dam at Fusina which separated the Brenta river from the Venetian lagoons. In its original form as a slipway it dates from 1437, but certainly by the end of the sixteenth century craft were being carried up or down the grooved single track on a wheeled cradle running on iron-strapped nutwood wheels. Motive power was a horse-gin.

Canals and pound-locks; Leonardo da Vinci

Europe's first summit-level canal, the Stecknitz, was built at the end of the fourteenth century. Previously, salt from Lüneburg had been brought down the River Ilmenau to the Elbe, then across to Lauenburg and the River Delvenau. This was unnavigable because of

***Axis in peritrochio.* A purchase consisting of a wheel and barrel, power being applied to the wheel, and weight being lifted by a rope coiled upon the barrel.

***Rouet à hérisson.* Wheel, with hand spokes like a ship's steering wheel.

mill-dams, so the salt had to be carted to the lake of Mölln and thus to the River Stecknitz, down which, thanks to three staunches, very small boats could work to the Trave, there to be reloaded for Lübeck, whence the salt was exported to Russia and Scandinavia, mainly to salt herring.

In 1390 Lübeck and the Duke of Saxony agreed to make the Delvenau and the Mölln lake navigable, and then the Stecknitz. The work involved the building of 15 flash locks (seemingly not with rising but some form of double-leaved gates) and also the cutting of a 13 km-long watershed canal. In July 1398 the first laden craft from Lüneburg passed from the Elbe to the Trave. Boats 19 m × 3.25 m, and carrying 12.5 tons, worked two together on the rivers, singly on the canal. Because the summit canal had little water, and because only flash-locks were used on the rivers, boat movement had to be by flashes, provided every other day by the millers. Moving downwards, a boat could with luck work through more than one lock on a single flash of water, but upwards it had to be towed against the flash, and often grounded for two days to wait for the next one. Not surprisingly, a voyage of some 100 km could take several weeks.

Probably after the original construction but before 1480, two pound-locks with wooden chambers were built at Hahnenburg and Buchhorst. With some further enlargements and improvements* over the years, the navigation continued until 1900, when, the voyage from Lauenburg to Lübeck still taking 8–10 days, it was replaced by the present Elbe-Trave Canal.

Much earlier in northern Italy, a water-supply and irrigation canal, 50 km long and with a fall of 33.5 m, had been built between 1179 and 1209 from an intake near Casa della Camera on the River Ticino, which runs from Lake Maggiore to the Po, southwards to Abbiategrasso, and then eastwards to end on the southern side of Milan. In 1269 it was enlarged, made navigable, and called the Naviglio Grande.

When construction of Milan's new cathedral began in 1386, using marble from quarries near Lake Maggiore, it seemed sensible to enable boats carrying it to rise about 2 m to the level of the city moat, along which they could be taken to the site. Therefore, at least inconvenient hours, the water supply of Milan along the Naviglio Grande was cut off by stop-planks, thus temporarily raising its level and enabling boats to enter the moat and so a short linking canal to the site. That done, similar planks were used to cut off the moat and reopen the Naviglio Grande. Before boats could return, the procedure

*A circular pound-lock chamber, dating from 1724, can be seen at Lauenburg.

Fig 4 The Conca di Viarenna, by the Via Arena in Milan, is built upon the site of Italy's first pound-lock, created in 1420 when two staunches were brought closer together to reduce water consumption.

had to be reversed. A staunch was then provided at Viarenna at the entrance to the linking canal, and later a second at the junction between moat and Naviglio Grande, to replace the stop-planks. By bringing the two nearer together to reduce water consumption, the engineers Filippo da Modena and Fioravante da Bologna in 1420 created Italy's first pound-lock. Before 1445 a second had been built in the old moat, now enlarged into the Naviglio Interno.

Navigation canal construction now began in Italy. In 1451 Bertola da Novate, engineer to the Duke of Milan, was asked to lay out a canal to Pavia near the junction of the Ticino and the Po. This became the Bereguardo Canal which ran southwards from the Naviglio Grande at Abbiategrasso to the village of Bereguardo, whence goods were transported by land to the Ticino for carriage to Pavia. Built between 1452 and 1458, 19 km long with 18 staunches for a fall of some 24 m, it was the first European lateral canal (one paralleling a river) to have its gradient controlled only by staunches.

More canal-building followed. Bertola himself built the Martesana, eastwards from Milan to near the River Adda at Groppello, then north for 8 km to an intake at Trezzo (the masonry wall separating river and canal on this stretch can still be found), 29 km long with two staunches. It has special interest because a small three-arched aqueduct carried it over the River Molgora, while another river, the Lambro, was culverted beneath it.

About the time Bertola da Novate was beginning work, L. B. Alberti wrote his *De Re Aedificatoria, Book X*. A flow of water, he said, could be held back in two ways, by a lifting gate, or one that was pivoted, in either case provided with masonry side-walls. If the first, wooden gear-wheels would make it easier to raise; if the second, the gate could be square or rectangular, but pivoted off-centre. Greater water pressure upon one side of the gate or the other, acting upon the longer side, would tend to equalise the pressures, and so make the gate easy to open. He does not mention an obvious disadvantage, that a boat could only use about half the total width of the gate in order to pass through.

What Alberti is describing is either a staunch or a single-gated watergate, but he goes on: 'Make the gates double, cutting the river in two places, with an intermediate space which comprises the length of a boat, so that if it is a boat moving upwards, when it will have been steered into it, the lower gate will be shut and the upper gate opened; if on the contrary it is going downstream, the upper gate will be shut and the lower gate opened. The boat, in this way, is carried with that part which is flowing, in following the course of the river.' Thélu writes: 'Here, clearly set out, is the principle of a pound-lock, even if he has not foreseen the need to equalise the levels on both sides of the gate before opening it; but the pivoting gate described earlier indeed utilises this difference of level to help it to open.'(9) It seems likely that Alberti had seen, at the Conca di Viarenna, that the water-level could be higher, first on one side, then on the other. This would

Fig 5 The Martesana Canal runs alongside the River Adda towards its intake at Trezzo.

explain his pivoted gate which could open either way, but it is obvious that his could only be a technically satisfactory solution for slight differences of level. The same principle of the pivoted gate was later reused by Leonardo da Vinci, but for paddle-gear and not for the gate itself. In spite of Alberti's description of a pound-lock, an illustrated manuscript of thirty years later by F. di G. Martini shows a river made navigable by vertically-rising staunches set at intervals.

Before considering Leonardo da Vinci's contribution to waterway technology between 1490 and 1515, let us move briefly to France.

The River Sèvre enters the Atlantic north of La Rochelle, and before the end of the fourteenth century had been made navigable up to Niort by six staunches. Afterwards wars and difficulties in collecting tolls caused their decay. However, a document of 1494, referring to the lock at Roussille not far below Niort where the Sèvre lost speed of current and entered a marsh, described repairs which were to be done at the cost of the owner of the tolls, among which was 'nectoyer le grenier et reffaire les murailles toutes neufves', and a reference to 'les portes' and not 'la porte'.(10) Later documents strengthen the likelihood that originally two staunches had been brought close together where the river gradient changed, and that later a masonry chamber was built which existed in a bad state in 1494.

Further east, near Vierzon on the River Yevre, a lock was built at Gourt, on the site of one of three existing locks, for which the contract of 1510 has been transcribed. Thélu concludes that 'The existence of two masonry walls and of a floor makes it clear that this is indeed a pound-lock, without one being able to know for certain whether the gates were simple or mitred; these incorporated paddles which were opened by making them slide in vertical slots and not by making them pivot around a vertical axis ...'(11) He thinks it a fair guess that the new work was to replace an earlier pound-lock with a wooden chamber, as on the Stecknitz Canal.

The gates and paddle-gear of the new lock, however, were to be built, the contract specifies, 'de la façon que l'ingénieulx de Millan a baillé la forme'. And according to the Bourges city accounts, 'l'ingénieulx de Milan et son fils' had already made their visit in 1505.

Which brings us to a consideration of Leonardo da Vinci's contribution to the development of the pound-lock, where again I have followed M. Thélu's account. Leonardo arrived in Milan in 1482, aged 30, having been sent by Lorenzo de' Medici of Florence at the request of Ludovico Sforza of Milan. At first he was concerned with the arts, but in 1498 he was appointed state engineer, with the duty of inspecting rivers and navigable canals, and held the post until

1503, though it is likely he had begun his water engineering studies earlier. His personal concern in Milan seems to have been the reconstruction of the San Marco lock at the junction of the Martesana Canal with the Naviglio Interno—and with his drawings. Later he moved back to Florence.

Two of the many drawings in the *Codex Atlanticus* are of special interest—one of a pound-lock with mitre-gates*, and one of the San Marco lock. These show construction details of a lock-floor, mitre-gates, and horizontally moving paddles set off-centre in the sluice opening on a vertical axis. There is a similarity to the lock at Gourt, but the Milanese engineer who went there in 1505 could not have been Leonardo, and also he had no son, though 'fils' might mean 'pupil'.

It is clear that Leonardo did not invent the pound-lock—we have seen previous examples; or the horizontally swinging gate, used in the Low Countries and elsewhere; or the horizontally moving gate set off-centre, which had been described long before by Alberti. As for mitre-gates, Thélu concludes that: 'one must remember also that mitre-gates could very well have been invented by one of the Milanese engineers of the last quarter of the fifteenth century, who were well spoken of.' Until such an example is found, however, we must surely award their invention to him.

Leonardo also pulled together for us the scattered constituents which together made a well-built contemporary pound-lock—two gates with a chamber between; that chamber built of stone founded upon piles, with a floor supported by cross-beams tied to them underneath the masonry; mitre-gates as an alternative to one or two horizontally moving gates, to one that is pivoted, or to one that rises vertically; paddle openings set in the gate, but controlled from above; and the opening of lock gates by windlass and chain (though gate beams had been used earlier).

In 1516, at the invitation of Francis I, who had become ruler of Milan in 1515, he went to France. The king installed him at the chateau of Amboise, where he interested himself in irrigation and other canals and works. He died in 1519.

This chapter began more than two thousand years before Christ. We may fittingly end it with Leonardo's death. Early times had ended; the Renaissance had arrived.

*Lock-gates which meet to form an apex facing the direction from which water pressure will come. Pressure therefore holds them tight against each other, and minimises leakage.

2
MERCHANTS AND PRINCES 1519–1759

In the quarter-millennium that followed Leonardo's death, Europe's great rivers were, as they still are, highways of waterborne trade. Almost all run roughly north and south: Volga, Don and Dvina; Dnepr and Bug, Niemen and Vistula; Oder, Elbe, Weser, Ems and Rhine; Seine. Danube flows east as well as south, Po east, Loire and Douro west. Of these, the Danube and Rhine were the greatest carriers of trade, as were the routes linking Rhine to Danube and Rhine to Mediterranean. Professor Glamann writes of the rivers at work as our period begins when the

> so-called 'great trading company of Regensburg' [on the upper Danube] had offices in Vienna, in Budapest, in Berne and Geneva, at Lyons, Avignon and Marseilles, in Milan and Genoa, in Barcelona and Valencia, at Antwerp and in Cologne and Nuremberg.(1)

The great rivers were served by tributaries, many partially navigable, and from early times these or the main streams had been linked by roads or packhorse tracks. Traffic moved both ways. The small seagoing ships of the time penetrated far inland before unloading, or transhipping their cargoes for carriage higher up. They brought imports, while down from each river's catchment area came wheat, barley, oats and, above all, timber. In the Middle Ages and through our period there was an enormous demand for timber for domestic and industrial use, for house- and ship-building and as a source of charcoal for cooking and heating. Hence the importance to Paris of Jean Rouvet, who in France is commemorated by monument and medallion as the inventor of 'flottages' in 1549. He organised woodcutting beside the Yonne, and then the collection of logs and their making into rafts to be floated down to Paris to be stored in depots until needed. Slowly coal supplemented and then largely replaced wood for fuel, as did iron and steel for construction; but timber rafts still move, notably on the waterways of Norway, Sweden, Finland and Russia.

It was not practicable to do much to develop large rivers. But something could be done for the smaller ones, and throughout our period piecemeal improvement went on, not usually as part of any larger plan, but by dogged local people interested in getting some transport betterment. Quite unusual determination was often needed to overcome the many obstacles—the canalisation of the Vilaine in Brittany between Rennes and Redon went on from the time Francis I granted letters-patent in 1539 to its completion with 15 pound-locks in 1585.

The drive of the time went into canal-building, from the sea inland, from trading town to trading town, more ambitiously from one river to another. Previously canal-making had been sporadic, but now, in France, the Low Countries and Prussia important systems were created, in Sweden, Russia and, as this chapter ends, Britain, began. Whereas, however, in France and Russia we can follow development straight through, in the Low Countries it must be divided vertically to separate the astonishing Dutch *trekvaarten* network from the rest, and in Germany and Sweden horizontally by the Thirty Years War (1618–48), during which canal-building was delayed or stopped altogether.

France: to the Canal du Midi and beyond

Early canal-building in Europe owes much to Renaissance princes who had visions of economic development, even if some were beyond the engineering and financial abilities of their subjects. We have seen Ludovico Sforza bringing Leonardo to Milan. It was Francis I who, when he returned to France in 1516, accompanied by Leonardo, discussed with him a watershed canal between the Mediterranean and the Atlantic. There were two possibilities: to join the Rivers Saône and Loire (the later Canal du Centre) or the Garonne and Aude (the later Languedoc or Midi Canal). The second, technically less difficult, was surveyed by Nicolas Bachelier, who reported in 1539, and later by Adam de Craponne, better known for his irrigation canals in Provence. Both chose a line very much that of the later canal.

Francis died in 1547, and though further surveys were made for both routes, it was only after the peace of 1598 that Henri IV and his treasurer Sully could turn their minds to economic improvements. Eventually Sully settled for an easier beginning, a Loire-Seine link, which became the Canal de Briare. The Loire had its own fleet of craft, and the country through which it ran yielded food that was needed in Paris. Thus began a long-lasting policy, that of linking the capital to the provinces by waterways.

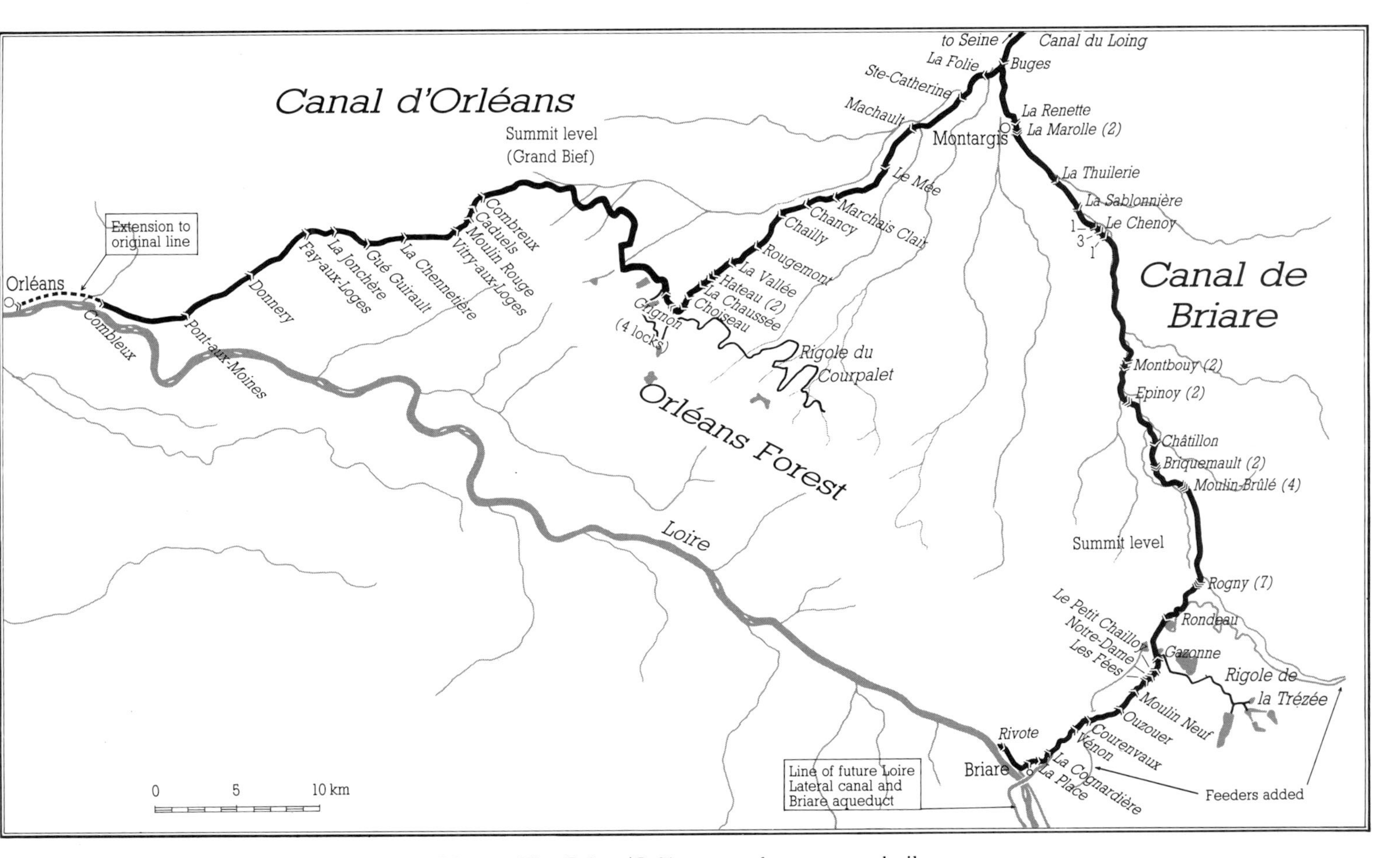

Map 2 The Briare/Orléans canal system as built

Hugues Cosnier, appointed contractor in 1604, worked out a scheme. With at least one eye on a possible link to the Mediterranean by subsequent construction of another watershed canal between the Loire and the Saône, the royal council approved it in 1605. Cosnier proposed to build a canal from the Loire at Briare that would climb for 11.25km up the Trézée valley to a plateau, have a 6km summit-level, then fall at Rogny to the Loing valley and so join that river at Montargis. Thence the Loing was navigable to the Seine and Paris. The canal rose 38m from Briare to the summit, then fell 79m to Montargis, using 41 masonry locks in all. Water supply was to be by feeder from streams, and also from a lake, the Etang de la Gazonne, itself given a 5.25km feeder from the upper Trézée. Cosnier also used a device that later was often adopted by canal engineers, the deepening of a 2.8km length of the summit-level to act as a reservoir during the winter months, with an intermediate lock at Rondeau that could be taken out of service when the water-level fell.

Sully sent troops to do the navvying. Henri and his queen visited the works in 1608, and by 1610 some three-quarters had been finished. Then the king was assassinated, Sully was forced to resign, and a commission of enquiry into the canal was appointed. From their report we learn what a remarkable job Cosnier had so far done: 42km of canal built out of 55, with 35 locks to take craft some 30.5m × 4.6m, incorporating the ground-paddles he had seen on the Brussels Canal (see p. 46). He had built lock staircases—a 7-rise* with a fall of 21m off the plateau to Rogny, 2, 3 and 4-rise elsewhere.

The government was then too preoccupied elsewhere to take action, and Cosnier left the area and died in 1629. The project was re-examined in 1628 by two engineers who recommended its completion, but also a second summit feeder to the Loing to provide water for the Rogny staircase. At last, in 1638, Louis XIII granted letters patent to Guillaume Boutheroue and Jacques Guyon, Sieur du Chesnoy, to finish the work and pay land debts still outstanding in exchange for ownership of the canal. They formed a company, and Guillaume's brother François then finished the canal in 1642, exactly, except for the second feeder, as Cosnier had planned it.

Thus the Canal de Briare, the first modern summit-level canal in Europe, was finished. Despite its design shortcomings, particularly regarding water supply, the canal proved useful and profitable. In the second half of the century it was carrying some 200,000 tons pa, and

*By-passed at the end of the nineteenth century, its ruins can still be seen.

so helping to provision Paris. The capital responded by starting to build quays on either side of the Seine, some of which are still used by barges. Modernised later, the Canal de Briare is still open.

Then between 1682 and 1692 the duc d'Orléans, Louis XIV's brother, who had been given the concession, built the 74km-long watershed Canal d'Orléans as a timber-carrier from the Loire at Orléans to the Loing at Montargis with 28 timber locks (rebuilt in masonry in 1726). The summit-level was again used to store water, a very long feeder channel, the Rigole de Courpalet, being built 32km long with a fall of only 4cm per kilometre. This astonishing piece of surveying and construction was the work of Sébastien Truchet, the canal's consultant.

The Briare and Orléans canals brought so much traffic to the Loing below Montargis that its 26 flash-locks were unable to cope. The lateral Canal du Loing, 53km long with 23 locks, was therefore cut to the river's confluence with the Seine at Saint-Mammès between 1719 and 1724 by Jean-Baptiste de Règemortes for the duc d'Orléans, who used royal troops for the purpose. Thus by 1724 an interlinked system of three canals showed what could be achieved by joining artificial to natural waterways. They were a portent.

The successful construction of the Canal de Briare and, not least, of its water-supply feeders to the summit-level, turned attention again to the old plans for a link between the Garonne and the Aude leading to the Mediterranean. A great engineer, Pierre-Paul Riquet (1604–80) with a highly competent assistant, François Andréossy, was found, while at the centre of government was the administrative ability of Colbert.

In 1661 Riquet worked out the water supply to the summit—it involved two feeders, one 42km long—and then proposed a line based on Bachelier's of a century before, for making navigable the upper River l'Hers from Toulouse to near Villefranche, the Fresquel from near Castelnaudary to the Aude and that river to the sea, with a canal between. A royal commission, however, decided not to depend upon rivers, but to build a canal the whole way from Toulouse to Sète. Work began in 1665 on the summit water supply. Once its practicability was confirmed, Riquet started on the canal, first from the Toulouse end, soon over the whole line, till some 8000 men were at work. Sadly, he died in 1680, seven months before the canal was opened in May 1681; it was his son, Jean-Mathias, who carried it to final completion in improved form in 1692.

From Toulouse on the Garonne the Languedoc (now Midi) Canal rose 63m with 26 locks in 51.5km. After a 5km summit it fell 189m in

185km with 75 locks to the Etang de Thau. Width was 19.5m and depth 2km, locks, being 35m × 5.5m with an average fall of 2.5m. A lock having collapsed early in the work, Riquet then adopted an oval shape for his walls to get greater resistance to earth pressure. This he achieved, though at the cost of considerable waste of water.

Here, then, was the greatest canal so far built west of China and Europe's finest seventeenth-century engineering work, perhaps her greatest since Roman days: 240km long with 101 locks, it had several special engineering features: a number of lock staircases, the greatest the 8-rise at Béziers; the first canal tunnel in Europe, 161m long at Malpas; three large aqueducts, one, the single arch over the Repudre, built by Riquet, the others, over the Orbiel and the Cesse, designed by 1686 by Sebastien Vauban the great military engineer and built by Antoine Niquet; the new canal port of Sète; and above all, its extraordinary water-supply system.

There was water in the Montagne Noire; too much in winter, too little in summer. Riquet therefore built the great St Ferréol earth-filled dam with masonry walls across the River Laudot, to create the world's first known artificial reservoir for canal supply: 780m long, it is 32m high above the river-bed at its maximum, with a greatest base thickness of 137m. Begun in 1667, it was finished four years later to hold 7 million cu m. With long feeder channels built by Riquet and Vauban to and supplementing the reservoir, the system still supplies the canal and is still a wonder of the waterways. With the opening of the Narbonne branch in 1776, more water was needed, and the Lampy dam, also still in use, was built during 1777–81 by Garripuy, then canal engineer. A dam similar to St Ferréol was built between 1788 and 1811 at Couzon near Saint-Etienne to feed the now disused Canal de Givors, branching from the Rhône.

Meanwhile, traffic on France's bigger rivers was growing. By the earlier eighteenth century, regular *coche d'eau* (packet-boat) services for passengers and small packages were working on French waterways. In 1747, for instance, boats on the Seine left Paris at 6am in summer and 7am in winter for Sens (Mondays), Montargis and Briare (Tuesdays and Thursdays), Auxerre on the Yonne (Wednesdays and Saturdays) and Nogent (Sundays). Others ran on the Oise and Aisne, on the Rhône from Lyon to Chalon (a 2-day voyage in winter, one in summer) to connect with coaches from Auxerre, and on the Saône from Chalon to Auxonne, usually with coach connections.

The end of the War of Spanish Succession in 1713, and the subsequent cutting of the Canal de Loing, encouraged interest in

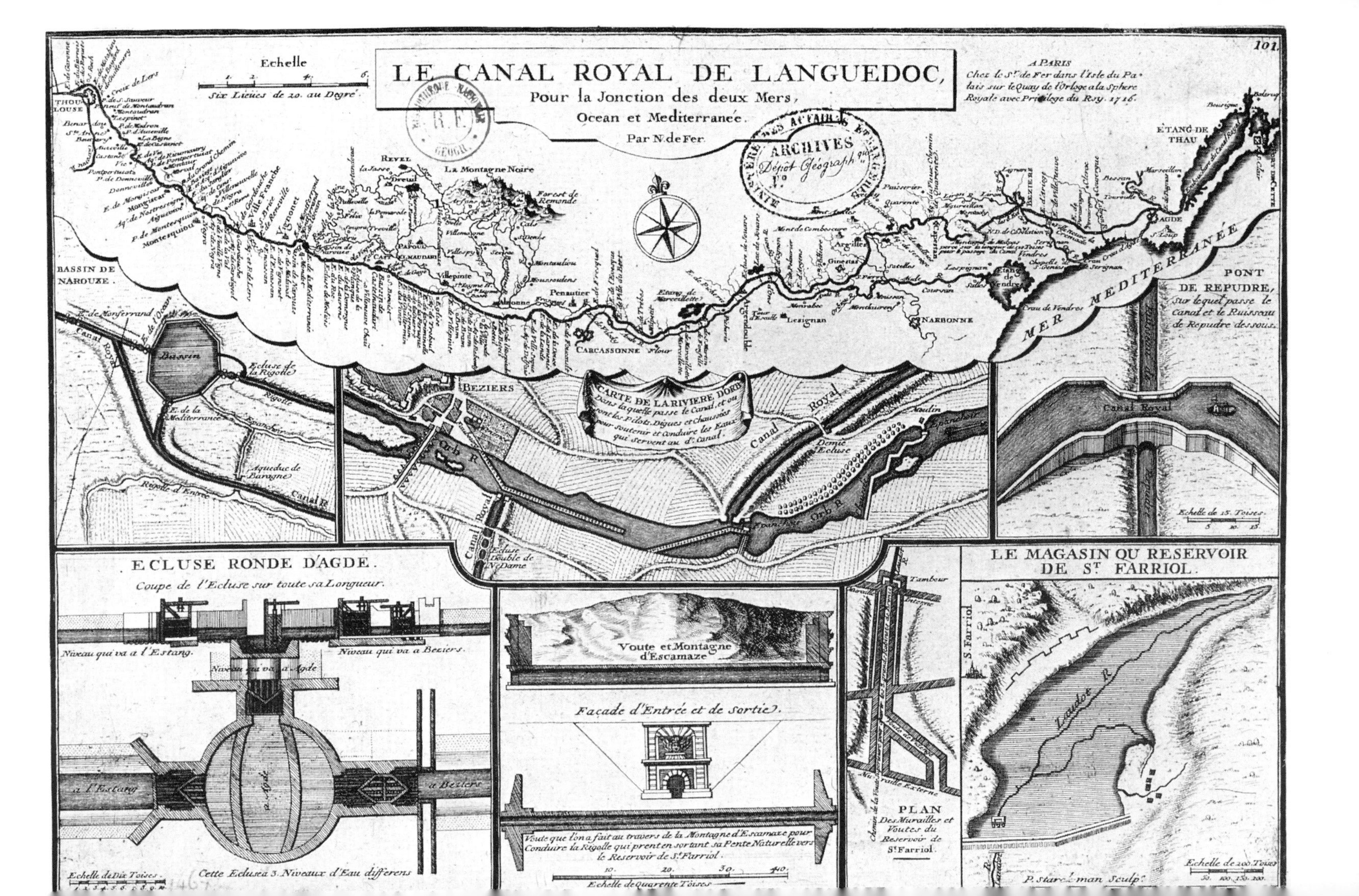
101
LE CANAL ROYAL DE LANGUEDOC,
Pour la Jonction des deux Mers,
Ocean et Mediterranée.
Par N. de Fer.
A PARIS
Chez le Sr. de Fer dans l'Isle du Palais sur le Quay de l'Orloge a la Sphere Royale avec Privilege du Roy. 1716.
Echelle
Six Lieues de 20. au Degré
ARCHIVES
Dépôt Géograph.que
MER MEDITERRANÉE
ÉTANG DE THAU
BASSIN DE NAROUZE
CARTE DE LA RIVIERE D'ORB Dans la quelle passe le Canal, et ou sont les Pilots, Digues et Chaussées pour soutenir et Conduire les Eaux qui servent au d.t Canal
PONT DE REPUDRE, sur lequel passe le Canal et le Ruisseau de Repudre dessous.
Echelle de 15. Toises
ECLUSE RONDE D'AGDE.
Coupe de l'Ecluse sur toute sa Longueur.
Niveau qui va a l'Estang.
Niveau qui va a Beziers.
Cette Ecluse a 3. Niveaux d'Eau differens
Echelle de Dix Toises.
Voute et Montagne d'Escamaze
Façade d'Entrée et de Sortie.
Voute que l'on a fait au travers de la Montagne d'Escamaze pour Conduire la Rigolle qui prent en sortant sa Pente Naturelle vers le Reservoir de St. Farriol.
Echelle de Quarente Toises
PLAN Des Murailles et Voutes du Reservoir de St. Farriol.
LE MAGASIN OU RESERVOIR DE St. FARRIOL.
Echelle de 200. Toises
P. Starckman Sculp.

building others. In France, as in Britain, the mid-eighteenth century was a time of great intellectual activity, as new possibilities seemed to open almost daily before humanity. In Britain, however, much of this activity went into invention, the organisation of industry and the expansion of commerce, as the Industrial Revolution (which we may date from 1709 when Abraham Darby first smelted iron with coke) got under way. But in France the main movement was towards intellectual freedom in religion, politics, science and economic theory as exemplified in the great *Encyclopédie* which appeared in 35 volumes between 1751 and 1776—rather than towards business.

One completion to be noted among several schemes was the Canal de Picardie (or Crozat) from St Quentin to join the Somme and then the Oise at Chauny. Authorised in 1724, the Sieur de Marcy's concession passed in 1732 to the rich M de Crozat, who succeeded in finding subscribers to 10,000 shares; the canal was opened in 1738.

In 1766 it was bought by the king from M de Crozat's heirs, at a time when its traffic remained small and its maintenance costly, because it needed to be part of a through Oise-Somme-St Quentin line onwards to the Escaut* – as it became later as the Canal de St Quentin.

Schelde and Meuse

When our period begins, the provinces of the Low Countries had one by one passed out of the control of their rulers into that of Philip of Burgundy, who married Juana, eldest daughter of Ferdinand and Isabella, joint rulers of Aragon and Castile. When Isabella died, Philip and Juana succeeded. Their possessions passed to their son Charles V and grandson Philip II of Spain, who found himself absolute monarch of Spain and its dominions, but also ruler of a loose, semi-self-governing federation in the Low Countries. In what we now call Belgium the liberties and energies of its great trading cities were eroded by Spain's policy of enforcing political centralisation and religious uniformity, until in 1713 the Spanish Netherlands were

*I use 'Escaut' in French contexts, otherwise Schelde.

Fig 6 This plan and illustrated guide to the Canal du Midi dates from 1724. The map shows its course from Toulouse to Sète (Cette). Beneath the map is shown (left) the basin at Naurouze (de St Ferréol) where Riquet's water supply channels were united before going to feed the canal; (centre) his ingenious level-crossing of the Orb at Béziers; (right) and his aqueduct over the Répudre. Below we can see (left) the three-way lock at Agde; (centre) Vauban's 121 m long Cammazes tunnel on the feeder system; (right) the layout of the St Ferréol dam.

transferred to Austria, who retained them until Napoleon's day.

In what we now call the Netherlands, a revolt under William the Silent led in 1579 to the formation of a union of provinces to resist Spain. Two years later the Netherlands proclaimed itself free, though fighting continued until, by the treaty of Münster, 1648, Spain recognised their independence.

In the provinces we now call Belgium trade followed the great rivers inland from the sea: the Schelde past Antwerp and towards France, the Meuse and its tributary the Sambre. The fearful storm of 1404–5 that permanently damaged Bruges as a seaport benefited Antwerp, for it had widened the Schelde's western branch and provided a direct route from the city to the sea, whereas previously ships had had to round the island of Walcheren.

Waterway development now had four objectives: to improve the communications of inland towns, notably Brussels, Louvain, Ghent and Bruges, with the sea; to complete a water line within the coast free from maritime dangers; to carry coal from fields near Liège, Charleroi and Mons; and to build up passenger services. We shall here be concerned mainly with the first two.

Navigation between Brussels and Antwerp was down the River Senne (upon which Brussels stands) to its junction with the Rupel, whence the Schelde and Antwerp could be reached. In 1531 it was decided to halve the distance by cutting a 30km canal from Brussels directly to Willebroek on the Rupel. Between 1550 and 1561 Jean de Locquenghiem built it with a cutting 9m deep at one point, and four octagonal mitre-gated locks 61m × 21.35m, designed each to take several of the local coasters. For them he put in the first known ground-paddles*. A stretch at the Rupel end of the canal he left tidal; it suffered from silting, and in 1570 he put in a lock with three pairs of mitre-gates**. After 1618 a passenger service, taking five hours in normal weather, was put on between Brussels and Antwerp, using a boat, partly horse-towed, partly sailed, called a *veer*, or *heu*, which included a *'hôtellerie'*. It ended in 1765, seemingly after a number of accidents. Eventually a road was built.

Later, between 1750 and 1753, a canal was made from Louvain (Leuven) to the River Dyle, and so also by the Rupel to the Schelde.

Ghent, higher up the Schelde than Antwerp, was becoming an important corn market. In 1561 a canal to Sas van Gent by-passed

*Paddle-controlled openings that admit water through a channel round the lock-gates, supplementing or replacing others in the lock-gates themselves.

**A third internal pair of gates saves water by enabling only part of a lock to be used when traffic is light.

Antwerp to give direct access to the Schelde estuary. A second outlet followed when in 1613 the broad and deep Ghent-Bruges canal was begun: 44km long, it was finished in 1623.

In 1626 the ruling Spaniards made a start on a canal still badly needed today: a Meuse-Rhine link. They began the Fossa Eugenia from Venlo on the Meuse to Rheinberg below and opposite Ruhrort. However, the Dutch, who thought it to be against their interests, sent troops to stop it. They eventually did so, and work ended in 1628.

Meanwhile, in the north-west of Flanders, developments were being based on the port of Nieuwpoort. In the sixteenth century a canal was cut to Furnes (Veurne), (whence it was to be extended to Dunkirk) and another, the Loovaart, thence south to the Yser, which was itself canalised back to Nieuwpoort. Later in 1640, the Nieuwpoort-Plassendale Canal was cut eastwards to the River Yperlee, giving access to Ostend and later to Bruges. By 1640, therefore, a waterway line, though often narrow and winding, had been completed inside the coast from Dunkirk to Bruges, as an alternative to the dangers of the sea passage during storms or war.

Antwerp, Ghent and Brussels were all badly hit by the Treaty of Münster (1648), whereby the United Provinces (Netherlands) acquired the southern side of the Schelde estuary and closed it to shipping. For exports and imports they had now to depend upon their outlets to Bruges. In 1722 she then sought to restore her old position by canalising the River Yperlee from the city to Ostend to admit seagoing ships. In 1753 this was extended to Ghent by the rebuilding of the older Ghent-Bruges canal, eventually to become the present Canal de Gand à Ostende. Not until 1792 was the navigation of the Schelde reopened, and then by the French revolutionary armies.

A happier situation existed in the south. Past Liège the River Meuse led upstream to Namur, whence it gave access to France. Its tributary the Sambre left it at Namur to continue past Charleroi and then eastwards also into France. An agreement of 1675 between the Spanish Netherlands (Belgium) and France allowed freedom of navigation on the Meuse for all except munitions traffic; thereby it encouraged the developing iron industry on both rivers.

The astonishing Dutch trekvaarten

In the Netherlands, early improvements included a mitre-gated lock of 1567 at Spaarndam near Haarlem, and another of 1578 at Gouda, which enabled small seagoing craft to move from north to south Holland. At Utrecht the difficulty of low-arched bridges was overcome by using the city moat—but at a high toll, because of objection by shopkeepers, who also made themselves felt at Gouda. In Amsterdam, the Amsteldam and the innermost ring canal, the Singel, date from about the early 1500s. Later, other semi-circular lines of canals were built, with their streets and walks, under the three-canals plan of 1612. Then mainly for commercial use, these are still characteristic of the modern city.

From the 1590s, the newly formed United Provinces were becoming, thanks to their development of sea transport, an industrialised society using imported raw materials and the mechanisms of international exchange. Though 'it was to fall short of the power-driven growth potential of Britain in the 1760s: it did nevertheless embody a number of mechanical inventions, labour-saving and labour-dividing processes that were to characterise the later Industrial Revolution proper'.(2) The province of Holland especially was a centre of economic growth for there agriculture, especially of specialised crops, flourished, yet industry and the growth of cities outstripped it. Much grain had therefore to be imported from the Baltic, France, England and the southern Netherlands, mostly using seagoing barges, cheap to build and run, but also some bigger craft. The causes of the Golden Age of the Netherlands in the mid-seventeenth century have often been studied. It seems likely that its economic base was due to the existence of widespread inland navigations, and the working of the country's stocks of peat. At a time when forests had been denuded by the Thirty Years War and wood was therefore expensive, these stocks gave the Netherlands an energy advantage. But to move peat required cheap transport, and this the waterways provided. Some 4000 craft of about 25 tonnes were used in peat transport, perhaps half those available. Many of these boats, carrying peat into the cities, also carried night-soil out, for use on the high-yielding farms and market-gardens; in the mid-eighteenth century 11 manure-barges were working from Leiden alone.

To these cargoes carried by sea, coasting and inland water transport were added from about 1600 onwards those created by busy Dutch industries like textiles, sugar-refining and brewing, and building.

Because the Netherlands was a federation and not a unitary state, canal- and lock-building could have been limited by the number of authorities interested: individuals, cities, provinces and drainage boards. Nevertheless, the impetus of an expanding industrialised society now produced Europe's first large canal system. Created quickly, it was in its main aspect unique, being built mainly for efficient internal passenger-carrying. Before 1631 the rivers, delta and

Zuiderzee were used by sailing boats carrying both freight and passengers: both *beurtvaren** or scheduled services, and private craft. These provided essentially local services, but as Jan de Vries** has shown, it would have been possible in 1650 to travel 578km by inland waterway and a few short road links from Dunkirk in the Spanish Netherlands to the furthest part of the province of Groningen in eight days. The *beurtvaren* were themselves supplemented by an extensive network of market-boat services worked between towns and their surrounding villages.

The natural waterways in the north Netherlands, however, themselves often circuitous, were made more so by the cities of Dordrecht, Gouda and Haarlem, who had rights to levy tolls, and the power to see that no new waterways were built to by-pass them. This created difficulties for passenger travel that were made worse by the navigation dangers of such waters as the Haarlemmermeer and the River IJ, dangers that were emphasised when in 1629 an accident on the IJ caused the death by drowning of the son of the Elector Palatine, his father himself having had a narrow escape.

This accident may have triggered action to create *trekvaarten*—waterways with towpaths, able to carry passengers and light freight regularly throughout the year independently of weather conditions† and, less important, carry them more quickly. To these real needs was added the very strong belief that such efficient passenger-carrying routes would stimulate commerce. Action was by cities, not the government, applications being made to the provincial assemblies for *octrooi*, powers to build waterways, acquire land, and levy tolls either by conversion of existing rivers or canals, or the building of new canals. The first move for a true *trekvaart* was in 1631, when Haarlem and Amsterdam proposed to build a line connecting their cities that would be independent of the IJ. Leiden and Delft followed in 1636, Amsterdam and Weesp in 1637, Amsterdam and Naarden and also Leiden and Haarlem in 1640.

So began a *trekvaart* fever which lasted until the late 1660s, which built 658km at a cost of over $4\frac{1}{2}$ million guilders, and which paralleled a rapid growth in the Dutch economy and population—under 400,000 in 1622, over 560,000 in the 1660s. By then the Dutch

*These were monopolies operated under contract to the terminal cities of each service, and by their skippers (*beurt*: a turn).

**Jan de Vries, 'Barges and Capitalism. Passenger transportation in the Dutch economy, 1632–1839, *A.A.G. Bijdragen* 21, 1978, to which I am greatly indebted for the following account.

†Except for an average of one month a year when *trekvaarten* were closed by ice. The operators then often provided waggons or ice sleds.

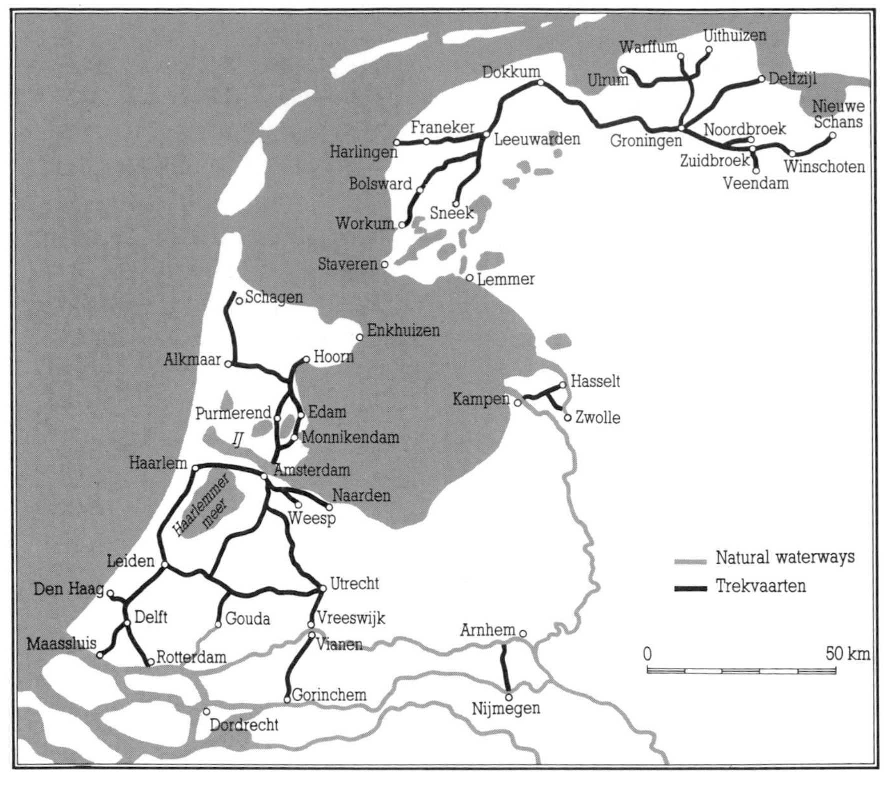

Map 3 *Trekvaarten* in the seventeenth century

system consisted of two networks, one south and one north of the Zuiderzee, the two being linked by numerous interchange routes across the Zee. Within the Dutch Republic the western provinces were far better served—indeed for the time uniquely so—than the eastern. A remarkable feature of the network was that in order not to interfere with established rights to take tolls and run *beurtvaren* services on the older waterways, the *trekvaarten* were largely confined to passenger craft by loading gauges—small-sized bridges at their entrances that prevented bigger barges from entering. Their creation did not affect the existing *beurtvaren* and market-boat services which continued, or indeed the building of better roads and the development of passenger and postal coaches in the last third of the seventeenth century, a development paralleled in different ways by the postal services in Germany of the Counts of Thurn and Taxis, the creation by Louis XIV of France of a civil engineering body, the Ponts et Chaussées, or the beginnings of a road turnpike system in England. Essentially as built in the forty years between 1630 and 1670, the *trekvaarten* lasted well into the railway age, and constituted Europe's first mass transportation system, and perhaps its first opportunity for leisure outings. Economically they were successful, especially in their early days. Most paid good interest upon their construction costs to their municipal owners.

Horse towpaths were already in use, as on the Rhine in Gelderland a century before. Passenger craft also ran earlier on other canals, as between Venice and Padua in the sixteenth century, or through the Willebroek Canal after 1618, and soon afterwards on the waterways connecting Ghent, Bruges and Ostend. What was new about the *trekvaart* system was the working of timetabled, frequent (sometimes hourly) craft carrying passengers at low and known fares all the year round. We in the nineteen-eighties, surrounded as we are by timetables, can hardly appreciate the innovative importance of punctuality, regularity and fare-reliability introduced into a world which had not long accustomed itself to public clocks. Many skippers were given sandglasses to check the passage of time, and passengers, were astonished that

> The hour for the Boat coming in, and oing out, is so punctually observed, that upon the Ringing of a Bell it goes off, without staying for any person whatsoever.(3)

As de Vries says, there was 'appreciation for the modern concept of time as a commodity'.(4)

Most newly cut *trekvaarten* were built straight from city to city, taking no account of intermediate villages. The Haarlem-Leniden line can be taken as an example: built 18.3m wide at water-level, and 2.4m deep, it had a small embankment on one side to protect the channel from floods, and on the other a larger embankment carrying the 6.4m-wide towpath, beyond which was a drainage ditch. Older lines were adapted to *trekschuit* working by giving them a towpath, which often involved the building of bridges over side channels.

Craft were long and narrow, from 10.2m to 15.3m in length and 1.8m to 2.6m external width, with passenger compartments 4.8m to 8.8m long. Most took from 24 to about 30 passengers on side benches in one compartment, at first tented, afterwards wooden with canvas or leather flaps that were later replaced by glass windows. It was called the *ruim*. Later, a *roef*, a separate first-class compartment aft, was often provided. This usually seated eight, but could also be hired for a private party. On some routes night boats were available, passengers sleeping on straw placed on the benches. An Englishman, in a book of 1743, wrote:

> This way of travelling is indeed far the most commodious, best regulated, and cheapest in Europe. (5)

The boats were worked by a skipper, an assistant, and a boy riding the towing horse at a light trot, giving an average speed of some 7 kph. Whereas in England on a canal with a single towpath, craft passed each other by one crew dropping its towline and allowing the other boat to pass over it, on the *trekvaarten* towing was from the masthead, boats going one way usually adjusting their masts to each other, though in some cases lines were dropped. At turns in the towpath, as later in England, vertical rollers were provided to guide the towline. At fixed bridges without towpaths Smeaton (6) describes the procedure.

> 'when the head of the Schuit, comes near the bridge a person at the head unhooks the line which hangs by a button and loop, and lets its run over the bridge and by such time as the rope is ready to drop on the other side of the bridge, the boat's head is also through, and the person ready to catch it, who hooks it on again, and away they go without the Schuit, or the horse stopping. (6)

De Vries considers that the development of the Dutch economy in the first half of the seventeenth century forged the relatively autonomous municipal economies into a single urban system, and that the intense demand for communications between these now interdependent cities was expressed in the construction of the *trekvaar* network. However, this integration lessened later. In the first half of

Fig 7 A Haarlem *trekschuit*

the eighteenth century the level of contact between the member cities diminished, while one city, Amsterdam, drew to itself much of the activity that had previously been spread throughout the urban system. Thereafter this partially dissolved urban system was further transformed by the rise of Rotterdam as a second central city.

These changes in emphasis were accompanied by a general decline of the Dutch economy from its late seventeenth-century levels until about 1740. The pattern can be seen in figures for the average annual passenger volume on two of the most used *trekvaart* routes:

	1660–70	*1700–10*	*1740–50*
Amsterdam-Haarlem	288,000	201,000	156,000
Haarlem-Leiden	136,000	83,000	60,000

Improved carriages travelling on better roads were a factor in lessening the use of *trekschuits*, but more pervasive reasons for economic decline were the gradual exhaustion of peat stocks, together with the increasing shallowness of harbours and rivers, necessitating more dredging, and competition from foreign coal, now increasingly available, especially from Britain. Yet even in relative decline, the traffic on the *trekvaarten* is still impressive.

Richard Castle, in his 'Essay on Artificial Navigation', 1730, the earliest substantial work on the subject in English, wrote:

> The Dutch have exceeded all nations we know of, in works of this kind, and the benefits they have and do receive thereby are evident to the most careless passenger that travels the country. The Province of Holland in effect is but one continued cluster of cities by means of their canals alone, which afford the most inland towns the same advantages with the most commodious seaports...(7)

He also notes 'The Dutch smack', which was developed in the 1680s to become a familiar sight on most Dutch waterways, which

> will carry fifty-five tons yet will draw but four feet of water, and to remedy the inconvenience of its flat bottom when exposed to broad lakes or arms of the sea, and to make the ship lie nearer the wind, they fix leeboards to her sides.

Berlin, a waterway city

Meanwhile in Germany, then a geographical expression covering many separate states, the lead was taken and kept by Brandenburg, which in 1618 was amalgamated with Prussia, and whose rulers in 1701 exchanged the title of Elector for that of King. These aimed to develop their possessions, and especially their capital, Berlin, intended by nature to be a waterway city. It lies on the River Spree, which flows past it from the south-east to join the Havel coming from the north at Spandau. The Havel continues past Potsdam to the Plauer See and then north-westwards to the River Elbe near Wittenberge. Eastwards of the sources both of the Spree and Havel, the River Oder runs north past Frankfurt to sea access at Stettin.

In the sixteenth century locks were built on the Havel and Spree to make easier the carriage of cereals, timber and wool to the Elbe and so down to Hamburg. Canal plans followed. Joachim II proposed a Havel-Oder canal in 1540, and in 1548 one between the Spree and the Oder above Frankfurt. Work on the latter did begin in 1558, but stopped in 1563. In 1605, however, the construction began of the Finow Canal, Joachim's first scheme. Money was scarce, but in 1620 it was finished, 40 km long with 11 locks. The canal ran level for 13 km from the Havel, then fell by 36.6 m down the Finow valley to Liepe on an Oder tributary. Its design was poor, and allowed Havel floodwater to enter it. This, and lack of care during the Thirty Years War, caused its dereliction and almost disappearance.

After the war, building began again. Joachim's old project for an Oder-Spree canal was revived by the Elector Frederick William. New plans were made and the canal restarted in 1662. With an Italian engineer doing the earthwork and a Dutchman building the locks and bridges, the Friedrich-Wilhelm was completed in 1669, 24 km long, rising 3 m from Neuhaus on the Spree and falling 20 m to Brieskow on the Oder, Europe's third summit-level canal after the Stecknitz and the Briare. It carried much Silesian trade to Berlin and onwards to Hamburg. Probably as a result, Frederick III converted the Spree locks from timber to masonry, and in 1694 issued a medallion to commemorate his work.

Stettin (now Szczecin) became Prussian in 1720, and Silesian in 1742. It then became national policy to improve the links between Berlin and the Oder which served them both. In 1744–51, therefore, a new Finow Canal was made, with two locks up from the Havel, a summit-level feeder, and 14 locks down to Liepe. It was quickly followed by the Plaue Canal (1743–6), 34.5 km long from the Elbe below Magdeburg to the Plaue See and the Havel. A waterway 'cross' had been built with Berlin at its centre.

A trans-Sweden canal is attempted

Another Renaissance prince initiated a great canal project. Canal building in Sweden has always been concentrated mainly round her four largest lakes, Vänern, Vättern, Mälaren and Hjälmaren*. Gustavus Vasa (1523–60) following earlier ideas, envisaged a navigable waterway across Sweden to Stockholm by way of the Göta river and Lakes Vänern, Hjälmaren and Mälaren, to avoid the tolls levied by Denmark who until 1653 controlled Skåne (southern Sweden) and the other dangers of the passage through the Sound. Work began in 1596 on an instalment, the canalisation of the Eskilstuna river between Lakes Hjälmaren and Mälaren with 11 timber locks, completed in 1610.

In 1607 Dutch engineers helped to construct a staircase pair of locks at Lilla Edet on the Göta river and also the Karls Grav** (Charles IX was then king) was dug between a small fjord leading off Lake Vänern and the great fall at Trollhättan to the Göta river, past which goods had to be portaged. Meanwhile, the Eskilstuna canalisation with its timber locks having proved unsuccessful, in 1628 a different route between the two lakes by way of the River Arboga was chosen. With ten masonry locks, a 605m rock-cutting and a fall of 22.9m, this Hjälmare Canal was opened in 1639 and superseded the older route. This time successful, it had to be rebuilt with 8 new locks 30.5m × 7.3m by the Dutchman Tilleman de Moll between 1691 and 1701.

In 1718 Charles XII revived the idea of a trans-Sweden canal, this time by the Göta river, Lakes Vänern, Vättern, Roxen and the Baltic, the route of the present Göta Canal, and Christopher Polhem, called Sweden's first engineer, agreed to build the whole line in five years. Two locks out of an intended three (with a total rise of 32.3m) were indeed cut in the granite at Trollhättan before the king died and work stopped: one can still be seen. In 1749 with Polhem, now 80, as adviser, a new start was made at Trollhättan, a shaft-lock† being built in 1754 on the Karls Grav with the then astonishing fall of 16m. Then work stopped after a flood, engineered, it was said, by those who stood to lose when the portage ended. The shaft-lock proved unusable in flood-time owing to insufficient height in the tunnel, and was replaced in 1768 by a staircase pair. For the end of the story we have to wait for the achievements associated with Nordwall and Count von Platen (see p. 81).

*Their areas is sq km and heights above sea-level are respectively 5500 and 44m; 1900 and 88.5m; 1100 and 0.3m; 500 and 22m.

**Much enlarged, it is now part of the Trollhätte Canal.

†Where a considerable lock rise would require very high lower gates, an enclosed shaft can be used for the upper part of the rise, with normal-sized gates placed at the entrance of the enclosed portion, which in this case was through a tunnel.

Peter the Great: canal-builder

In Russia and Siberia, the rivers were main transport routes. The commonest barges were flat-bottomed and decked or half-decked. The largest measured about 25 m × 8.5 m, drew up to 2 m loaded, and carried up to 150–70 tons. Upstream they were horse-towed, bow-hauled, rowed, sailed or warped by anchors. Additional downsteam traffic was by raft.

Peter the Great, who died in 1725, towards the end of the seventeenth century determined to bring Russia into the civilisation of western Europe. Having seen the waterways and locks of the Low Countries during his travels, he caused a start to be made upon a through route from the Neva to the Volga, and from that to the Don, using where necessary imported engineers and craftsmen. So Russian waterway engineering began as for a long time it continued. Because stone was often scarce, river dams were built of timber, as were locks also, probably on a piled foundation and with sides built up from timber cribs or boxes filled with clay puddle.

From several possible Neva-Volga lines, the Vychene-Volotski* route was chosen, leaving the Volga near Tver by the Tvertsa river and then via the Msta to Lake Ladoga and the Neva. A summit canal from the Tvertsa past Vychene-Volotski, 3.2 km long and 15.5 m wide with two locks, was begun by Dutch workers in 1703. However, the route suffered from silting, and between 1719 and 1723 a second short but deeper and wider canal, the Tsna, was built nearby. Supplemented by two dams to increase water supply, the route now became practicable. Thereafter it was steadily improved, notably by adding a locked by-pass canal to Lake Ladoga and later to Lake Ilmen. In 1757, it was carrying 194,000 tons pa of merchandise to St Petersburg. Between 1710 and 1714 an English engineer, Capt John Perry, surveyed for a second Volga-Neva route—that which, from 1799, was to become the Mariinski line (see p. 66).

The Sultan of Turkey, Suleiman I, had begun a canal some 20 km long through difficult country to connect river tributaries of the Volga and the Don, but left it unfinished. Peter the Great, after conquering Azov from the Turks in 1696, ordered building to start again along almost the same line, using troops as navvies under John Perry. However, after some years a Swedish war broke out, he needed the troops, and work stopped. A second Volga-Don link was also begun by way of Lake Ivanovsko, the source of the Don, and the River Oka to

*In this book, I have until 1917 mainly used older Russian names and spellings.

the Volga at Nijni Novgorod. By 1707–24 locks had been built, and some 300 craft had used the route. Then in 1711 Azov was returned to the Turks, and building ended.

English river engineers foreshadow her canal age

And so we come to England, whose earliest-known pound-locks had been those on the Exeter Canal with vertically rising gates built by John Trew between 1564 and 1566, that at Shelford on the Trent of 1576 and the timber mitre-gated lock also of 1576 on the River Lea at Waltham Abbey. Thereafter England's engineers built locks and made cuts so successfully that by 1760 1127 km of engineered rivers had been added to 998 km of natural navigation.

In the early eighteenth century increasing agricultural wealth caused a rise in demand that encouraged industrial growth. So, encouraged by important inventions, the process began that we call the Industrial Revolution. Towns grew, and needed coal; factories and mills multiplied and needed iron, raw materials, and transport for their products; urban populations needed food and warmth. Out of the beginnings of these demands came the Canal Age, personified by the Duke of Bridgewater, who in 1759 began the canal that bears his name from his Worsley collieries towards Manchester. His scientific bent was in his own nature, but the development of his interest in canals must have had two origins: abroad, his tour along the Canal du Midi and the Garonne from Sète to Bordeaux in 1753, and the building near Worsley of the Sankey Brook Navigation, authorised in 1755 and partially opened in 1757.

The duke's canal initiated the rapid building of a canal network. Whereas, however, the *trekvaarten* over a century before had been for passenger-carrying and been financed mainly by cities, that of Britain carried industrial raw materials, manufactured products and food, and was privately financed. With it arrived modern times.

3
BRIDGEWATER TO VON PLATEN

The Bridgewater Canal and its effects

The opening in 1761 of the first part of the Bridgewater Canal energised the start of Britain's canal age. Thereafter entrepreneurs, mostly colliery owners and industrialists, built local canals in England, Wales and Scotland to help feed and warm the people of the new industrial towns, to bring raw materials to and carry away the products of the mines and factories where they worked. Most were before long linked to long-distance routes connecting the country's main rivers and ports, serving as many inland cities as possible, and forming a largely interconnected whole. These longer canals were usually promoted and financed by big landowners and industrialists, supported by middle-class people looking for profitable investments. In Ireland, however, which had few mines and little industry, more public and less private money went into canal-building and river improvement mainly intended to promote future agricultural and economic development.

The results were spectacular, given the limited resources of manpower, capital and engineering skill then available. In 1760 England and Wales had 2387km of inland waterways, most of them rivers: in 1790 they had 3718km, the greater part of the increase being artificial canals. These were small, usually with 2.15m locks and boats carrying some 30 tons when James Brindley was the engineer; or 4.3m carrying about 60 tons when, slightly later, they were engineered by William Jessop. They were economical in capital and water, and well served the immediate purposes for which they were needed. By 1790 Trent had been joined to Mersey, Severn and Thames to both, and Forth to Clyde; in 1804 the Pennines surmounted; in 1804 Ireland was crossed from Dublin to the Shannon, in 1810 the Bristol Avon linked to Thames. A pattern had been completed in fifty years. Additions were then made, including a third type of waterway, the Caledonian Canal, built by the state to enable what were then large seagoing ships to transit Scotland.*

*The story of British canal development is told in my books *British Canals* and *The Canal Age*.

Because of Britain's industrial and commercial lead, which had been helped by her organisation as a unitary state, and her orderly social, legal and political system, her canal and road network grew quickly, and themselves helped further development in town and country. She became a showpiece. Engineers from all over Europe visited her to learn what they could, and then went home to apply the lessons learned. In the New World also her influence was felt.

Piecemeal navigation building continued for a time on the Continent: I have in this chapter instanced a sea-to-sea route across Denmark, and examples of local industrial waterways in France, Sweden and Germany. Then country after country initiated canal ages of their own: notably France, but also Spain, Prussia and Russia, until the French Revolution and afterwards Napoleon interrupted most of the continuities. Throughout the turmoil of the times, however, Chinese canals remained uninfluenced, epitomising the past as did Britain's the future.

Sea to sea across Denmark

Denmark needed an extended *diolkos* or a Corinth Canal. Indeed, the Vikings are said to have used horses to drag boats across the peninsula on rollers. In medieval times goods from the North Sea moved from Tönning up the Eider river, to be carried onwards, by road to the Baltic. In 1777, however, the Danes began to improve the winding and shallow Eider, tidal to Rendsburg. They built a lock there and, 10 km upstream, cut a 32 km canal with five more locks to near Kiel. An edict of 1785 declared the new waterway, some 161 km long, open to all the nations of Europe. By 1840 it was being used annually by some 3000 vessels of 200–50 tonnes. The canal changed hands after the Danish-Prussian war of 1864, to be replaced in 1895 by the present Kiel Canal.

Waterways for coal

France needed a waterway along which Belgian coal from the Mons-Charleroi area could be brought to the industries and people of the north-east, and especially to Paris. A good answer would be a link between the Escaut (Schelde) and the Oise leading to the Seine. The earlier building of the Canal de Picardie (see p. 45), joining Oise to Somme, led to proposals by Devicq and Laurent for its extension from St Quentin through a ridge of high ground to the Escaut near Cambrai.

Laurent's line was agreed in 1769 on the duc de Choiseul's initiative. It included an enormous 13,682 m tunnel which, if it had been built, would have been the world's longest. Only a specimen section was dug to the full intended dimensions of 6.5 m broad and high. Beyond, a pilot tunnel 3.9 m high and 3.25 m wide was cut unlined through the chalk. When in 1787 Arthur Young climbed down 134 steps to it, he saw a plaque saying that in 1781 Joseph II, Holy Roman Emperor, had visited the works and gone by boat from that spot to shaft No 20, and had said:

> Je suis fier d'être homme, quand je vois qu'un de mes semblables a osé imaginer et executer un ouvrage aussi vaste et aussi hardi. Cette idée me lèvel l'ame.*

Yound tells us that 5000 toises (9745 m) out of the total of 7020 toises (13,682 m) of pilot tunnel, with only a few inches of water in it, had by then been cut through the chalk, but that the work was at a standstill for lack of money, 1,200,000 livres having been spent, and 2,500,000 livres more being needed. Resumption on an altered plan had to wait for Napoleon (see p. 70).

Iron ore and timber were being moved from Sweden's Dalarna mining region overland to Stockholm or ports on Lake Mälaren. Then in 1764 Johan Ulfström, a director of the College of Mines, proposed to make the Kolbäck river navigable with short canals by-passing rapids, to create a 100 km waterway with 26 locks from Smedjebacken on Lake Barken to Mälaren. Building began in 1777, and the Strömsholms Canal was ready in 1795.

Frederick II of Prussia's territories included much of the coal-bearing Ruhr. Interested in industrial development, he authorised a Dutch-financed company to start sending coal from around Witten down the Ruhr river to the Rhine at what was to become the port of Ruhrort. They began in 1772 by transhipping cargoes over the mill-dams from barge to barge, a plan that had around 1760–1 been tried on the River Stroudwater in England, but soon suggested the building of locks instead. Frederick, advised by his chief mining official, F.A. von Heynitz, who had visited England, agreed, building some of the 17 timber locks himself and allowing millowners to build others, a few of which, indeed, were not on Prussian territory. They took broad, shallow-draught barges carrying some 30 tons. In 1780 the first barge-load of coal from Witten passed down the 75 km navigation to Ruhrort. Much of what followed was to be exported to

*'I am proud to be a man, when I see that a fellow man has dared to imagine and execute such a vast and bold work. The idea lifts up my soul.'

the Netherlands. Thanks to a visit by Reichsfreiherr von Stein to England in 1787, Ruhr colliery owners were then encouraged to build horse-railway lines to carry their coal to the river.

Navigation maintenance costs were financed by lock-tolls, wharfage charges at Ruhrort, and a subsidy from the mining administration. After 1815, with Prussian control of the whole river, differential lock charges were introduced, higher for locks nearer Ruhrort, lower for those further away. Thus production from upriver collieries was stimulated. The original locks were rebuilt larger in stone in 1837–42, and again twenty years later, now to take 180-tonne boats. River coal tonnage reached its peak in 1860. Twelve years later a parallel railway was built, and commercial traffic ended. Afterwards, and further north, the Dortmund-Ems and Rhine-Herne canals were to restore water transport for much Ruhr district coal.

In England the Duke of Bridgewater's engineers had built a system of underground colliery canals connected to the Bridgewater Canal that eventually extended to 74 km on three levels. Other colliery canals were also tunnelled in Britain, and at Waldenburg in Silesia an underground coal canal was cut that had been copied from Worsley. It was later, as at Worsley, to load coal into the boats in small containers for ease of handling.

Canal-cutting in Spain

These had been enterprises directly concerned with mining or industry. Waterways were also being built in Europe as longer-distance transport routes, many either paralleling difficult rivers or joining two rivers across a watershed.

In Spain, the Imperial Canal of Aragon, a by-pass to the Ebro between Saragossa and Tudela 85 km higher up, had been built for irrigation. Between 1770 and 1790 it was enlarged and made navigable for 100-ton craft, a celebratory moment being reached when six barges arrived at Saragossa on 6 October 1784. To the west the Castile Canal, taking barges up to 75 tons, had originally been intended to provide better transport between the port of Santander on the north coast and the central Castilian plain, and at the same time to support irrigation and waterpower for mills, for which reason the canal was built on a slight gradient. In fact, the line was never made through the mountains. Its northern branch began at Alar del Rey on the River Pisuerga, to end 74.5 km and 24 locks away in the River Carriòn at Calahorra de Campos. After a level crossing of the river the canal continued for 78 km with 7 locks and 3 aqueducts to Medina de Rioseco.

From this later section at El Serrón the southern branch left for Palencia and Valladolid, 54km and 18 locks away. Between 1753, when work began, and 1791, the northern section had been built from Alar del Rey to Calahorra. Then pressures of war at home and abroad stopped work until 1831. In 1835, the first barge reached Valladolid, that to Medina de Rioseco in 1849.

Louis XVI of France: young man in a hurry

In 1774 Louis XVI came to the French throne, a hopeful young man of twenty with big ideas of economic development—by the next year drawings of the canals of France were hanging on his drawing-room walls, Then in 1778 a remarkable assessment of waterway development throughout the country was published by Delalande, a professor of mathematics; it may well have influenced royal policy.

There had been almost two centuries of often heated debate among promoters and engineers upon the best route for a canal to link Paris with the Saône and thence the Mediterranean. The scheme first proposed under Francis I (see p. 39) was for a link, then called the Canal du Charollais or Bourbonnais, now the Canal du Centre, between Loire and Saône. Indeed, once work had started on the Canal de Briare under Henri IV, there was a strong argument for building the Charollais as complementary to it, and three times during the seventeenth century plans were considered. However, under Henri IV it was also realised that a Saône-Yonne link through Burgundy (and serving its capital Dijon) would be more useful, since it would provide a more direct line though the kingdom, with the important advantage of better navigable conditions on the Seine and Yonne than on the Loire, always a difficult river. The estates of Burgundy preferred the Canal de Bourgogne, but had for long been frustrated by the technical difficulties of the project, involving a very high summit-level. The debate was finally settled when the engineers Perronet and Chezy reported in 1765, declaring the Canal de Bourgogne to be feasible and more profitable than the Canal du Charollais (although twice as costly). Royal edicts authorising the canal and committing limited funds were promulgated in 1773 and 1774. Work began in 1775, with four companies of soldiers sent to dig sections in the Armançon valley at the Yonne end. However, the next year only 140,000 livres were allocated, or less than 1 per cent of the total cost estimated by Perronet and Chezy.

Soon, war with Britain intervened. When it was drawing to a successful end, the king, in his mind also the advantages of large-scale

Fig 8 French medallions, four of them showing the burst of canal building of Louis XVI's reign: (top), Canal de Briare, 1617; (second row left), Canal du Midi, 1667, issued to commemorate the start of work at Toulouse; (second row, right) Canal de Picardie (St Quentin), 1785, with an inscription referring to the originally-planned 13,772 m tunnel; (third row, left) Canal de Bourgogne, 1785, probably to commemorate the re-start of work; (third row, right) Canal du Charollais (Canal du Centre), 1785, upon which work had begun in 1784; (bottom) Canal de Frańche-Comté (part of future Canal du Rhône au Rhin), 1783, upon which work started in 1784.

public works upon which his disbanded troops could be employed, caused Louis XV's great roads to be repaired after several years of neglect, sought to have ports improved, and also to push forward with navigation building as a precondition for industrial and commercial growth.

The time was right for the estates of Burgundy to begin upon three river-linking projects. Work restarted on the Canal de Bourgogne mainly at the Saône end, but by 1793 only 50km had been completed from St Jean de Losne past Dijon to Pont-de-Pany. The first stone of the Canal du Charollais from the Loire at Digoin to the Saône at Chalon was laid in Chalon in 1784 by the Prince de Condé, governor of Burgundy, the works being partly financed with funds raised by sale of the citadel at Chalon. With a relatively easy line and a competent engineer, Gauthey, the canal, 118km long with 81 locks, was completed in 1792 at a cost of 11,420,000 livres, and opened to navigation in 1793. The third project was the 17km Canal de Franche-Comté, from St Symphorien on the Saône to Dôle on the Doubs, which avoided the lower course of the latter river and its influential millowners, who had succeeded in prohibiting navigation. This canal, forming the first section of the future Canal du Rhône au Rhin, was started in 1784 and opened to navigation in 1802.

Another north-south watershed canal which had been considered in the seventeenth century was the Nivernais, from the Loire at Decize to the Yonne at Auxerre. Work began in 1784 at the Loire end, but was barely a third completed by 1793; complete opening had to wait for 1842.

To the south, the estates of Languedoc had been creating the important waterway link between the Canal du Midi and the Rhône. The first section, the Canal des Etangs from Sète to Aigues-Mortes, was indeed completed, but the second, the Canal de Beaucaire, though begun in 1774, was unfinished at the Revolution in 1789. Napoleon completed it in 1808.* Not far away, the Sieur Zacharie of Lyon had been given letters patent in 1761 for a 16km canal to run from the Rhône at Givors to the coal-mining centre of Rive-de-Gier, with 28 locks rising 82m, his ambition being to continue over an exceptionally high watershed to link with the Loire near St Etienne. Work could not start immediately because the owners of mules engaged on coal transport to Givors opposed registration of the letters patent. When building did begin in 1763 Zacharie misguidedly adopted an untried design for the locks, with bottom-hinged

*Both canals together are now called the Canal du Rhône à Sète.

collapsing gates at both upstream and downstream ends. Other problems arose from inadequate planning, and Zacharie had spent all his fortune when he died in 1768, with only 5km completed. Later a company was formed, and work restarted on a sounder basis in 1774, to be completed in 1781.

In then remote Brittany, the estates had put forward a plan for making Rennes the centre of a waterway system. The Ingenieurs du Corps Royal des Ponts et Chaussées worked on the idea along with a commission appointed by the Académie des Sciences, and in 1786 came up with sensible recommendations to make a start on the River Vilaine. One of the nobles, speaking in the estates in December 1784, said what might have been repeated in any outlying part of Europe by someone who knew what waterways were achieving for others better placed.

> It is then solely to the lack of an inland transport system and of easily-reached outlets, the bad state of our main roads, still more the total decay of our cross-country roads, that we owe this general torpor, this agricultural apathy and finally our deprivation of all industry. Let us then dare, with an inland navigation, to make fruitful all the happy lands that it will pass through. The carriage of their produce, the example of activity that the continuing sight of a great trade will give them, by creating happiness and removing misery, will encourage them to work, and will tempt industry from neighbouring districts nearer and nearer.(1)

In the north, too, canal links were being built round Lille, around Dunkirk, and—most important—the Canal de Neuffossé, joining the River Lys (a tributary of the Escaut) at Aire with the River Aa at St Omer. It had a 5-rise lock staircase at Fontinettes near Arques. This canal, one of several designed by Vauban in the seventeenth century, was built under military administration and completed in 1774.

Meanwhile, rights to run *coches* and *diligences d'eau* (passenger-boats that sometimes also carried mail) on rivers and canals were farmed out, the fares charged and services provided being regulated by royal decree. They ran, for instance, from Paris to Auxerre, Sens, Nogent, Montereau, Briare, Montargis, Nemours, Moret, Melun and Corbeil in 1784. Previously, in 1778, a service had also been started on the Seine between Paris and Rouen.

Arthur Young, perambulating France in 1787, 1788 and 1789, was surprised to find how good many of the roads were. Himself in favour of waterways, he approved the Briare, Neuffossé, and Midi Canals. 'A noble work!' he says of the last-named at Béziers. 'The port is broad enough for four large vessels to lie abreast; the greatest of them carries from 90 to 100 tons. Many of them were at the quay, some in motion, and every sign of an animated business. This is the best sight I have seen in France.'(2) He notes traffic on the Oise, the corn trade down the Aisne to Paris and Rouen, food passing down the Garonne to Bordeaux and, on the Loire at Orléans, the many locally built craft that were being loaded with timber, brandy, wine and other cargoes for transport down the Loire to Nantes, where they would be broken up and sold with the cargo. We shall find such disposable boats in use elsewhere, as in early days on the Ohio and Mississippi. When the monarchy fell in 1793 and old France died, 1770 km of artificial canals had been authorised and 1000 km built and opened, as well as many river improvements made by individuals or groups given by the king the right to levy tolls.

Prussia builds eastwards

In Prussia the linking of rivers by canals had gone on. In 1774 the Bromberg* Canal had been completed from the Netze (Noteć), a navigable tributary of the Oder, to the Vistula (Wista). Whereupon, using the Warthe (Warta), Oder and Spree to the west, the Vistula and its linking waterways to the east, it became possible to take a boat from Berlin to the limits of Prussia beyond Königsberg and, indeed, into Russia. Southwards, the Klodnitz Canal from the Oder just below Cosel to the Silesian coalfield beyond Gleiwitz (Gliwice) was begun in 1788, to be completed in 1806.

Progress in Russia

The canal age also reached Russia, and resulted in three important long-distance systems—the Mariinski, Berezina and Oginski.

The Mariinski route linked the Volga with the Neva and the capital, St Petersburg, by way of a river, lake and canal line that left the Volga at Rybinsk, and for 433 km followed its tributary the Sheksna as far as Lake Bielo. Across this, it entered the River Kovja, and by two locks ascended it to the entrance of the Mariinski Canal. This took it across a watershed to the River Vytegra which runs into Lake Onega, whence Lake Ladoga could be reached by the River Svir—and so to the Neva and St Petersburg. This important route, deeper, shorter and more easily navigable than that by Vychene-Volotski, was decided upon in 1802, opened in 1810 and steadily

*Now Bydgoszcz in Poland.

became busier. An alternative route between the Volga near Rybinsk and Lake Ladoga, by way of the River Mologa and the Tikhvine Canal, which only took craft of 32 tonnes, was opened in 1811. Less successful, it nevertheless carried some 6000 barges a year, many of which took valuable goods to and from Nijni Novgorod fair.

As the Mariinski and Tikhvine lines joined the Neva and Baltic to the Volga and Caspian, so links were needed between Baltic and the Dnepr river leading to the Black Sea. Three resulted, to the Western Dvina, Niemen and later to the Bug (see p. 88), of which the first, the Berezina, was preferred as the only one entirely within Russian territory. This river-lake-canal route ran from the Berezina river, a tributary of the Dnepr, through a canalised section some 157km long built between 1797 and 1804 to the Western Dvina, and so to the port of Riga. Its northern end to the Baltic was mainly used to carry timber.

The Oginski route from the Dnepr ran along its tributary the Pripet and smaller rivers to a short canal section, then down the Niemen and across the Prussian frontier to Memel. Michael Oginski of Vilna began the 48km canal section in 1768, the government finishing it between 1799 and 1804. In our period traffic to Memel was mainly grain and timber, though later railway competition was to take much of the grain. In 1893, however, 20,000 timber-rafts (some also cargo-carrying) floated down the Niemen, and 2000 other craft carrying some 100,000 tonnes.

The stillness of China

Contemporaneously, in 1793–4, Lord Macartney was sent on an embassy to China, and travelled for some days over the whole length of the Grand Canal, apart from a disused section at the Peking end. It was then a country whose government's policy was to discourage novelties, partly because they had no very high opinion of them, partly to discourage 'their subjects as much as possible from entertaining a higher opinion of foreigners than of themselves.'(3)

His canal observations showed that Chinese engineering had in fact gone backwards in the preceding centuries, helped by an amplitude of labour that made it unnecessary to save it. The canal had no pound-locks. Flash-locks took the form of

> flood-gates thrown across the canal wherever they were judged to be necessary, which was seldom the case so near as within a mile of each other... They consist merely of a few planks let down separately one upon another, by grooves cut into the sides of the two solid abutments... of

stone that project, one from each bank, leaving a space in the middle just wide enough to admit a passage for the largest vessels.... Some skill is required... in order to direct the barges through them without accident. For this purpose an immense oar projects from the bow of the vessel, by which one of the crew conducts her with the greatest nicety. Men are also stationed on each pier with fenders, made of skins stuffed with hair to prevent the effect of the vessels striking immediately against the stone...(4)

His chronicler, Sir George Staunton, says

The flood-gates are only opened at certain stated hours, when all the vessels collected near them in the interval, pass through them on paying a small toll... The loss of water occasioned by the opening of the flood-gates is not very considerable, the fall at each seldom being many inches ... (5)

He passes over embankments—

The earthern embankments in this part of the canal, were supported by retaining walls of coarse grey marble, cut into large blocks, and cemented together ... Those walls were about twelve feet in thickness; and the large stones on the top were bound together with clamps of iron.(6)

Barges and yachts usually had to lower their masts at bridges:

The height of the arches, and the steps upon them, prevented the passage of wheel-carriages over them, the number of which was small, and the use infrequent, as all heavy articles and most passengers, are carried upon the rivers and canals with which the country is intersected...(7)

Where one canal joined another at a different level, slipways were still in use, and must have been testing to travellers' nerves: '...in two instances, the travellers were launched in their barges with prodigious velocity down the stream'.(8) A wooden beam or gate held back the water of the upper canal. When it was swung aside, the boat plunged down the slipway, perhaps 3m vertical and at a slope of 45°, by its own gravity. To be hauled up, a boat

requires sometimes the assistance of near a hundred men, whose strength is applied by the means of bars fixed in one or more capstans placed on the abutments, on each side of the glacis. Round the capstans is a rope, of which the opposite extremity is passed round the vessels's stern ...(8)

The interruption that was Napoleon

Meanwhile, back in Europe, the world was changing. On 17 October 1784 the 15-year-old Napoleon Bonaparte changed at Nogent from the mail coach that had brought him from Champagne to a cheap, four-horsed passenger barge that took him to Paris. He arrived on the afternoon of the 21st. Some fifteen years later, Napoleon had another contact with waterways. In March 1789, at Seurre on the Saône, a barge was being loaded with wheat bought by a businessman of nearby Verdun. The locals, fearing that their food was disappearing, stopped the barge leaving. Some soldiers were sent to restore order, Napoleon being among the officers.

Before that, however, the confusion that followed the Revolution of 1789 had stopped not only canal construction, but most maintenance as well, though a decree of 1790 did exempt rights to charge for canal improvements from the general abolition of old grants. When Napoleon came to power, it was urgent to develop France's—indeed, the Continent's—internal communications to compensate for Britain's command of the sea. Canal passenger services were started or restarted, and postal services speeded up. Because of the armies' need for money, finance for new canal-building had to be found by selling some of those existing, the ownership of which had partly or wholly passed to the state at the Revolution. Thus the Canals d'Orléans and du Loing were sold to one company, and the state's share of the Midi (the Riquet-Bonrepos family retained part) to another. Later, an imperial decree of 1810 amalgamated them. The cash raised went towards paying for the extension of the Canal de Franche-Comté from

Fig 9 A copper badge issued under Napoleon's Empire to the master of a postal boat on the Canal du Midi. Such boats, like present-day post-buses, also carried passengers.

Dôle to the Rhine as the Canal Napoléon*, the Breton canals, and the proposed Canal du Nord.

Part of the Canal Napoléon was built in his reign, and the Canal de Bourgogne extended further north. In Brittany, mainly because of the British naval blockade, Napoleon promoted with company help a considerable network of canals—the Canal de Nantes à Brest and its branch to Lorient, the Canal du Blavet, the Canal d'Ille-et-Rance from Dinan to Rennes, and the navigation of the Vilaine thence past the Nantes-Brest Canal to the sea. The Canal d'Ille-et-Rance (87 km) was started in 1804, the Canal de Nantes à Brest (374 km)—in fact, a series of canalised rivers with canal links—in 1806 with the River Aulne, and in 1811 the main line itself from Châteaulin, though in Napoleon's time work was only done on two locks and the first pound.

The Canal du Nord, to link the upper Rhine to the Meuse and then to the Schelde and the port of Antwerp, was one of Napoleon's international and imperial ideas. Planning for this great concept began in 1803 upon the section from Venlo on the Meuse (the site of the earlier attempt of 1626) to the Rhine nearly opposite Düsseldorf, 53 km long with 20 locks. Plans were approved in 1806, but no work done. However, on the corresponding section from the Schelde to Venlo, a portion was built, the Noordvaart, 34 km long with one lock. It is a still navigable branch of the Zuid-Willemsvaart.

A related through route was from the Seine to the Schelde. This required the completion of the Canal de St Quentin. Napoleon personally ruled out Laurent's line with its 13,682 m tunnel and in 1802 substituted Devicq's. The canal, linking Oise, Somme and Schelde, was completed in 1810 with its two tunnels of Bellenglise (Bony, Riqueval), 5670 m and de Lesdins (du Tronquoy), 1098 m, and allowed coal from the northern mines to reach Paris. More was later to reach the St Quentin Canal from a branch off the Schelde that was mainly built in Napoleon's reign, the Canal de Mons à Condé.

Between 1789 and 1813 only 200 km of canal had been opened for traffic. Yet they meant much to Napoleon. Even at Elba, when told of the poor communications between its two good harbours of Porto Ferraio and Longone, he had said: 'I will dig a canal.'(9)

Boat-lifts and inclined planes

Two problems sometimes arose in canal building—scarcity of water on summit-levels to supply flights of locks on each side, and how to

*It is now the Canal du Rhône au Rhin.

build a canal through hilly country without spending uneconomic sums on locks. From the 1780s answers to both were found in vertical boat-lifts and in inclined planes down which boats could be carried, usually on wheeled cradles. These used little water, and each could raise a boat several times the height of an ordinary lock.

A tiny vertical lift was built on the Churprinz, a works canal in the Freiberg mining district near Dresden, and used from 1789 to 1868. It is the first known, and remained in itself unique, for 2½ ton boats were raised 7m dry by hand tackle without using the later universal tank or caisson. In the 1790s several experimental lifts were built in England, but none were considered robust enough for regular use. Not until 1838, on the Grand Western Canal, did seven, using counterbalanced water-filled tanks each taking an 8-tonne, begin operational service. The biggest raised boats 12.8m vertically.

We have already noted early forms of inclined plane, with rails, as at Fusina, or slipways, perhaps fitted with rollers, as in the Low Countries. The first successful modern plane was that of 1788 on the Ketley Canal in England, with a vertical lift of 22.25m. It had double-railed tracks, up and down which boats were carried dry and level in cradles. Because loaded traffic was downwards, working was by counterbalance. It was soon followed by three others on the neighbouring Shropshire Canal, built like Ketley by the Reynolds family of the Coalbrookdale ironworks, and opened in 1791–2. These also were double-tracked and carried small boats level in cradles, but were steam-powered. The greatest had a vertical rise of 63m.

In 1795 a small plane clearly based on Ketley was opened at South Hadley (see p. 281) in Massachusetts, and in 1796 the American engineer Robert Fulton, who had studied the Coalbrookdale planes, published his book *Improvements in Canal Navigation.* A French translation was fathered in 1799 by the military and canal engineer François de Récicourt. It advocated small tub-boat canals for hilly country, using planes or lifts. His ideas were to be partly carried out later on England's Bude Canal, upon whose six railed planes ran boats fitted with wheels. One plane on this canal, Hobbacott Down, had at 68.5m the world's highest lift until the American Portage Railroad (see p. 307) started carrying canal boats. Fulton was in France from 1797 to 1801 and, his ideas seemingly having been taken up and improved by the French engineers Bossu and Solages, in 1801 a canal was authorised from the Rigole de Torcy to the Creusot ironworks. This was to have three locks, three inclined planes and three lifts. By 1806 one lift at Creusot basin, and one inclined plane, had been built on Fulton's lines. The incline, waterwheel powered, carried boats on cradles. This seems not to have saved as much water as expected, and further construction was stopped.

In 1808 Solages proposed a new type of inclined plane which, he thought, could replace eight locks to overcome the difference of land level of 22 m near. Etreux on the Canal de la Sambre à l'Oise, then being built by de Récicourt. It would use a caisson to carry boats floating longitudinally instead of, as previously, cradles to carry them dry. This caisson was to be of wood, carried on ten wheels running on cast-iron rails, and counterbalanced by two stone-and-water-filled 6-wheeled trucks. For rope cables 7 cm in diameter were to be used. This incline was not, however, built. Instead, Etreux got conventional locks. The first working caisson-carrying planes were to be those on the Chard Canal in England, opened in 1841–2.

Britain's canal age

Meanwhile in Britain canal construction had gone on throughout the French wars. Whereas Napoleon had three parallel motives for building waterways—national development, military efficiency, and imperial unity—broadly Britain had only one, to facilitate her industrial and agricultural revolution. The burst of construction that followed the Bridgewater Canal had added some 1331 km of waterway in England and Wales. Then in the early 1790s came a second burst,

Fig 10 Jessop's and Telford's iron trough at Pontcysyllte, on Britain's Ellesmere (now Llangollen) Canal, 307m long and 38.4m maximum height above the valley floor, opened in 1805.

the first years of which, from about 1789 to 1796, are called the canal mania. With William Jessop as its principal engineer, it had by 1810 added some 1931 km more. Thereafter the pace slowed as from 1830 competition from railways and steam locomotives increased, another 885 km being added up to 1850, to give England and Wales (Scotland and Ireland accounted for several hundred more) over 6437 km of navigable waterway.

Britain had also built some 58 km of canal tunnel—far more than at that time existed in the rest of the world. Her three longest were Standedge (4989 m as built), Strood (3608 m) and Sapperton (3490). She had also constructed such notable acqueducts as the iron troughs of Longdon-on-Tern (Reynolds and Telford) and Pontcysyllte (Jessop and Telford), and the superb masonry structures at the Lune (Rennie), Dundas (Rennie) and Marple (Outram). Of these Pontcysyllte, with its height of 38.4 m above the River Dee, is still the loftiest navigable canal aqueduct ever built. Finally, Jossop and Telford designed and built the great 8-rise lock staircase at Banavie on the Caledonian Canal. As France had been in the seventeenth century, so Britain was in the later eighteenth and earlier nineteenth centuries the leader in canal engineering, until her place was taken first by the United States, then by France, then Germany.

Europe after Napoleon

Soon after Napoleon's abdication, a thoughtful European concerned with water transport could have forecast a range of developments that would spring from not always compatible motives.

Over Europe as a whole, the need to free the great rivers from the man-made obstacles of a multiplicity of tolls would strike him, for rivers were water roads along which seaborne commerce could pass, and also the best means of internal bulk freight transport. Two other likely activities would be to improve river ports' access to the sea, and to continue linking rivers together by canals to form a waterway network, to which branches to important places could be added as necessary. He would realise also that after Napoleon's empire there would follow a rebuilding of nation states, and consequent efforts to improve their economies and bind together their peoples. Two recent inventions he would think likely to complicate the scene: first the steamboat which in one form could carry passengers with greater speed, regularity and comfort than any sailing or towed vessel, and in another enormously increase barge tonnages; second, the railway locomotive.

In the remainder of this chapter we shall glance at what happened in France, the Low Countries and Sweden; at the beginning of the next at the steam revolution.

The forty-four years between Napoleon's abdication in 1815 and the start of work on the Suez Canal in 1859 were to be for our ancestors a challenging yet confusing, hopeful yet worrying, time. At the end of this period, an observer might be forgiven for thinking inland water transport other than by ship canal to be obsolete: but he would have been wrong.

The Congress of Vienna, ending the Napoleonic period, introduced the future by its decision that rivers running through more than one state or forming a frontier between states—then a common enough situation in German-speaking areas—should be free to the citizens of all such states from the upper limit of navigation to the sea. This was a vital step in freeing rivers from a network of medieval tolls which had long hampered traffic development. In this decision lies the origins of the Rhine Commission, said to be Europe's oldest economic body, set up to bring together riverside and other states to regulate the river. Its powers modified by the Convention of Mannheim (1868) and the Treaty of Versailles (1919), it still functions effectively.

The end of the Napoleonic Wars saw traffic on the great rivers begin to increase. On the Rhine, for instance, trading and passenger craft, some carrying mails, were sailed where possible, otherwise towed—on the middle Rhine by horses, above Speyer by gangs of men.

Within the Germanic Confederation formed in 1815, and still more after the Zollverein or customs union of 1834, agreements were made to carry out the Congress decision on the Elbe (1821), Weser (1823), Rhine (1831) and Ems (1843). Thereafter other negotiations simplified or abolished the tangle of tolls, transit dues and local duties, as on the Rhine (1851, 1854, 1866 and 1868, when the last toll disappeared), Weser (1856), Elbe (1863 and 1870) and, under the Treaty of Paris (1856), the Danube. Between 1861 and 1863, notably, tolls were reduced on the Neckar, Main, Lippe, Ruhr and Ludwigs Canal, and abolished on the Moselle. Treaties of 1863 enacted free navigation for all on the Schelde.

After Napoleon: France

In France, planning had to begin again. Though road transport was active, the previous twenty years had left a pent-up demand for better water carriage. There were two schools of thought on techniques, two on finance. On the former, one group, foreshadowing the later Freycinet (see p. 146) wanted a network built to uniform dimensions to allow barges to move widely; another to copy the Britain of before the war and encourage economic growth by building small canals quickly. On finance, some looked to state money and direction, others to private enterprise.

Two canals were indeed built on the British system. The Canal du Berry had been begun in 1808, planned to run from Montluçon, near which were useful collieries, down the Cher valley to St Amand-Montrond and the valley of the Auron, and then by Bourges, Vierzon and Noyers-sur-Cher to the navigable Cher and so the Loire. Bottom width was to be 7 m and navigable depth 1.5 m, with locks of the then normal size of 34 m × 4 m. Ten such were built before work stopped. When it restarted in 1819, however, it was as a 'British' canal to reduce costs, with locks 27.75 m × 2.7 m (they took a special 60-tonne barge, the *berrichon*) and a bottom width of 5 m. At these dimensions, and with an additional branch to the Loire at Marseilles-les-Aubigny it was finished by 1839, 261 km long with 96 locks, and a fine 9-arched masonry aqueduct over the Cher near St Amand-Montrond. However, overdependent on the Montluçon coal traffic and too small to take barges from other canals, the Canal du Berry was never a success. Closed in 1955, it has now gained the cachet of a preservation society.

The second was the Canal de l'Ourcq. In 1802 it had been decided to divert water from the Ourcq river (a tributary of the Marne) into a canal that would provide Paris with a water supply, and would also be navigable from its intake to la Villette basin in Paris, where it would divide into two others, the Saint-Martin and Saint-Denis, to give access in different directions to the Seine. The city of Paris began to build it. Then, unfinished, the works were taken over by a company, the Canaux de Paris, formed in 1818. It was eventually completed in 1825 as 11 km of canalised river from Port-aux-Perches, and 97 km of small canal making 70–75 tonne craft*. In the early 1880s the-short-lived Beauval inclined plan was built between the then busy Canal de l'Ourcq and the River Marne (see p. 137).

Eventually, in 1820 Becquey, director of the Ponts et Chaussées, drew up a plan to provide France with a comprehensive system of navigations: he reported that 2760 km were under construction, but 10,800 km more were needed. This first effort at planning on a national scale resulted in laws of 1821 and 1822. They authorised a start on three new routes, the Canal des Ardennes (a heavily locked connection between the Aisne and the Meuse), the canalisation of the

*The Canal de l'Ourcq has 10 locks, the Saint-Martin 9, the Saint-Denis 7.

Map 4 The French canal system grows

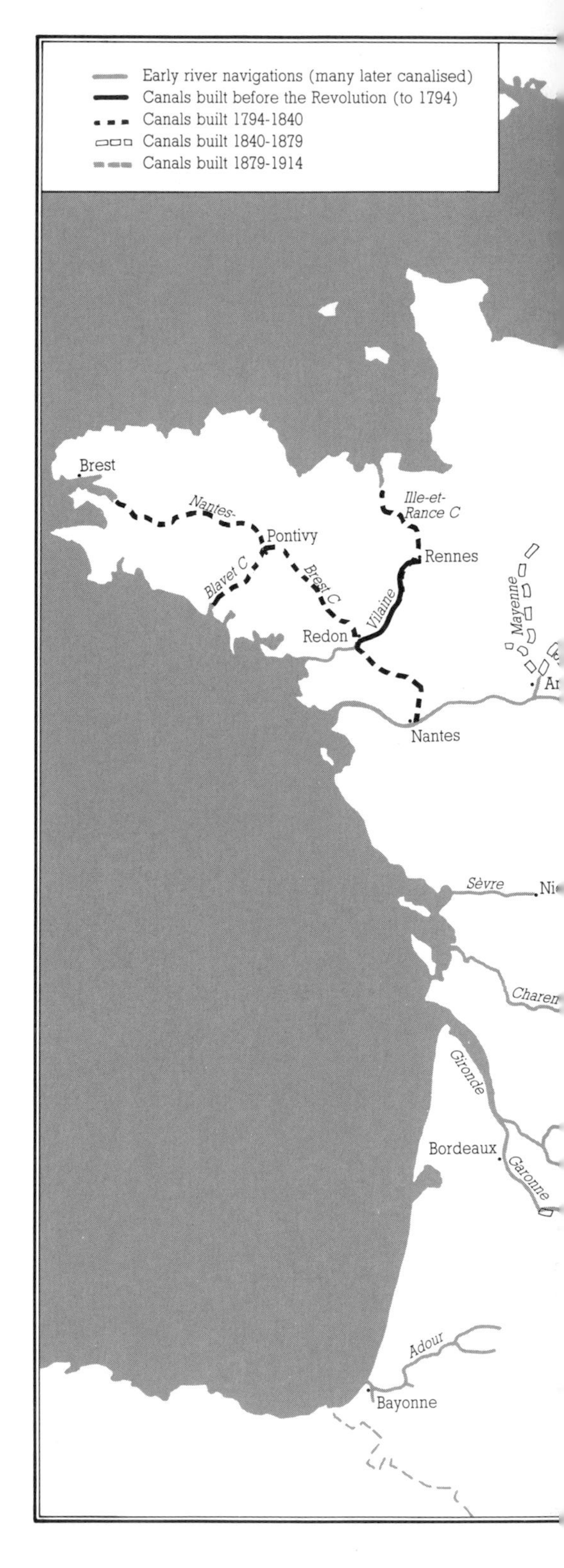

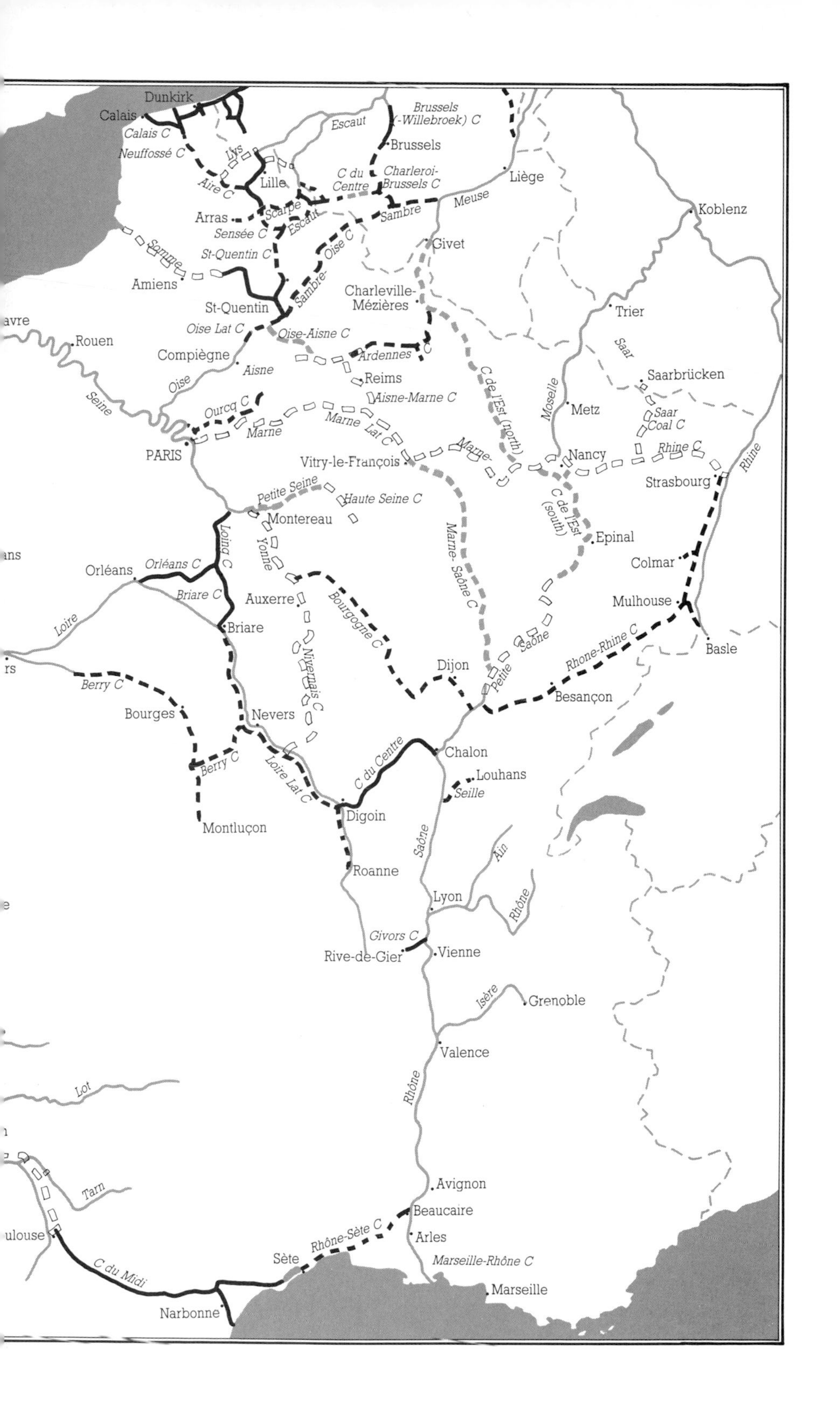

Dunkirk
Calais
Calais C
Neuffossé C
Lys
Aire C
Lille
Arras
Scarpe
Escaut
Sensée C
St-Quentin C
Somme
Amiens
St-Quentin
Oise Lat C
Rouen
Compiègne
Oise
Aisne
Seine
Escaut
Brussels (-Willebroek) C
Brussels
C du Centre
Charleroi-Brussels C
Liège
Meuse
Sambre
Oise C
Sambre-
Givet
Koblenz
Charleville-Mézières
Trier
Oise-Aisne C
Ardennes C
Reims
Aisne-Marne C
C de l'Est (north)
Moselle
Saar
Saarbrücken
Metz
Saar Coal C
Ourcq C
Marne
Marne Lat C
PARIS
Vitry-le-François
Marne-
Nancy
Rhine C
Rhine
Strasbourg
Petite Seine
Haute Seine C
Montereau
C de l'Est (south)
Epinal
Loing C
Yonne
Marne-Saône C
Colmar
Orléans
Orléans C
Briare C
Auxerre
Bourgogne C
Mulhouse
Loire
Briare
Petite Saône
Rhone-Rhine C
Basle
Nivernais C
Dijon
Berry C
Bourges
Besançon
Nevers
Chalon
Berry C
Loire Lat C
C du Centre
Louhans
Seille
Digoin
Montluçon
Saône
Ain
Roanne
Lyon
Rhône
Givors C
Rive-de-Gier
Vienne
Isère
Grenoble
Valence
Lot
Rhône
Tarn
Avignon
Beaucaire
Rhône-Sète C
Arles
Sète
Marseille-Rhône C
C du Midi
Marseille
Narbonne

Oise upwards to the St Quentin Canal junction (the main route between the Seine and the Escaut), and the Canal latéral à la Loire, to by-pass that river between the Canal du Centre and the Canal de Briare. The first two lines would much improve industrial links with north-east France and Belgium, the third the route between the north and south of France. Eight other part-built canals were to be completed, among them the former Canal Napoléon (now the Canal du Rhône au Rhin), Canal de Bourgogne, Canal du Nivernais (now to link the Loire Lateral and Bourgogne routes), and three in Brittany. The programme was commemorated by a medallion issued in 1822 which celebrated *Canales undique versum effosi* (Canals are cut in all directions), and showed Geometry measuring a map of France and giving orders for navigation works.

Five companies, under which the canals were grouped, were over ten years to raise the money for construction. Tolls were then to be fixed and remain unaltered; out of them maintenance costs were to be met and payments made to amortise the capital. The balance was to be shared between the lenders and the government. Additionally, during the Restoration (1815–30) period, some concessions were granted to private enterprise, notably for the Canal de la Deûle to link the Rivers Scarpe and Lys, and to improve the Canal de St Quentin.

As a result, between 1814 and 1830, 901 km of new canal were opened and others enlarged. Soon afterwards, in 1834, the watershed Nivernais Canal followed, though its summit-water feeder carried over a spectacular aqueduct was not to be ready until 1842*. However, because little attention had been given to river improvement—and many canals joined rivers—the resultant traffic tended to be local rather than long-distance.

After Napoleon: the Dutch-Belgian kingdom

The years between 1815 and 1830, during which Belgium and the Netherlands were united as a single state under the latter's King Willem I, are remarkable for three ship canals that much improved the sea access of Amsterdam, Ghent and Rotterdam.

During this time of peace the great ports of the Low Countries realised that, as sailing ships grew bigger, they either had to lose their trade or build canals big enough to take them. Britain, a maritime state, had already given a lead in building two large-dimension canals, though only one was to an inland port, the other being

*The 25 km Rigole d'Yonne brings feeder water from a dam at Pannesière across the 30.5 m high Montreuillon aqueduct into the canal's summit-level.

isthmian: the Gloucester & Berkeley, begun in 1793 to by-pass a difficult river navigation to Gloucester, and the Caledonian, partly naval in origin, begun in 1803. When the first of the Low Countries' canals, the Great North Holland, was started in 1819, neither had been completed: the Caledonian was opened in 1822, the Gloucester & Berkeley in 1827.

Though the port of Amsterdam had over 12 m depth of water, her access route through the Zuiderzee had been shallowing since the seventeenth century, thus progressively preventing larger ships from reaching it. Dredging had been begun in 1738, and by 1790, with as little as 3 m over the shoals, camels (water-filled pontoons that could be lashed on each side of a ship and then pumped dry to raise her in the water, so lessening her draught) were being used.

Therefore the Great North Holland Ship Canal was begun, 82 km long, 38 m wide at surface, and 5 m deep, with double towpaths, to run from the IJ at Amsterdam up the North Holland peninsula past Alkmaar to Den Helder, which had a good harbour and deep water close inshore in the Nieuwediep, making use of as many existing waterways as possible. Completion in 1825 with a tide-lock at each end and three intermediate locks enabled 800-tonne ships to reach the port. In 1847 it was said that 'The time which vessels require, to make the passage from Amsterdam to the Helder... varies... Fly-boats, with six relays of four horses, make it in ten hours. A large East Indiaman requires two, three, and four days, according to the wind.' (10) The enterprise was only partially successful. Some ships continued to use the Zuiderzee route; others just avoided Amsterdam. Later it was to be both shortened and enlarged to 2000-tonne standard before being replaced in 1876 as Amsterdam's main artery to the sea by the North Sea Ship Canal.

Its construction has interest, partly because, as it lay through alluvial soil, fascine bundles of willow were used when necessary as a basis for making the bed watertight, a technique more often used for river banks and embankments of polder; partly because of the eleven timber floating bridges, each of which had an extending section over the central portion of the canal carried on a boat hauled by a windlass and rope. Partly also because at the smaller of the two Willem locks, a form of the modern sector gate*—then called a fan gate—was used.

*'Each fan gate consists of two gates rigidly connected at an angle of 70°, one the actual gate, and the other a wing with a rather greater width, revolving on the heel-post into and out of a recess in the side wall formed like a quadrant. By causing water, admitted through sluices in the lock wall, to press upon one or the other side of the wing, the wing is turned to the front or back of the recess, closing or opening the gate even against a current of water'. (L.F. Vernon-Harcourt, *Rivers and Canals*, 1896, 577.)

A second Dutch-Belgian enterprise was the Ghent-Terneuzen Canal. Till then, small craft bound for Ghent went from the Schelde on the tide up an inlet to Sas van Gent, where a small canal ran for 21 km to the city. Begun by a company in 1824, and taken over by the state upon its completion in 1827, this new enterprise was intended for land drainage as well as navigation. Taking 1000-ton ships, the works included a lock at Terneuzen on the Schelde estuary about 61 km below Antwerp and a 30.8 km canal thence to Ghent. Thereupon, in 1828, Ghent achieved its first dock, 1700 m × 60 m, which also linked the new canal to the Ghent-Bruges canal, the upper Schelde and the Lys. After Belgium and the Netherlands were divided in 1830, the canal began in the first and ended in the second: nevertheless, other expansions followed, and it served both countries well until its rebuilding in the 1870s.

The Kanaal door Voorne was a third of great importance. Rotterdam's original approach to the sea, only 30 km away, had been by the Nieuwe Maas and the Brielse Gat (south-west of the present Nieuwe Waterweg), but by the beginning of the nineteenth century this had become semi-blocked owing to the increasing size of the island of Rozenburg. Large vessels had therefore to reach the sea by way of the Oude Maas and Dordste Kil, and then the Haringvliet or Volkerak—maybe 100 km, routes themselves made difficult by shoaling. Therefore in 1822 a canal from the Nieuwe Maas below Rotterdam to Hellevoetsluis at the seaward end of the Haringvliet was proposed. The project was approved, though with sailing ships mainly in mind. Willem I's wish that the needs of the new steamers should be considered was not accepted, it being thought that they could use the roundabout routes. The canal, with sea-locks at either end, was opened in 1830. Highly successful, it had 1650 transits in 1831, 3000 in 1847, 6600 in 1863 and 9600 in 1872. Thereafter it was replaced by the Nieuwe Waterweg.

Other notable inland waterways were also fruits of the Dutch-Belgian state: the Canal de Maastricht à Bois-le-Due* (now the Zuid-Willemsvaart) was intended to by-pass the Meuse between Maastricht and 's Hertogenbosch (where it joined the River Dieze and so the Bergsche Maas) by a line then in a united country, but soon to start at Maastricht in the Netherlands, cross into Belgium to run to beyond Bree, and then to cross back into the Netherlands. Begun in 1823, the Belgian section was completed in 1826. It was enlarged in 1864–70.

Child of a different border change was the Antoing-Pommeroeul

*'s Hertogenbosch.

Canal. The Mons-Condé had been begun by Napoleon in what was then France, but was finished in 1818 only after difficult negotiations between post-war France and the new Dutch-Belgian state. However, because of the dues charged by the French at the border near Condé, Willem I ordered the building of the Antoing-Pommeroeul Canal to link the Mons-Condé to the Schelde within Dutch-Belgian territory. It was opened in 1826. Lastly, the Sambre upwards of Namur to the French border was being improved by concessionnaires*, and what was to become an important coal-carrier, the Canal de Charleroi à Bruxelles, built by a company for 35-tonne barges. It was opened in 1832 by Leopold I, king now of the independent Belgians. Rebuilding to 300-tonne standard began in Leopold's own reign, being slowly achieved between 1854 and 1914.

Swedish achievement

Meanwhile in Sweden, a company was formed about 1790 to try once more to connect the Göta river below the Trollhätte cataract with the level of Lake Vänern. They asked Erik Nordwall (later Nordewall), born 1754, who had earlier engineered the Hjälmare and Strömsholms Canals, to check their plans for a by-pass canal. Boldly, he proposed instead to build a cut with 8 locks in two staircases, one 3-rise and one 5-rise, directly down through the rocky side of the river gorge. The company agreed, he moved to Trollhättan in 1795, and the through communication was opened on 14 August 1800.

Among his supporters had been young Count Baltzar von Platen, born in 1766, endowed with energy, perseverance and influence—all of which he was to need in abundance. Trollhättan locks opened, von Platen now determined to push forward the old idea of a waterway right across Sweden, this time based upon a survey made in 1781–4. King and administration having agreed, von Platen, who had heard of the somewhat similar trans-Scotland Caledonian Canal, then being built by Jessop and Telford, wrote to the latter to ask for a visit. Telford agreed, sailed from Leith on 28 July 1808, arrived at Gothenburg six days later, and spent about six weeks surveying and laying out a route and siting locks. He then wrote a report and left again for Britain on 1 October. He recommended certain changes to the 1781–4 line, along with more and shallower locks to take craft up to 32m × 7m, dimensions a little less than the Caledonian.

Construction began in 1810 under Samuel Bagge as engineer. With

*These were French (the French branch of Rothschild's and the Société Generale), since the Sambre was the principal route for coal moving to France.

troops made available by the government as navvies and drawings provided by Telford, work went steadily ahead, and the whole Göta Canal was opened on 24 September 1822, a month before the Caledonian. Sadly, von Platen had died nearly three years before.

Begun of course before the steamboat era, the Göta proved too small to be commercially successful, especially after railways added their competition, and tolls through the Sound were ended in 1857. However, visitors today can still pass through in passenger ships that run a summer service. The whole line is 558 km long from Gothenburg to Stockholm, of which 97 km is man-made. After the Göta river and the Trollhätte Canal comes Lake Vänern; then the Göta Canal proper from Sjötorp on the far side of Lake Vänern past Lakes Vättern and Roxen to Mem on the Baltic, whence ships pass through an archipelago into a fjord, whence the 3.2 km Södertälje Canal (opened 1819) leads into Lake Mälaren, and so to Stockholm. There are 65 locks in all, and a summit-level of 93 m. The 7-rise staircase at Berg can be compared with the 8-rise at Banavie on the Caledonian, and the earlier 8-rise (now 6-rise) at Fonsérannes (Béziers) on the Canal du Midi.

At Trollhättan in 1844 Nils Ericson cut a new flight of locks out of the granite, large enough to take craft able to pass the now completed Göta Canal. He used more locks with smaller rises than had Nordwall—eleven, in two staircases of four and one of three—which served until in 1916 they in turn were replaced by the present four large ship locks in a staircase of three and a singleton. A new lock of similar size has also been built at Lilla Edet on the Göta river.

Already, in our consideration of the Dutch-Belgian state and of Sweden, steamboats have been mentioned. Let them lead us to see how steam changed the scene.

Fig 11 Nordewall's 1800 locks at Trollhättan as shown in a lithograph by Eugène Ciceri, published in the 1840s.

4
TRANSFORMATION BY STEAM

For the fast-moving period that extends roughly from the 1820s to the opening of the Suez Canal in 1869, we have to hold in our imagination four worlds; those of roads, carrying passengers and goods; steamboats, mainly on the bigger rivers, as passenger-carriers and tugs; emergent railways, first to carry people, then merchandise, then some bulk traffics; and inland navigations, whose development, because they took so long to plan and build, tended to lag behind general transport changes. In different countries, all these worlds were appearing, changing, increasing, diminishing, at varying speeds, because of differences in geography, civilisation, fuel resources and economic development.

Changes in passenger transport

Good roads were well developed before our period began in economically advanced countries or in those with a strong central government which might need them for military reasons. Britain indeed then had a good turnpike system. By the 1820s, even in rural England, a country dweller had at his disposal an efficient personal transport system by public coach or hired carriage, and a postal service comparable to, though more expensive than, today's. Speed was the limitation, for even in a fast private carriage with good horses, 24 kph was rapid travelling. Long-distance freight transport was also well organised and efficient, but not for bulk goods, where over distances of more than a few miles it became prohibitively expensive. Hence a rapid growth of river, canal and coastal transport.

What was true of English roads applied even earlier to France: Arthur Young in 1787 wrote: 'If the French have not husbandry to show us, they have roads ...'(1). After the Napoleonic Wars, the canal passenger services on the *trekvaarten* revived, especially on the Friesian routes less liable to competition from road carriages. On the bigger waterways of what we now call Belgium, however, the large horse-drawn *trekschuits* working between important cities offered comfort and efficiency to English tourists able again to visit the

Continent, and so forced road vehicles into acting as their ancillaries. As departure time approached, coaches would set down passengers and baggage; a carriage might even be taken on board. On arrival, porters and cabs would be waiting to take passengers to hotels and homes.

In 1815 Robert Southey left Ostend for Bruges:

> The Trekschuit, being flat-bottomed, is much more roomy than would be supposed from its size. The best cabin is somewhat splendidly fitted up with cut crimson plush, a seat covered with the same material running round it. There are cabins, both at the head and stern, and in the middle a large apartment full of market-women returning from Ostend ... (2)

Writing rather later, an English engineer was to say:

> The boat on which I was conveyed between Ghent and Bruges ... was commodiously fitted up with separate state-rooms, containing one bath in each, and was, in other respects a most comfortable and agreeable conveyance. (3)

And yet road services were quickly to make the *trekschuits* seem slow and old-fashioned even though they were comfortable and regular. In 1829 a voyager wrote:

> ... when I am travelling I am eager to *feel* that I am travelling. The gliding, murmurless movement of a trekschuit bores me; I much prefer to sit in a galloping postal coach. (4)

Nevertheless, *trekschuit* fares cost only a quarter to a third of coach fares. One answer was to speed up the boats. The *vliegende schuit* or flying barge, using two horses, was introduced on the Amsterdam-Utrecht run in the 1820s, and in Britain fast services using light boats and galloping horses were provided on such canals as the Bridgewater, Lancaster or Forth & Clyde. Then, from 1830 when the Liverpool & Manchester was opened, came railways, to defeat canal and road passenger services alike. There was an intermediate moment when the older mode co-operated with the newer. In the mid-thirties passengers reaching Carlisle from Newcastle by rail transferred to a swift boat on the Carlisle Canal, thence to change again at Port Carlisle for the Liverpool steamer. And in 1839 the Rev F. E. Witts left his Gloucestershire rectory in his own carriages for Northleach, where the caught the London coach running via Oxford. At Maidenhead, however, to which the construction of the Great Western Railway to Bristol had then reached from London,

> The coach drove into the station yard where the passengers alighted, leaving their luggage with the coach which proceeds a little onwards and

> reaches the level of the railway by a road constructed for that purpose. There the carriages are placed each on a railway truck, ready to be hooked on to the train ... At the Paddington station the coach is met by a pair of horses, the passengers resume their seats, and the journey is continued ... (5).

But soon the horse was to admit defeat in the business of long-distance passenger-carrying, and give way to steam: steam on rails, but also on water.

Steam changes the waterway balance

We have seen building up in France, Germany and elsewhere, as in the British Isles, a single waterway network. Rivers, like canals, came to be given paths for towing by animals or men (bow-hauling); the same barge might be sailed, occasionally poled, sometimes towed, on a river and then towed on a connecting canal. Occasionally on a big river, as also for ferry-boat use, a powered pre-steam unit had appeared, the horse-machine (see p. 90).

Slowly, too, old-established flash-locks on larger rivers began to be replaced by pound-locks such as canals already had, but on lesser ones by new devices, movable dams, intermediate between the two. These were important in encouraging the canalisation of, eg the Saône, Lot, Marne, Seine and Aisne in France, and of rivers in Russia and elsewhere. The flash-lock provided a navigable opening in a fixed dam. Then in 1834 near Clamecy on the Yonne in France, the engineer Poirée put across the channel a number of metre-wide iron frames. Having a footbridge above, they were linked by a bar above and a sill below, against which rested a row of paddles (needles in France) that could be raised or lowered vertically from the bridge. The frames could then be lowered by chains, and the bridge removed in sections.

The Poirée needle-weir enabled the opening for boats to be enlarged according to the water-level. In good water, they had a clear run; in low, they worked through flash-lock equivalents. Developments followed, notably the type built by Chanoine on the upper Seine at Conflans in 1858. This used shutters and a framework supported by props which, when released, enabled it to fall flat to the river bed. The Chanoine weir was adopted in the United States (where it was called a wicket dam) on the Kanawha and Ohio. A third and more sophisticated type, the drum-weir, dating from 1857, was to spread from France especially to Germany.

Steamboats changed the waterway balance, for they developed mainly on the great rivers, which were thus helped to hold their own, always better and often successfully, against the coming railways. Therefore a separation began between great rivers and other waterways which was to widen and continue.

In the Old World, we may perhaps date paddle-wheeled steamboat development from William Symington's *Charlotte Dundas* of 1803, first useful steam-tug, and Henry Bell's *Comet*, which in 1812 began the first public passenger service in Europe between Glasgow, Greenock and Helensburgh on the Clyde in Scotland. Soon afterwards the *Margery*, which had in 1815 worked on the Thames, was transferred to French owners, renamed the *Elise*, in 1816 crossed the Channel, and started work on the Seine. The year before, Charles Baird had put a steam engine into a wooden barge, the *Elizabeth*, and begun a service on the Neva between St Petersburg and Kronshtadt. In Prussia the *Prinzessin Charlotte*, German-built but engined by Boulton & Watt, began work on the Elbe, Havel and Spree in November 1816, and in 1817 the *Die Weser*, also with a Boulton & Watt engine, started on that river. In 1822 Charles Manby took the *Aaron Manby*, first iron vessel to put to sea, direct from London to Paris, Aaron Manby having in 1819 started an ironworks and shipbuilding yard at Charenton near the French capital.

The steamboat's arrival had little effect on canal-building. The post-Napoleon period saw the revival and completion of many unfinished canals, as well as the filling in of gaps in the existing network. It saw, too, the planning and starting of new trunk lines, either lateral canals to by-pass difficult rivers or lakes, or much needed links over watersheds between existing river navigations.

Waterway development in Russia and Finland

Let us begin with Russia and Finland (the latter mostly Swedish till 1809, then ruled till the end of the century by the Czar, but as a separate state), broadly considering canal development first, then the great rivers and especially the Volga.

The early nineteenth century saw the building of new waterway lines using short canals to link rivers or lakes: the Duke Alexander of Württemberg system, the Dnepr-Bug route, the August river-lake-canal system, and the improvement of others.

The Duke Alexander river-lake-canal line, built between 1825 and 1828, linked the Volga by way of the Sheksna river, which left it at Rybinsk, with the Northern Dvina leading to the White Sea, 1750km in all from Rybinsk to Archangel. Once opened, however, the main traffic became timber moving from the White Sea by this and the

Mariinski line to St Petersburg. The line included five short canals and 11 locks.

The third Black Sea-Baltic waterway route to be developed was the Dnepr-Bug. It followed the Oginski line (see p. 67) up the Dnepr and Pripet before turning off up the Pina to Pinsk and a 76.5 km summit canal, whence it ran by a canalised tributary, the Monkhovets, to the Bug at Brest-Litovsk, and so to the Vistula for Danzig. Begun at the end of the eighteenth century, it was reconstructed between 1837 and 1843. Because ample water was available, the canal was exceptional in having only flash-locks set in Poirée dams, each altering the water-level by 1 m to 1.5 m. There were ten of these, with another 11 on the Monkhovets and one on the Bug between the former's junction with the latter and Brest-Litovsk. Because flash-lock navigation caused such slow changes of water-level, barges worked in groups of 20 or more. When railways came, canal traffic was an easy victim; by the end of the century it consisted largely of log-rafts.

The beginning of our period saw Lithuanian grain being barged down the Niemen to Prussia (merchandise returning) and Polish down the Vistula, also to Prussia. The August line from near Augustów to Grodno, 102 km long, with 18 masonry locks (they included one 3-rise and one 2-rise staircase), taking 180-tonne barges, was built between 1825 and 1837 to avoid Prussian tolls at the Vistula's mouth by taking cargoes instead to the Russian Baltic port of Ventspils. By completion, however, the political situation had changed, and the future of the August line was to be mainly as a timber carrier.

Special difficulties hindered Russian canal construction and operation: organising a sufficient work-force was one; incompatibility between a canal-sized barge and those used on connecting rivers or the rough water of a large lake was another. This latter difficulty required effort to be put into canals by-passing lakes that could have gone to connecting canals. A third was, of course, winter closures. Canals were not emptied. Masonry locks were covered over, water being allowed to run beneath through both sets of paddles, while timber locks were built independent of the ground on which they stood, and so free of expanded clay or earth.

In Finland where the first steamboat started work in 1833 the Grand Duke, Nicholas II, Czar of Russia, in 1844 authorised the Saimaa Canal—first link for the country's timber trade between its inland lakes and the sea at Viborg*. Completed in 1856, 60 km long

*Now in the USSR. The rebuilt canal (see p. 260) crosses the border.

with 28 locks,* a fall of 76m, and several miles of masonry construction across bog, it took 260-tonne barges and proved highly successful. The other notable Finnish canal and river line, the 64.5km Pielis-Elf, carried steamboats as well as other traffic from Joensuu northwards through 10 wooden locks to Lake Pielis.

No canal had yet been cut from the Volga to the Don (itself 40m higher at the convenient junction), but in 1852 Laurence Oliphant noted a tramroad from Duborka on the Volga to Kakalinskaia on the Don, which carried iron, timber, Siberian produce, and manufactured goods going to south Russia, and the other way, produce from Turkey and southern Europe. Goods were also carried in bullock-carts, along with knocked-down barges to be reassembled for the Don voyage and then sold for firewood; or complete barges moving on rollers, sometimes hauled overland by oxen. This variegated portage was later replaced by a railway from Tsaritsyn on the Volga to Kalach on the Don.

The Volga and its tributaries provided very long navigation routes uninterrupted by rapids or locks, but with beds mainly of loose sand with ever-shifting channels and sandbanks. The great lines were that of the Volga itself, navigable for 3564km, but most important from Astrakhan for 2688km through the grain-growing lands between Saratov and Kazan to Rybinsk, and its great tributary the Kama, joining it about 72km below Kazan. Navigable during summer months for 906km upwards to Perm, the Kama and its own tributaries were a main route for salt, timber and metal from the area between the Urals and the Volga, and also for Siberian and Oriental products. These rivers bore primitive but large barges that were often broken up at the end of the voyage: an example from the Kama that carried firewood and salt to the lower Volga was 64m × 21m, drew about 1.5m, and carried 1400 tonnes. Others drawing, and carrying, more could be used when water was plentiful.

The workhorse type carried up to 400 tonnes, often on the grain run north. It might sail in two months from Samara on the lower Volga to Rybinsk, where the cargo would be stored for the winter before being carried onwards, perhaps to St Petersburg. It usually had one huge square sail hung from a yard about as long as the vessel (maybe 11m to 32m), and could take a 60-man crew an hour to hoist. The mast on bigger craft might be part of a whole tree. Oliphant thought it

> a pleasing object, with its elaborately-carved triangular stern, and spacious deck that projected like a stage over each bow, on which a sort of

*Including as single locks three staircase pairs and five 3-risers.

wooden pedestal, also painted, and sometimes decorated with flags, was erected. Here, six or eight feet above the deck, stood a booted and sheep-skinned figure heaving upon the long tiller ...(6)

When not being sailed, these craft were worked downstream by large sweeps, upstream by poling, bow-hauling by the *burlaki* or Volga boatmen, or, above Samara, horse-towing. Below Samara warping (kedging) was also used, though it was expensive in manpower and only gained about 10km a day: two rowboats alternately dropped anchors ahead, each on some 1300m of warp, up to which the vessel was hauled. Then a French emigré, Jean-Baptiste Poidebard, in 1810 invented and built a horse-machine boat. It still used anchor boats, but hauling up on the warp was done by horses walking round a capstan. A large horse-machine in the 1840s would draw up to 2.15m

Fig 12 Volga barges seen at Kazan in the 1870s. Note that those being towed are also using their sails.

loaded. It could itself carry 800–900 tonnes and haul three others of some 900 tonnes, speed over long distances being some 12 km a day. In 1846 some 200 horse-machines were moving nearly half of all Volga Traffic. By then, however, capstan boats, similar to horse-machines but steam-powered and serviced by steam anchor boats, were being introduced. With a smaller draught and cheaper to run, some by the late 1850s could haul about 8000 tons. Such craft from the Kama, Sura or Oka could reach Rybinsk in one navigation year; those from lower down the Volga usually wintered at Nijni.

Here is the great annual six-weeks fair at Nijni Novgorod, itself largely a product of water transport, held on a peninsula where the Oka joined the Volga, as seen in 1852:

> Both rivers are covered with every conceivable shape and description of boat and barge; some from the distant Caspian, laden with ironware, Persian shawls, Georgian carpets, and Bukharian skins, or dried fruits: these vessels, of square, unwieldy construction, are elaborately painted and ornamented, and on their decks are erected curious wooden habitations, from the peaked roofs of which flutter gaudy flags, while out of the carved windows peep Eastern maidens. Others, rude and strongly built, have come down the Kama with Siberian iron or tea*; while the more civilised appearance of a few denotes their Western origin, and these have threaded their way from the shores of the Baltic, laden with the manufactured goods of Europe.(7).

Road-building, like the Moscow-Siberian highway, began to take traffic from the rivers. Then from 1817 came a few steamboats, but rapid growth was to follow a government decision in 1843 to allow unrestricted private enterprise in providing steam cargo or passenger services, subject only to safety and navigation regulations; and the starting in service of the *Sokol* in the same year. She had been designed on American western steamboat lines, longer, broader, drawing less water yet carrying more cargo than older types. Over the next twenty years American practice was to prove decisive.

The Volga Steamboat Co was founded as a freight-carrier in the same year of 1843 by an English merchant, Edward Cayley, with Russian colleagues. Progress was slow until experience had been gained, but by 1849 five steamboats and 14 barges were being worked, and 22,000 tonnes had been carried. Competent boat- and engine-builders were scarce, but in 1849 the Nijni Novgorod yard opened, and quickly became a major supplier of excellent products. Other companies followed: the Mercury in 1849, the Kama-Volga (an

*From Kiahta the tea was carried overland to the Kama.

expansion of the shipyard) in 1854. The opening in 1851 of the St Petersburg-Moscow Railway, which crossed the Volga at Tver, itself caused the foundation in 1854 of the Samolet company, carrying passengers and express goods only down to Rybinsk and Nijni Novgorod, and also the Pol'za company, primarily for freight-carrying not only down to Rybinsk, but also along the Sheksna river, the first section of the Mariinski water line.

American-type side- and sternwheelers usually had cargo, crew and machinery in the hold (machinery on deck in the sternwheelers) third class on the sheltered main deck, first and second class in cabins on the upper deck, with the pilot house above. Some passenger steamers towed barges, but a class of tugs increased rapidly, some cargo-carrying, towing barges with 1000 to 4000 tonnes of cargo. By 1856 the Volga had 61 steamers other than capstan and anchor boats, though they only carried or hauled a small proportion of the total traffic; by 1884 the figure had become 491. The fuel was wood, reloaded every other day, with anthracite occasionally on the lower river. In the 1860s the residue of petroleum refining at Baku began to be used; by 1892, 747 out of 922 steamers consumed it.

All this activity required on the one hand a stronger and better designed barge able to withstand steam-towing and carrying more, on the other hand, channel improvement. The earliest improved barges on the Volga were copied from some first built on the Dnepr in 1845 by Prince Paskievitch. By the 1890s, such barges were carrying from 100 to over 2000 tonnes.

On the upper Volga some dredging and rock clearance had been done on the Tver-Rybinsk section before the first major improvement. This came in 1843, a reservoir, water from which could raise the river level at Tver (198km below) by 28cm for 64 days. In 1848 channel marking was begun, and in 1854 dredging of shallows. Then from Tver down to Rybinsk, systematic rock clearance and channel deeping was commenced, including the building of training walls to confine the current. From 1880 the marking of shallows and sandbanks was begun, together with supervision of vessels to prevent over loaded craft running aground and obstructing traffic.

Astrakhan on the lower Volga was the entrepôt for the trade of the Caspian Sea (in fact a lake). Steamer services worked thence, eg to Persian ports. From Baku some also worked across to Krasnovodsk, whence others ran to Russian and east or west Persian destinations.

Steamboats in India

Whereas in Europe passenger and fast freight steamboat services would slowly give way to railway competition, in the world of Asia and later of Africa, though some succumbed, others were to expand and flourish up to comparatively modern times. They were, for instance, a godsend to the development of the Indian sub-Continent, where in 1819 William Trickett built at Lucknow India's first, with an engine imported by Henry Jessop from the Butterley works in England. They provided comfort and regularity for passengers, and reliable and comparatively fast transport for goods, so supplementing the numerous cargo and passenger-carrying sailing country-boats of 30–75 tonnes that have always predominated there.

It was in 1834 that the East India Company began a service on the Ganges from Calcutta to Allahabad: 'side paddle steamers, each with a tall thin funnel and a single mast with two square sails ...'(8) they carried only cargo, but towed an accommodation flat also for passengers. Should the steamer ground, the flat was held off by a wooden beam secured on each vessel by a socket and pin. The journey took 22–28 days depending on river depths.

In 1844 the India General Steam Navigation Company was founded, and soon established a service to Allahabad. Coal supplies were a problem in the days of coal-eating simple engines: they had to be taken in sailing boats to upriver depots. By 1859 the company had ten steamers and ten barges. Hitherto barges had been towed, until the captain of the *Madras* experimented with lashing one on each side of his steamer. It proved practical, and the custom survived.

Profitable business brought competition from other companies, notably the Rivers Steam Navigation Company (which in 1889 came to a working agreement with the India General), and then from railways. A line reached Allahabad in 1864, and soon, as the great irrigation systems of northern India were begun, the Ganges began to lose water, silting and shallowing, so by 1874 the India General had closed the Allahabad route. The Ganges run was replaced, however, by one up the Brahmaputra to Assam: but that is a later story.

India had had some early small coastal canal and lock systems: the Nulland Canal (Calcutta), the Midnapore Coast Canal (Orissa), the sea-water Buckingham Canal (Madras), the Malabar backwaters (Kerala) and the Yhana/Calyah Creek (Bombay), though the obvious need for a Hooghly (Calcutta)-Ganges Canal, often considered, could not then be answered. Large-scale canal-building had to wait upon irrigation needs.

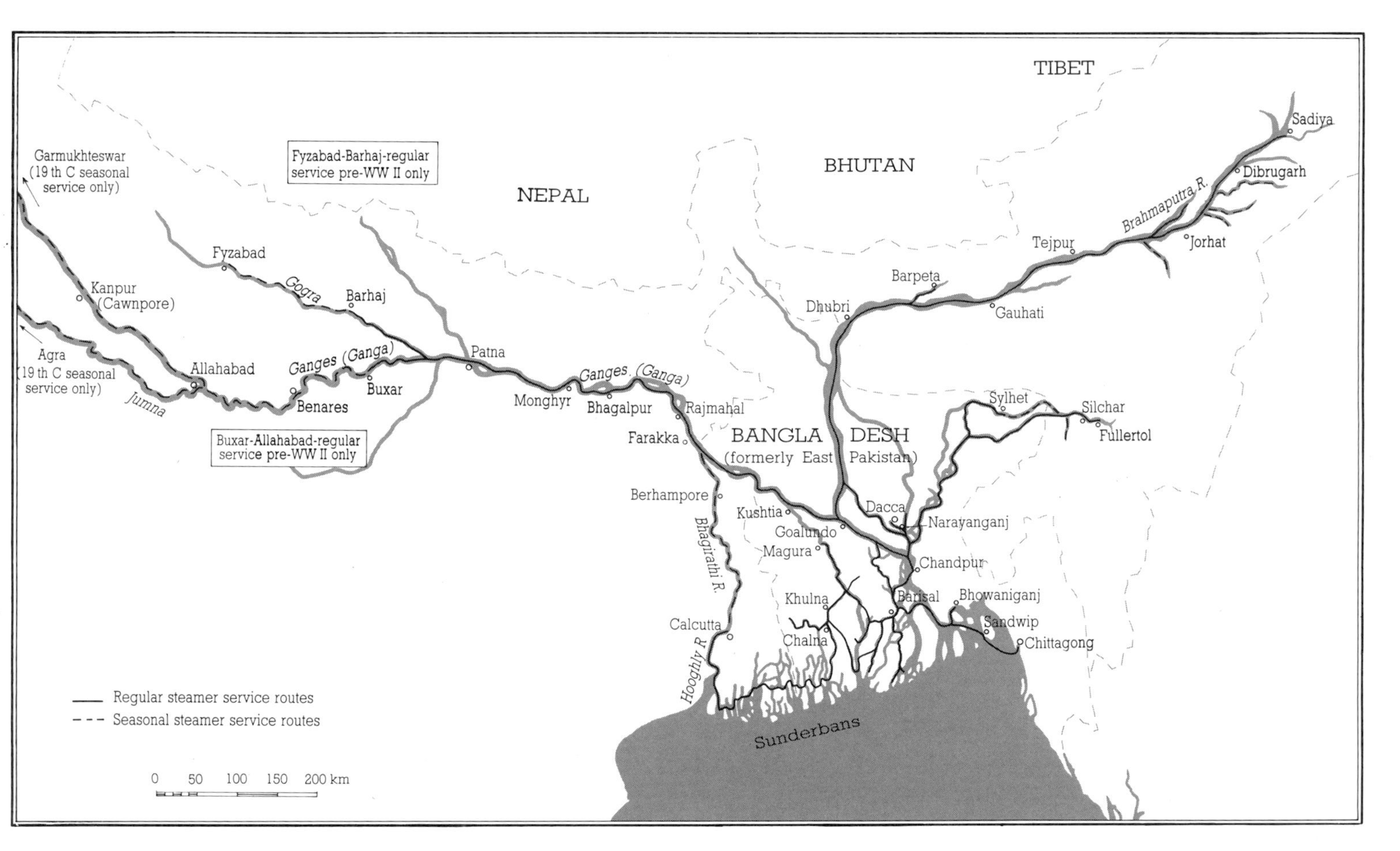

Map 5 Indian steamer routes in the early twentieth century

British Army and civil engineers gave nineteenth-century India the world's then greatest artificial irrigation system. Between 1817 and 1901, 5483km of main channel and 29,282km of distributaries were built in upper India alone, many of the main channels being also made navigable. Let the Ganges Canal represent the earlier part of the century. It was built by the British Army officer Proby (later Sir Proby) Cautley between 1842 and 1854, took water from the upper Ganges at Hurdwar, and ran to Cawnpore (Kanpur) lower down, with several branches. The Governor-General, Lord Ellenborough, in 1842 only agreed to work being begun provided that it should be primarily a navigation, and only secondarily an irrigation canal. He also wanted it brought down to Allahabad, not Cawnpore, since steamboats came up to the former. Because the canal had to be designed with a current*, it was as built little used for navigation, though provided with locks.

The Ganges Canal was the world's longest irrigation, and also navigation, canal, with 827km of dual-purpose waterway from the Ganges to Nanoon, including its two branches, one to Cawnpore, the other to Farrukhabad and a further 740km of branches for irrigation only. With its locks and towpaths, its deep cuttings, one over 20m, and 15-arched Solani aqueduct, itself 335m long and 7.3m above the river, approached by embankments 3265m long on one side and 830m on the other, the Ganges Canal was an extraordinary achievement. It remains, along with others, as much a memorial of the British in India as do their roads of the Romans in Europe.

Steamboats in Western Europe

Britain herself was an exception to what was happening all over the world. With a busy coasting trade and rivers small by Continental standards, steamboat services of any importance developed only in Yorkshire, along the Rivers Trent, Aire and Ouse.

Things were very different elsewhere in Europe. In France, for instance, steamers were soon working on many rivers, company after company being formed, many with British capital, a good deal of which was lost in over-hopeful expansion. Let us take as examples the easily navigable Saône and the difficult Rhône.

A service on the Saône between Chalon and Lyon began in 1827 with *La Chalonnaise* and *La Lyonnaise*—small wooden flat-boats 27.6m long by 5.28m wide, drawing 0.8m, powered by a 14–16hp beam

*The slope was 45.7cm to the mile.

engine, and carrying up to 100 passengers. They took nine hours down for the 136 km, and fifteen back, operating for six months of the year. So fast was progress however, that in March 1843

> ...nine boats belonging to three different companies are at this moment used...six of which provide the regular service between Chalon and Lyon. Consequently three boats go down and up every day, leaving at fixed hours, with a half-hour interval between each and in an order laid down by the municipal police, in order to avoid any racing between them. These boats stop at different wharves fixed by orders of the prefectures ... These same boats are inspected at Chalon and Lyon by authorities, who satisfy themselves that safety measures laid down in the rules have been properly fulfilled... It is remarkable that during fifteen years of steam navigation on the Saône, there has never been an accident caused by the use of this power-source.(9)

The finest boat was then the paddle-wheeler *Hirondelle No 5*, built in 1842. Her iron hull was divided into three, passenger accommodation being at each end and the engine-room amidships. The rear compartment included a dining saloon, lounge and ladies' room, 'le tout parfaitement meublé et décoré dans le dernier goût'. (10) Fitted with two 60 hp Watt engines built by Murray, Jackson & Fenton, 48 m × 5 m, and drawing 0.63 m loaded, she could carry 300 passengers, and perform the round trip from Chalon to Lyon and back in one day.

The first serious efforts to provide a service on the far more difficult Rhône began in 1827, and in 1829 the 50 hp *Pionnier* managed the upstream voyage with the help of capstans, oxen and horses. By 1843, 28 steamboats with a total of 2288 hp were available, though not all in active use, for passenger- and cargo-carrying between Lyon and Arles. The downstream from Lyon to Beaucaire now took 10½ hours, and the ascent 32 to 34 hours. Four then existing boats could do the upstream run in two days during the good water season, and that from Avignon to Lyon all the year round. Slowly the steamboats killed off the horse barges, on the Rhône as on many other rivers: no one had more vividly described the struggle between them than Bernard Clavel in his novel of the Rhône, *Le Seigneur du Fleuve* (translated as *Lord of the River*).

The first Rhine steamboat, the British *Defiance*, worked up to Köln in 1816, to be followed in 1817 by another, *Caledonia*, commanded by James Watt Jun, that reached Koblenz. In 1823 a Dutch steamboat company was formed to provide a passenger and express goods

Fig 13 A waybill of 3 April 1852 for a consignment from Marseille to Lyon, to go first by rail and then, probably from Avignon, by steamboat.

AGENCE GÉNÉRALE DES BATEAUX A VAPEUR

DU RHÔNE.

N°

TAFFE fils de Jacques
Rue Paradis, 17

Voiture F.
Frais
Assurance
Rembours^t
Total

NOTE D'EXPÉDITION

L'EXPÉDITEUR EST SEUL RESPONSABLE DE L'ENVOI DES PIÈCES DE DOUANE

Marques	Numéros	Poids	Colis
LN	1	40	1

LE VOITURIER NE RÉPOND PAS DE L'EMBALLAGE NI DES AVARIES, DÉCHETS OU MANQUE PROVENANT DU VICE D'EMBALLAGE.

Marseille, le trois Avril 1872

A la Garde de Dieu vous recevrez par Chemin de fer et par Bateau à vapeur les colis ci-après désignés dont l'assurance est couverte moyennant la prime de un quart pour cent sur une valeur déclarée de trente francs exigible avec le prix de voiture, et ce, aux clauses et conditions générales, stipulées dans les formules imprimées des polices adoptées actuellement par la réunion des Négociants assureurs de Marseille.

Un Ballot sac vide

marqués et numérotés comme en marge pesant brut quarante Kilos qu'ayant reçus en temps illimité et bien conditionnés devant le port d'une des compagnies de Bateaux à vapeur vous en payerez la voiture à raison de trois francs soixante & port plus cinquante centimes pour frais et rembourserez en outre cinq centimes d'assurance

Dires de Messieurs
Noilly Prat & Cie

à Messieurs Noilly Prat & Cie
à Lyon

NB. Le transporteur n'est pas responsable de la rupture des choses fragiles, du coulage des liquides, du déchet naturel à la marchandise ni du manque ou avaries provenant du vice d'emballage. Il n'aura également aucune retenue à éprouver pour détérioration d'emballage.

pour L'Agent général
Robinard

service between Rotterdam and Köln, and in 1825 another began a Mainz to Speyer run on the upper Rhine. In 1826 came the Preussisch-Rheinische concern of Köln, the ancestor of the present Köln-Düsseldorfer company. By 1827 *Concordia* and *Friedrich Wilhelm* were constantly working through the gorge between Köln and Mainz, in that year carrying 33,352 passengers. In 1832 the *Stadt Frankfurt* succeeded in getting to Basle in Switzerland.

Journey time reductions and the voyage's pleasantness compared with road travel made steamer services increasingly popular. Much later Cecil Torr was to write:

> A steamboat was nicer than a diligence; and that really was the reason why people were always going up the Rhine. It was much the easiest way of getting to Switzerland and Italy. Going by the Rhine in 1855, my father notes that it was the seventeenth time that he had gone that way, either up or down the stream...he never wished for anything more rapid than the steamboat on the Rhine, whereas I have found it tedious, and gone up by the train.(11)

On the freight side, steam-towing was introduced after 1825 between Rotterdam and Mannheim: passenger steamers sometimes towed barges, and steam paddle-tugs began to replace towing horses or gangs of men, and to haul strings of barges neither could ever have managed. Steamboat-building had hitherto been done in Britain or the Netherlands, but in 1836 Friedrich Harkort built a steamer at Duisburg that pioneered local steamship-building. By 1850 Ruhrort by Duisburg had become a centre of steamer construction.

Most lakes of any size in central Europe had steam passenger services calling at lakeside ports throughout the nineteenth century. Boat services still run on many today, geography often making them more convenient than road or rail alternatives. Freight was then carried in sailing or tug-hauled craft, as now in motor barges.

Wilhelm in 1824 was the first paddle-steamer to run a service on Lake Constance. Of Württemburg origin, she worked from Friedrichshafen to Rorschach across the lake in Switzerland. Next year a second Württemburg craft began to call at other ports, including Konstanz, and also went down the Rhine to Schaffhausen. In 1831 came *Leopold* and *Helvetia* from Baden, in 1838 *Ludwig* from Bavaria, and in 1850 *Schaffhausen* from Switzerland. All these and others were privately operated, and by 1846 competition had made necessary an agreement to share services. Between 1855 and 1863, however, they were taken over by the four states' railway administrations. Austria, the fifth lakeside country, did not begin a railway-administered

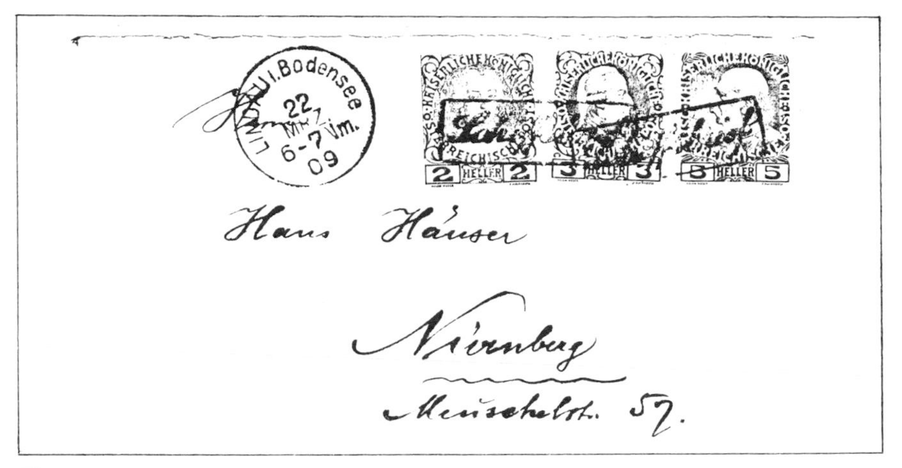

Fig 14 The international character of the Bodensee (Lake Constance) is shown by this envelope of 1909. The letter originated in Austria, and its Austro-Hungarian Imperial stamps are cancelled 'Schiffsbrief' (ship-letter). It was then postmarked just over the border in Lindau, Bavaria, for onforwarding to Nuremberg.

service till 1884 with *Kaiser Franz Josef I* from Bregenz.

In the early 1850s, the northern part of Lake Maggiore was in Switzerland, its western in the kingdom of Sardinia, its eastern in Austria (until 1859, when it became Sardinian or, as it was now called, Italian). At its southern end the Ticino river ran to join the Po at Pavia. At this time a service of Austrian government paddle-steamers served the eastern ports, and also those in Switzerland. In October 1852, however, the Austrian Lloyd company began a service across the Adriatic from Trieste to the Po estuary. There passengers, mail and freight changed to passenger steamers or freighters to be taken upriver to Pavia, and thence up the Ticino to Maggiore, where they worked up the Sardinian side. Both Austrian government and Lloyd lines on the lake ended with war between Austria and Sardinia in 1859. Afterwards, new Italian services began.

Lake Garda was an oddity. The northern tip ran into the Tyrol, and remained Austrian until 1914. Most of the rest was Austrian till 1859, then Italian, while the eastern side stayed Austrian until 1866 before becoming Italian. Here Austrian warships kept on the lake ran a passenger and mail service crewed by civilians from Austrian Lloyd, until the latter date.

Steamboats and improvements on the Danube

In the late 1820s and after, the rulers of the Austro-Hungarian Empire developed traffic on the Danube, not only within their own frontiers of Passau with Bavaria and Orsova with the Turks, but also below, as a means of increasing their influence in that decaying

Fig 15 European lake transport: (above) the Lake of Geneva (Lake Léman); (centre, left) Lake Como; (centre right) Lake Garda; (below) the Vierwaldstättersee (Lake Lucerne). *Uri* entered service in 1901, and is still steaming.

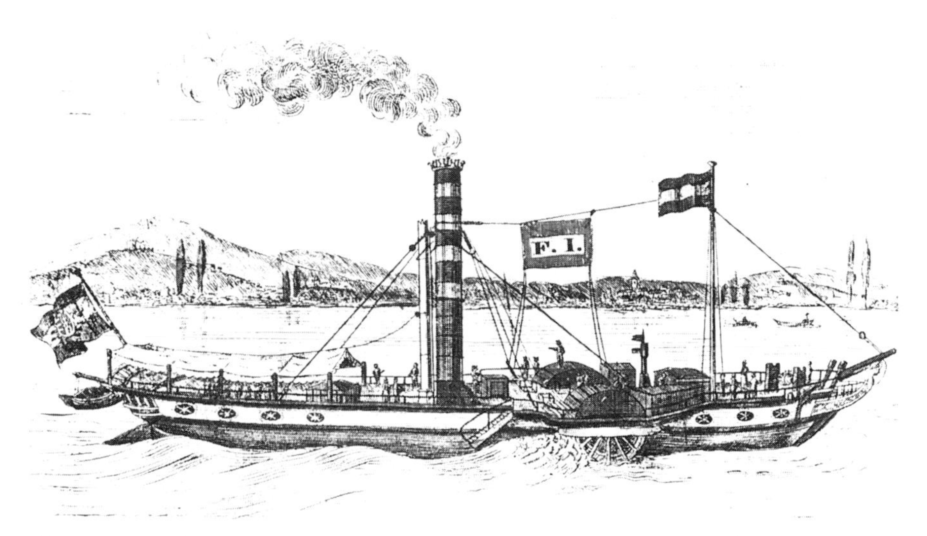

Fig 16 Steamboats develop: DDSG's *Franz I*, 1830, the first successful steamer on the Danube.

empire. They were helped by two Britons, John Andrews and Joseph Pritchard, who in 1829 were given a 15-year grant of exclusive trading on the Austro-Hungarian Danube, and then started the still existing Danube Steam Navigation Company (DDSG)*. It began with the 60 hp *Franz I*, which was soon working experimentally from Vienna to Pressburg (Bratislava) and Budapest. In 1834 the *Argo* passed the Iron Gates and reached Galatz. A year later the company's river ships met others of theirs operating from Black Sea and eastern Mediterranean ports. By 1837, when the *Maria Anna* first worked upwards from Vienna to Linz, the DDSG had seven river and three seagoing paddle-steamers. This year, the first in which river services were worked as regularly as weather and water conditions permitted, is celebrated as that of the company's true beginning. As new steamers were delivered, express boats calling only at important places supplemented local services. Later, steamers also ran up the River Tisa to Szeged, and the Sava to Bosanska Gradiška. As the century moved on, passenger traffic was slowly lost to railways, but freight remained. Before long the company had organised agencies along the river from Linz to Sulina near the Black Sea, with regular passenger and commercial services, while above Linz the Bavarian-Swabian shipping line was the developer. By 1857 the DDSG had 101 steamers and 359 tugs.

Laurence Oliphant, travelling on the DDSG's *Boreas* from Galatz to

*The Erste KK pr Donau Dampfschiffahrt Gesellschaft, or First Imperial Royal Privileged Danube Steam Navigation Company.

Orsova in 1852, did not think much of it. There were few private cabins, and most of the passengers lived in common:

> If it was difficult to sleep on board the Boreas, it was still more difficult to wash. The only basin supplied by the company was required at nine o'clock AM by the stewards, so that the crockery might be washed in it immediately after the passengers. It was therefore necessary for some of the party to begin their ablutions before daylight—as we had scarcely done fighting for the basin when we began to fight for places at the breakfast-table: and then the food was so atrocious... (12)

The lower Danube exhibited fierce rivalry between Turkey, in possession, and Russia, who coveted the Balkan lands. War broke out, and Russia dictated the treaty of Adrianople in 1829. Before it, the Turks had controlled the Sulina channel to the Black Sea, and by the primitive means of compelling every outgoing vessel to tow a large rake which stirred up the silt and enabled the current to carry it further out, had maintained some 5m of depth. The treaty took control from Turkey, but did not lay down responsibility to dredge. Silting increased. In 1840 Austria-Hungary agreed with Russia that the latter would dredge in exchange for a levy on ships. She did not do so, though she collected the levy, and was suspected of wanting the Sulina channel to silt, so that traffic would revert to the northern, Kilia channel, Russian-dominated, and once the deepest. Then came the 1856 treaty of Berlin and the problem's solution (see p. 167).

The lower Danube at Cernavodâ is only 64km from Constanta on the Black Sea. Because the voyage to Constantinople (Istanbul) could be much shortened were a canal to be cut from Cernavodă to Constanța, and because this would be free of Russian influence, a line was surveyed in 1837, but not enough water was found to supply the 50m-high summit. By 1840, therefore, the DDSG had organised an overland freight route between Cernavodă and Constanța. Then in 1856 the British-promoted Danube & Black Sea Railway was incorporated to build a line; it was opened in 1860. Thereafter passengers and freight from the DDSG could easily reach Constanța and Austrian Lloyd ships. The canal project was revived in 1949, and the canal opened in 1984 (see p. 234).

Early railways influence river improvement

Steamboats on larger rivers transformed passenger travel, and opened astonishing, though temporary, possibilities during a time when road travel was dying and railways were being born. In 1842 Thomas Waghorn published a guide to routes from Britain to ports whence

Alexandria could be reached, and so India, One, across France to Marseille, used road diligences except for Saône-Rhône steamers from Chalon to Avignon; a second, steamers all the way from Rotterdam to Basle, then road to Venice and ship to Trieste; a third, in the summer only, from Köln to Mainz by steamer, rail and road to Regensburg, and by steamer down the Danube to the Black Sea and so Constantinople.

But within thirty years of their success, long-distance and many intermediate services had conceded victory to the railways, though surviving for local services. For freight, steam-tugs, able to haul larger and larger barges as their power increased, and supplemented by a few self-propelled craft, needed improved river channels and navigable depths to be economical. Rail trucks might divert their better-paying freight, but coal, iron ore, grain and similar cargoes remained, and produced track and dock improvements.

On the Rhine the lower river had been given a navigable channel in Frederick the Great's time, but steamboats and increased freight traffic forced engineers to consider how to improve channels and equalise variable water-levels, so that services were not made seasonal by lack of water. Improvement went on from 1818. Notably, Frederick William III of the Prussia in 1832 had the Binger Loch made in the gorge—a channel 30 m wide through a rock barrier which lowered the river 3 m in 1000 m. Before, cargoes had often been portaged.

When railways came, the lower Rhine traffic was so well established that the first rail lines were built as feeders rather than competitors: Heidelberg-Mannheim (1840), Minden-Deutz (opposite Köln) (1847), Kaiserslautern-Ludwigshafen (1847). But competition soon followed. Freight traffic indeed increased on the lower and middle river, but contracted above, where fast currents and shallow channels made navigation difficult. Freight-carrying virtually ended to Basle after 1847 and to Kehl (opposite Strasbourg) after 1855, whereupon for half a century Mannheim-Ludwigshafen became the head of navigation, dock facilities being enlarged at both ports. Then from 1851 middle- and upper-river channel improvement began. Here are figures for goods on the Rhine in thousands of tonnes:

	Emmerich (Dutch frontier)	*Koblenz (Prussia)*	*Kaub (Hesse-Cassel)*	*Mainz (Hesse-Darmstadt)*	*Mannheim (Baden)*
1836	329.3	154.3	145.3	124.4	104.2
1840	381.8	291.5	274.7	215.4	91.7
1850	573.2	585.3	567.5	450.2	142.6
1860	1045.5	1063.7	936.1	791.8	221.4

The winding navigation channel of the Seine above Paris had a depth of under 0.89m for a third of the year, the banks were unprotected against floods, and towpaths in poor shape. Yet along it came timber-rafts and cargoes of coal, corn, wine and much else the capital needed. Under laws of 1837 and 1846, work was first done between Paris and Montereau. Locks (51.2m × 8.6m) and weirs were built, the channel dredged to give a minimum depth of one metre, and towpaths raised. At Paris itself, more barge quays and basins were built. Below, the river, though it had been improved from time to time, was obstructed by shoals and bridges, so that large gangs of men and horses had to be maintained at the worst obstacles. Then lock-building was started, together with dredging and improvement of towpaths and bridges. Below Rouen, embanking began, so that larger ships could reach Rouen from the sea.

There had in the 1840s been a flourishing passenger-steam boat service between Paris and Le Havre, while over 550,000 tons of goods pa were carried by water. Then came the Paris-Le Havre railway, the first section, Paris-Rouen, being opened in 1843, the remainder in 1847. With it went the long-distance passenger traffic, and some of the freight. Nevertheless, the programme continued. By the time of the Franco-Prussian war of 1870, eight locks had been built along the 240km between Paris and Rouen (the smallest 76m × 12m), giving an average depth of 1.7m except near Port Villez, where one lock was unfinished.

Have canals a future?

We have seen steamboats bring an exciting, though brief, new life to towns and villages on the larger rivers, and steam-tugs facilitating carrying in much larger barges. What now of the intricate water network made up of canals and smaller rivers with which many of them connected?

Canal-building has always been a slow business; slow to plan, start, and complete. When the railway age began in 1830, and became obvious to all by the mid-1840s, it caught the network just getting its second wind after its major upset by Napoleon. Whereupon Britain went one way; everyone else another.

In Britain, canals being almost entirely privately owned, new construction slowed down, and except to supplement what already existed, stopped. The new railways, also privately owned, bought over a quarter of the canal mileage, and were able to discourage traffic on it to their own benefit. Waterways thus tended to lose long-distance traffic to railways, and fall back upon local. Only in the north-east did the Aire & Calder and in Cheshire the Weaver, the one solidly based on the coal trade, the other on coal and salt, fight back with modernised weapons.

On the Continent, however, public ownership was widespread, though not universal, and public responsibility for waterway maintenance widely felt. The result was the creation of a railway system alongside that of the canals, though as rail went ahead, so canals tended to fall back.

In the Netherlands, *trekschuits* on main routes had fought back against road passenger carriages. But in 1839 the railway between Amsterdam and Haarlem had been opened. It was soon extended to Leiden and in 1847 to Rotterdam, third-class fares being set the same as the *trekschuits*—and less than elsewhere in Europe—for much faster services. Soon afterwards most *trekschuit* services ended, to continue for another two generations only for local carriage away from railway-served areas. The slow growth of a railway system in the Netherlands—a system relying heavily on passenger receipts—reflects, of course, not *trekschuit* opposition so much as the strong competitive position of the Dutch freight-carrying waterways—one they still hold today. On the Canal du Midi in France, the transit time taken by boats carrying passengers and mail, to compete with faster coach services before the railway age, had been reduced from 80 to 36 hours. Against trains even faster speeds up to 17 kph were obtained, but nevertheless by 1860 the mail boats had gone.

All over Europe, from France to Russia, Scandinavia to the Balkans, railways were now being built, at first tentatively, as feeders to waterways or using waterways as parts of a through line. For instance, in the early 1840s it was thought that a railway from Paris to Chalon could be extended by steamboat services from Chalon to Lyon and on to Avignon or Beaucaire, whence another railway would run to Marseille. But the obvious disadvantages of transhipment soon ended such schemes, and longer rail routes were planned and built.

However, the future of railways was then unknown and the extent of their spread not realised, whereas a half-built canal and river navigation system already existed in most countries, much of it publicly owned. If extended, notably by building lateral canals beside difficult rivers, and others over watersheds to link those already navigable, it would probably continue to carry bulk cargoes cheaply even though losing passengers and fast freight to rail. A presupposition, however, was central planning and control of construction and working. Let us take as examples what happened in Belgium, Prussia, Bavaria and finally France.

Canals in the Low Countries

The waterway history of independent Belgium between 1830 and the sixties is one of ambivalence. She and the Netherlands, when united, had developed on a large scale. But separation gave the Netherlands a long stretch of the Meuse, and the approaches to Antwerp on both sides of the Schelde. When railways came, the post-1830 Netherlands with good waterways and access to the sea, good roads and little heavy industry, was not enthusiastic about them. Even in 1865 she had only 865km of lines, most built by the state and leased to private companies. But Belgium, helped by her nearness to England and her situation as a European junction, led all Europe in turning to the new means of transport, and between 1835 and 1843 constructed a planned system.

Nevertheless, while her railways were being built, Belgium first reorganised and then for twenty years expanded her waterway system, after which she paused for thought. In 1830 most of her waterways had been rivers. The trunk was the Schelde upwards past Antwerp and on to Ghent and the French frontier, together with its principal tributaries the Dendre (to Ath), Lys, also to the French frontier, and Rupel with its tributaries the Nethe and Dyle. Further south the Meuse was navigable south of Maastricht to Namur and on to France. Canals were few. Most important was that from the Schelde at Ghent past Bruges to Ostend, and its branch the canal parallel to the coast past Nieuwepoort towards Dunkirk. Two old canals also ran from the Rupel, to Louvain and past Willebroek to Brussels.

She first made a major organisational change by starting to transfer her principal waterways from provincial adminstrations (and to a small extent from concessionaires) to the state. Of her total of 1618km in 1830, the state controlled only 156km, the provinces 1034km, the communes 111km, and concessionaires 317km. Twenty years later the state controlled 1180km, chiefly by the transfer of the Schelde, Lys, Meuse, Sambre, Dendre, Dyle, Demer, Rupel and Petite-Nethe, and of the Brussels-Charleroi, Ghent-Ostend and Mons-Condé Canals. More were to follow. Otherwise little was done in the decade except the enlargement of the waterway lines from Brussels and Louvain to the Rupel and so the Schelde.

The 1840s saw a lower Meuse by-pass built between Liège and the canal from Maastricht to Bois-le-Duc ('s Hertogenbosch). A new cross-country line was also begun, to run from the Nethe (and so the Schelde) to the Bois-le-Duc Canal near Bree, and so within Belgian territory give access north or south to the Meuse. This was completed in 1856 past Herenthals to Antwerp—it is now partly the Albert Canal and partly the Canal de Bocholt à Herenthals—with two branches, one opened north to Turnhout in 1846, the other south to Hasselt in 1858. Otherwise the 1850s saw only one notable improvement, a canal link between Lys at Courtrai and Schelde at Bossuit that was begun in 1858 and opened in 1863.

The 1850s also saw some river improvement. Early in the decade two dams were built on the Schelde, two also on the Dendre, and a lock each on the Dyle and Demer. Then the state paused, wondering whether railways would make further waterway improvements useless. Only on the Meuse above Liège did work continue throughout the decade, and there only on building wharves, barge harbours and cuts.

Canals in Germany and Sweden

In Prussia round Berlin, there was a foreshadowing of the great building era that followed the Franco-Prussian War of 1870. The coming of railways to the city and the consequent large increase in its population encouraged canal-cutting to help carry building materials, fuel and food. Notable was the Landwehr Canal (1845–50), 10.5 km long with two locks and a harbour at Schöneberg to by-pass a section

Fig 17 A masonry-walled cutting in the central section of Ludwigs Canal.

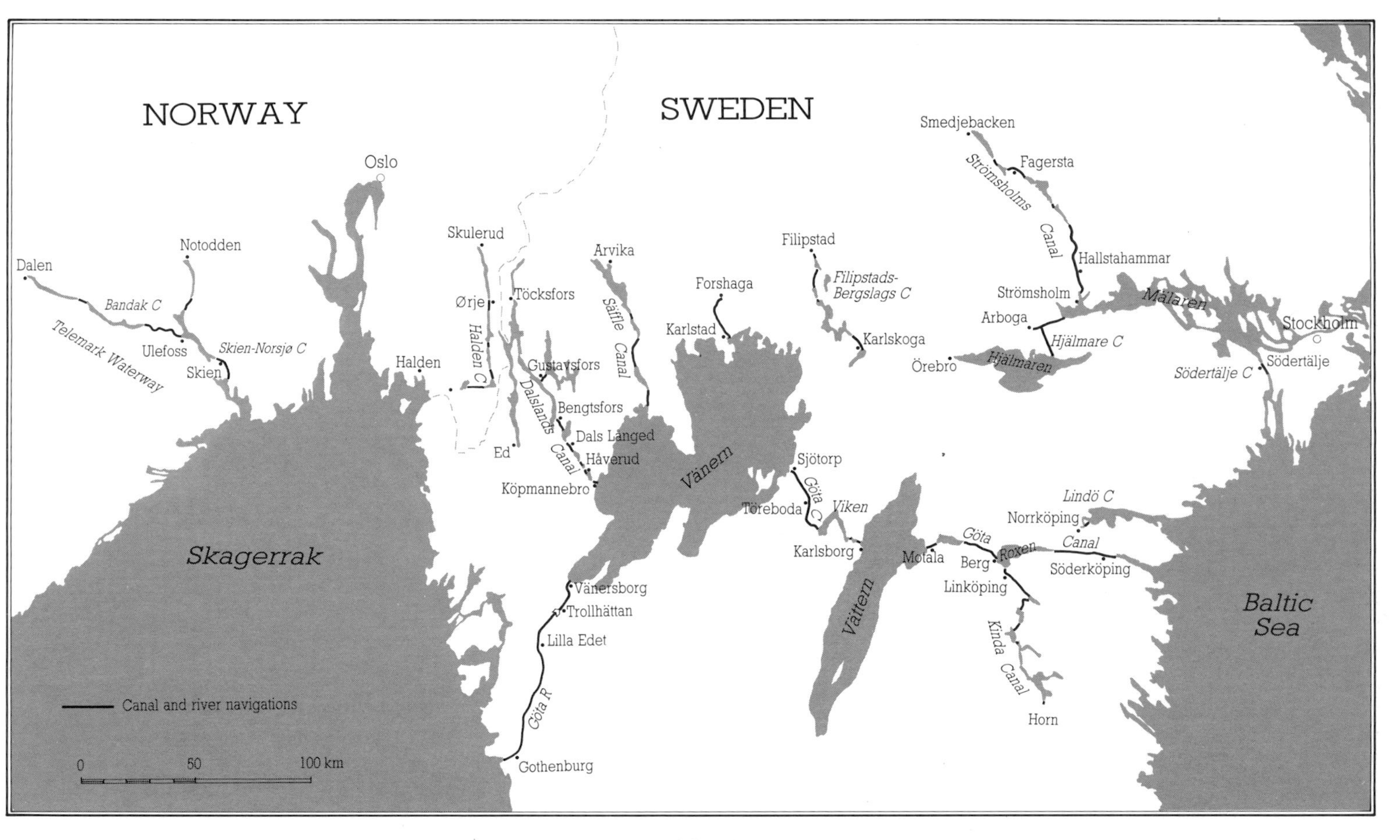

Map 6 The canals of Sweden and Norway

Fig 18 King Ludwig I of Bavaria's aqueduct over the narrow, rocky bed of the Schwarzach river, on Ludwigs Canal near Nuremberg, together with Herr Haas' modest restaurant, is commemorated on a 'Gruss' or greeting card.

of the Spree. It was followed by a linking branch back to the Spree, the Luisenstädtische Canal (1852) and the Berlin-Spandau Canal to join Lake Tegel (part of the Havel route) and the Spree (1859). Later, in 1872–5, the Charlottenburg Junction Canal linked the Landwehr and Berlin-Spandau lines to give a direct Oder-Spree to Havel waterway.

To the south King Ludwig I of Bavaria realised the ancient dream of connecting Main and Danube by building the 172 km waterway that bears his name, in my choice Europe's most delightful canal. It began at Bamberg in the Main's tributary, the Regnitz, and ran by Nuremberg to Neumarkt on the summit, some 110 m above the Regnitz at Bamberg and 457 m about sea-level. Thence it fell to Dietfurt to join the now canalised Altmuhl and continue to Kelheim, where a basin was provided for barges to await reasonable navigation conditions in the Danube.

The king had ordered a survey in 1828, and received it in 1832. Four years later Bavaria began construction with 6000 men, an early steam shovel being used on the deep cutting near Dörlbach. The line was opened between Bamberg and Nuremberg in May 1843, and throughout in August 1845. Operated by a company with 25 per cent state ownership, Ludwigs Canal took 100-tonne barges through its

100 locks and substantial aqueducts and cuttings. In the brief years before railway competition, traffic grew to a peak of 196,000 tonnes in 1850, though working was hampered by water shortage in dry seasons and winter frost closures. Also, in the canal's earlier years, the navigation of the Main was poor, while that of the upper Danube was often downright dangerous. Therefore through traffic was scarce, and the canal had to depend too much on local. Yet it was a brave attempt by an enterprising monarch.

The engineering at Håverud makes the Swedish Dalslands Canal noteworthy. Opened in 1868 from Lake Vänern to the Norwegian border to serve industry and carry timber, it was built by Nils Ericson. At Håverud he had to cross a steep river gorge and also provide four downhill locks. There were built on one side, the gorge then crossed slantwise by a single-span 32.3m-long iron aqueduct and the canal taken round a sharp turn to run beneath the cliff to the fourth lock.

Canals in France

In France under the July Monarchy (1830–48), while the Napoleonic canal programme was still being completed, notably with the 238-lock Canal de Nantes à Brest in 1836, the Corps des Ponts et Chaussées continued to put forward needed trunk waterways, while at the same time, under a general railway Act of 1842, planning national railway main lines from Paris to the frontiers worthy to set beside the existing royal roads and great canals present and to come. But perfectionism, along with an active debate on state versus private enterprise, left French railway expansion well behind Belgian and German.

The canal programme included a great lateral canal and two important river links. The Canal latéral à la Garonne was authorised in 1838 to by-pass that river's difficult navigation between Castets above Bordeaux and Toulouse, where it joined the Canal du Midi. It was 193km long with 53 locks, and included the Agen masonry aqueduct (539m) over the Garonne.

Much more commercially important, however, was the Canal de la Marne au Rhin, to link the centre and north of France to the Rhine at Strasbourg and the important industrial areas of Alsace-Lorraine. Work began in 1846, the canal and its accompanying railway being built together, in some cases running alongside each other, in curious resemblance to the earlier parallel building of the Chesapeake & Ohio Canal and the Baltimore & Ohio Railroad in America. In France, however, both were finished in 1853. The resultant canal, a difficult one to construct, was 310km long, with 171 locks 38.7m × 5.13m and two summit-levels, one 281 m and the other 267 m above sea-level. Originally 1.7 m deep, it was later given 2.2 m. Also in 1846 the shorter but important Canal de l'Aisne à la Marne was begun to the same dimensions, to give the new cross-France waterway a link with the north without going round by the Seine. It had one outstanding feature, Mont-de-Billy tunnel, 2300 m long.

On the side of rivers, a preliminary was settled in 1835, when an ordinance defined which river lengths were to be considered navigable and which 'flottable', ie raftable. A few French rivers had locks, but weirs were fixed, and did not allow regulation of the flow. When in 1834 Poirée built his first movable weir, it became possible to give a river a uniform navigation depth; thus a unified canal-river network became possible. Money was therefore allocated for river improvement, especially on the Saône, Lot, Marne, Seine and Aisne.

The period of the July Monarchy had added 2000 km to France's inland waterways, and had seen their reorganisation begun. Under the laws of 1821 and 1822, the tolls of canals built under them could not be altered. But, given railway competition, these were now too high, and industrialists allied themselves with boatmen to protest. The government was therefore compelled to reduce tolls by decree, after which it set up a national budget for public works, including waterways, and from 1845 also began to buy back the waterways from their owners. A law of 1853 authorised the repurchase of the Rhône au Rhin, Quatre Canaux and Bourgogne companies, and thereafter most canals reverted to state ownership, though one considerable exception was made when in 1858 the Midi railway company was allowed to lease the Midi Canal, suffering badly from rail competition, and also the Canal latéral à la Garonne. The fifties saw national energies absorbed by railway-building, but in 1862–6 the still-important Canal des Houillères de la Sarre (Saar Coal Canal) was cut from the Saar river for 63 km to Gondrexange on the Marne-Rhine Canal. Because it helped the sale of Prussian coal in France, its cost was shared between the two countries—perhaps the earliest such arrangement in Europe. By 1902 it was carrying over ½M tonnes a year.

Just as the Aisne-Marne Canal had been begun to link the east-west Marne-Rhine to the north-east, so the Canal de la Marne à la Saône in 1862 was to join it to the southern rivers; 224 km long, with a summit-level 340 m above sea-level, four reservoirs and two tunnels (one, Balesmes, 4820 m), it was not to be completed until 1907. Let us regard it as the last of the old or the first of the new, for during its building a man named Ferdinand de Lesseps had changed the waterway scene.

5
WHAT RAILWAYMEN WOULD HAVE THOUGHT IT?

In 1875 W. J. C. Moens took his steam yacht *Ytene* through the French and Belgian waterways from Le Havre to Calais by way of Paris, Brussels, Ghent and Bruges. He passed 143 locks, travelled 1243 km, yet in all his voyaging mentions no improvement works except the resumption of construction on one Seine lock interrupted by the Franco-Prussian war. In general, he found the waterways of France and ever more of Belgium busy, congested and old, and remarks that 'it must be the policy of Belgium to discourage the use of its ancient waterways and to throw all traffic possible on the railways'. (1) More than once he meets a horse-drawn *trekschuit*, and the only sight of modernity is chain-towing on the Seine and the approach waterway to Brussels. Almost arrived at that city, he passes the end of the coal-carrying canal from Charleroi: 'The locks' he says, 'are only nine feet wide and seventy feet long'—in other words, they took 35-tonne craft. Today, after its second enlargement, the canal takes 1350-tonne barges. What has happened since he wrote?

In fact, Moens was seeing waterways about to change radically. In this chapter we shall look at some broad trends, in the next more closely at events in the principal countries of Europe.

Ship canals become the rage

Several forces had been at work since the late 1850s, but before businessmen would invest, legislatures pass laws or governments vote money, men's imagination had to be caught by the idea of waterways as technologically advanced and nationally prestigious. De Lesseps did this, first for France and then for Europe, when he planned and built the Suez Canal. The first large isthmian ship canal, it used the most modern excavating machinery that could be made available, including huge steam bucket dredgers working from water and land, and it was being built by France. Britain had begun and powered the railway age and, though France had ably seconded her upon the Continent, was building railways widely beyond the seas. France would initiate an age of great isthmian canals, and in doing so, would get in a pleasantly public dig at Britain. For did not the Suez Canal supersede a railway, and had not the great British engineer Robert Stephenson, who had planned the railway, pronounced the canal impracticable?

Britain on the whole opposed the idea of a Suez Canal. William IV was an exception when he wrote to Lord Palmerston in 1833 that 'It has been shown lately that Communication by Canal and by Steam Vessels might be established from Cairo to Suez and the Red Sea' (2), and that therefore it was Britain's interest, as ruler of India, to be the closest ally of Egypt's ruler. The king was thinking of the steamship route for mails and passengers between Britain and India via Alexandria and the Red Sea which was then being pioneered by Thomas Waghorn. The overland route began at Alexandria, using horse-hauled passenger- and luggage-boats on the old Mamoudieh Canal, rebuilt in 1819–20. Some 17 km long, it joined the Nile at Atfeh. Thence Waghorn's line ran to Cairo, using sailing boats or steamers, and on across the desert to Suez by camels or horse-drawn vans. The route was partially replaced when the Alexandria to Cairo railway was opened in 1856, and wholly with its extension to Suez in 1859.

After de Lesseps began to promote the idea, Britain disliked it even more, politically because of the increase in French influence it might bring, economically because she did not yet realise the cost savings in shipping to India and Australasia that would follow, and also technically, though Sir John Hawkshaw was an exception to the view of Stephenson and others.

The story has often been told. Enough to say that de Lesseps, his imagination caught by Napoleon's dream of a canal across the isthmus, had got to know Prince Saïd of Egypt while French consul at Cairo in the 1830s. De Lesseps had gone on to plan, think and dream, with little hope of fulfilment until, retired from foreign service, he heard in 1854 that Saïd had come to the throne. Two months later, in Egypt, the prince and he came to an agreement. No engineer, but a superb organiser, diplomat and public relations man, de Lesseps and 20,000 workmen thereafter created a 164 km lockless ship canal from Port Saïd to Suez*. Its triumphant opening by the Empress Eugénie of the French on 17 November 1869 in the imperial yacht *Aigle*

*A side-product was the Sweet Water Canal, built to provide fresh water to Suez, Ismailia and Port Saïd, but also locked, and providing a barge route from Cairo to the Canal at Ismailia.

Fig 19 Dredges and elevators at work during the last stages of making the Suez Canal. Most of the canal was cut by such machines, used as soon as enough water could be introduced along the line, though portions, as at the Chalouf rock cutting near Suez, had to be entirely hand cut.

crowned his achievement, even though the ship that followed, the British P & O liner *Delta*, could be taken as a portent. Men, especially Frenchmen, saw their canal as epitomising the age. Had not the seemingly impossible been performed? Was their canal not an achievement so gigantic that it rivalled the building of the Great Pyramid, yet one that superadded the ideal of benefiting all mankind by a waterway freely open to the ships of every nation?

Two years later, France had been humiliated by Prussia and the empress was a refugee in, of all countries, Britain. But her canal triumph persisted, though success came slowly. For two years, Suez Canal receipts were less than operating expenses, and the company could not pay the statutory minimum of 5 per cent on its shares. With difficulty, bonds were placed to provide interim finance. However, in 1872 receipts exceeded costs, and thereafter rose quickly. In 1874 Britain bought Saïd's shareholding and with it the influence upon the canal company that earlier she could have had for the subscribing; so she turned from hostility to co-operation in seeking the canal's success. Profitability was helped also by decision in 1883 to put on the board seven directors representing user interests.

Depth and bottom width were increased, and night passages began, made possible by using compressed-gas illuminated buoys, and providing portable generating equipment that could be put on transitting craft to give a searchlight ahead and smaller lights on each side and astern. First through at night was the P & O's *Carthage* in 1886. By 1909, fifty years after opening, bottom width, originally 22 m, was 30 m, there being also 23 passing places additional to Lake Timsah and the Salt Lakes. Curve radius had been greatly improved, and depth, once 8 m, increased to 10 m. So the 486 ships totalling less than $\frac{1}{2}$M tonnes that transitted the canal in 1870 had by 1908 increased to 3795 and 19,110,981 tonnes. Yet receipts, £374,680 in 1870, were only £1,200,000 in 1908, so greatly had charges been reduced. The share price, however, 205 francs in 1873, was over 6000 in 1912.

From the beginning of work to opening and then, after its first difficult years onwards to increasing success, the Suez Canal was a dominating influence. Ship canals proliferated. The Dutch authorised their North Sea Canal from Ijmuiden to Amsterdam in 1862 and New Waterway from Hook of Holland to Rotterdam in 1863, and opened them respectively in 1876 and 1872; and the Russians in 1875 began

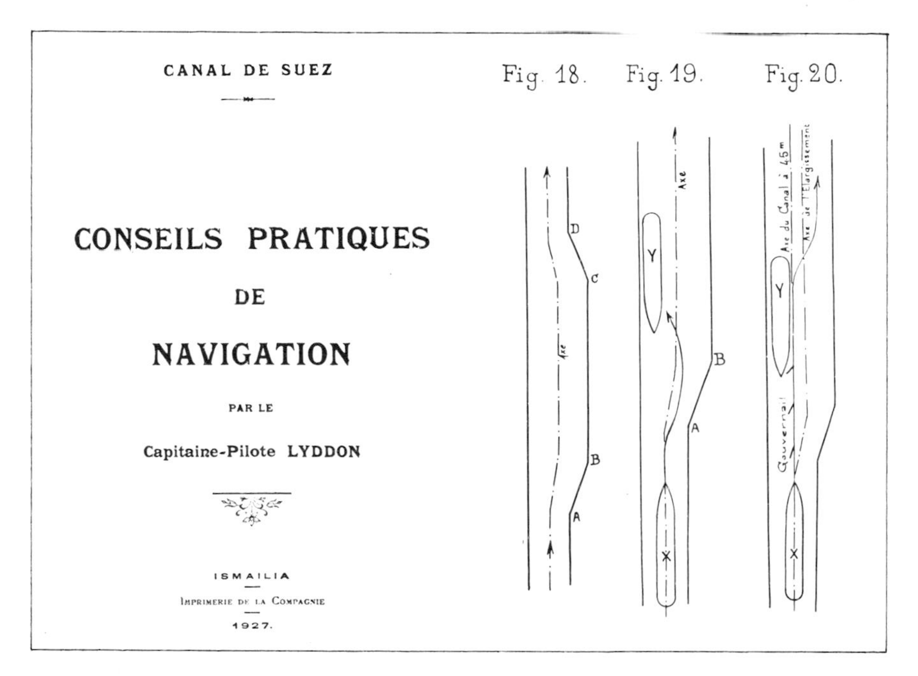

Fig 20 The work of a Suez Canal Company's pilot. In 1927 Capitaine-Pilote Lyddon wrote a handbook to pilotage, which was printed at the company's works at Ismailia. Its careful instructions and drawings included advice (right) upon how to use the channel's passing places. Later the company initiated the convoy system, by which a line of ships transited the canal one way, to be succeeded by another going the other, thus making passing places unnecessary. Finally, canal widening made passing practicable along most of its length.

40 ENGINEERING. [JAN. 26, 1894.

JOHN H. WILSON & CO., LTD.

BANKHALL LANE,

SANDHILLS, LIVERPOOL.

LONDON OFFICE - - - - 6, DELAHAY STREET, WESTMINSTER, S.W.

STEAM CRANE EXCAVATORS.

SIMPSON & PORTER'S PATENT.

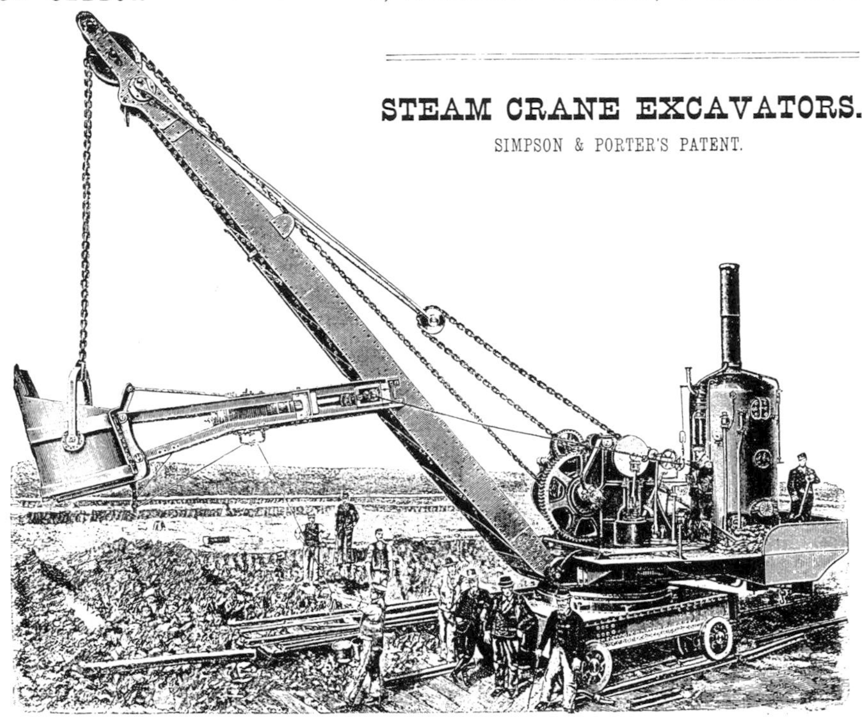

VIEW TAKEN ON MANCHESTER SHIP CANAL AT IRLAM.

We Supplied to the Contractors of the **MANCHESTER SHIP CANAL:**

150 LOCOMOTIVE STEAM CRANES.

7 STEAM CRANE EXCAVATORS.

12 Carey & Latham PATENT CONCRETE MACHINES.

Also for Mr. Walker's Buenos Ayres Contract:

50 STEAM CRANES.

6 MUD PUMP DREDGERS.

THEIR QUICK TRAVELLING CRANES WERE MUCH APPRECIATED.

Telegrams—"ENGINEERS, LIVERPOOL." "DRAQUE, LONDON."

Fig 21 Modern technology helps the contractors for the Manchester Ship Canal. An advertisement appears in the month in which Queen Victoria had opened the canal.

Fig 22 Emperor Wilhelm II's Kiel Canal is celebrated by the German equivalent of the Navy League. A lavish gold-embossed head of the Kaiser on the left is surrounded by a gold wreath. Above is the German eagle carrying a gold crown, and beside it the legend 'Our Future lies on the Water'. A warship passing through the canal and the naval gun, at bottom left, emphasises where that future is to lie.

their part ship canal, part dredged channel, the Morskoy Canal, between Kronshtadt and St Petersburg, and finished it in 1884. Meanwhile de Lesseps himself had begun to work and plan for a greater canal than Suez, at sea-level across the isthmus of Panama (see p. 353). Work on it began in 1880.

A fourth, smaller and shorter, but offering a considerable engineering challenge, was tackled in 1882 by a French-promoted company, one through the isthmus of Corinth. Avid for fame, French engineers in 1883 also surveyed for a maritime canal across the isthmus of Kra from the Bay of Bengal to the Gulf of Siam, though here no action followed. In 1887 the British began their Manchester Ship Canal and opened it, 57 km long, on 1 January 1894, with triple entrance locks, the largest 183 m × 24.4 m, and such notable engineering works as its Mersey embankments. Our British forebears were fascinated by the construction plant used, which included 75 steam excavators, 124 steam cranes, and seven land dredgers working from the banks, four of them French and three German Lübeckers. Designed to link the Lancashire cotton industry to its Indian and Chinese markets, the canal was given the same depth as Suez.

In 1887 also, Germany, soon to be led by her ambitious Emperor Wilhelm II, began the Kiel Canal that was opened in 1895. Its naval effect in enabling Germany easily to transfer warships from the Baltic to the North Sea was soon to influence the United States, with a two-seas fleet, to built her Panama Canal; it was to be opened a few days after war had begun in 1914. Two great canal events, therefore, exactly span our period and provide its background, Suez's opening in 1869, Panama's in 1914.

PIANC a symbol of engineering advance

Men now understood that great canals were achievements worthy of late Victorian and Edwardian promoters, engineers, investors and governments, not only because of their size and benefits, but because of the technological advance represented by their equipment. A feeling soon spread that waterways were desirable subjects for technological betterment, and advances were made over a wide front in the engineering of port and harbour installations, river and canal channel improvements, structures like locks, aqueducts, vertical lifts or inclined planes, and of craft and their means of traction or propulsion.

Symbolic of the time was the establishment of an international body where waterway engineers could meet. The Permanent International Association of Navigation Congresses (PIANC) met for the first time in Brussels in 1885. In 1985 it celebrated its centenary, again in Brussels.

Disenchantment with railways

If ship-canal building and the application of technology to waterways were positive influences of the period, then overreaching railway administrations were another. Railways had spread over Europe. But they had overplayed their hands by raising freight rates too fast and too high whenever they had a monopoly position. So doing, they drew public attention to the lower rates they charged where waterway competition existed. Investors, too, were becoming disenchanted as they found railway maintenance and renewal costs higher than they had expected, so leaving smaller margins for increased dividends. One result was to encourage a move by heavy industry towards sites near the coast or a river navigable from the sea, in order to avoid or reduce rail carriage. We shall examine others later.

We can therefore watch ship-canal building and port and estuary improvement interreacting with betterment of the lower and middle

Fig 23 A steamboat threads her way past an assortment of river craft at Bremen on the Weser in 1862.

reaches of rivers to bring ships further inland, and also improve ship-barge cargo interchange at ports. Up to 1866 at Hamburg, seagoing vessels had moored to dolphins, and then transferred their cargoes either to river craft to ascend the Elbe, or to barges which took them, by way of the many canals that intersected the city, to warehouses. Between the 1860s and 1880s, however, several basins and their accompanying warehouses were built. As a result, the number of Upper Elbe craft reaching Hamburg increased from an average of 6081 in 1871–80 to 15,978 in 1896, and tonnage carried from 492,000 to 2,080,000. Upstream tonnage rose from 434,000 to 2,969,000. On the lower Weser, the increasing size of ships prevented many from reaching the ancient tidal port of Bremen. Here from 1887 a triple policy was followed, of increasing the river depth to Bremen itself from 2.75m to 5m, building new dock accommodation there, equipped with the latest in hydraulic cranes and warehouse lifts, and also developing another port, Bremerhaven, 68km further down the Weser, for bigger ships. In 1893, four ships drawing 5m, and 51 drawing 4.5m, had reached Bremen itself; in 1897 the figures were 122 and 511.

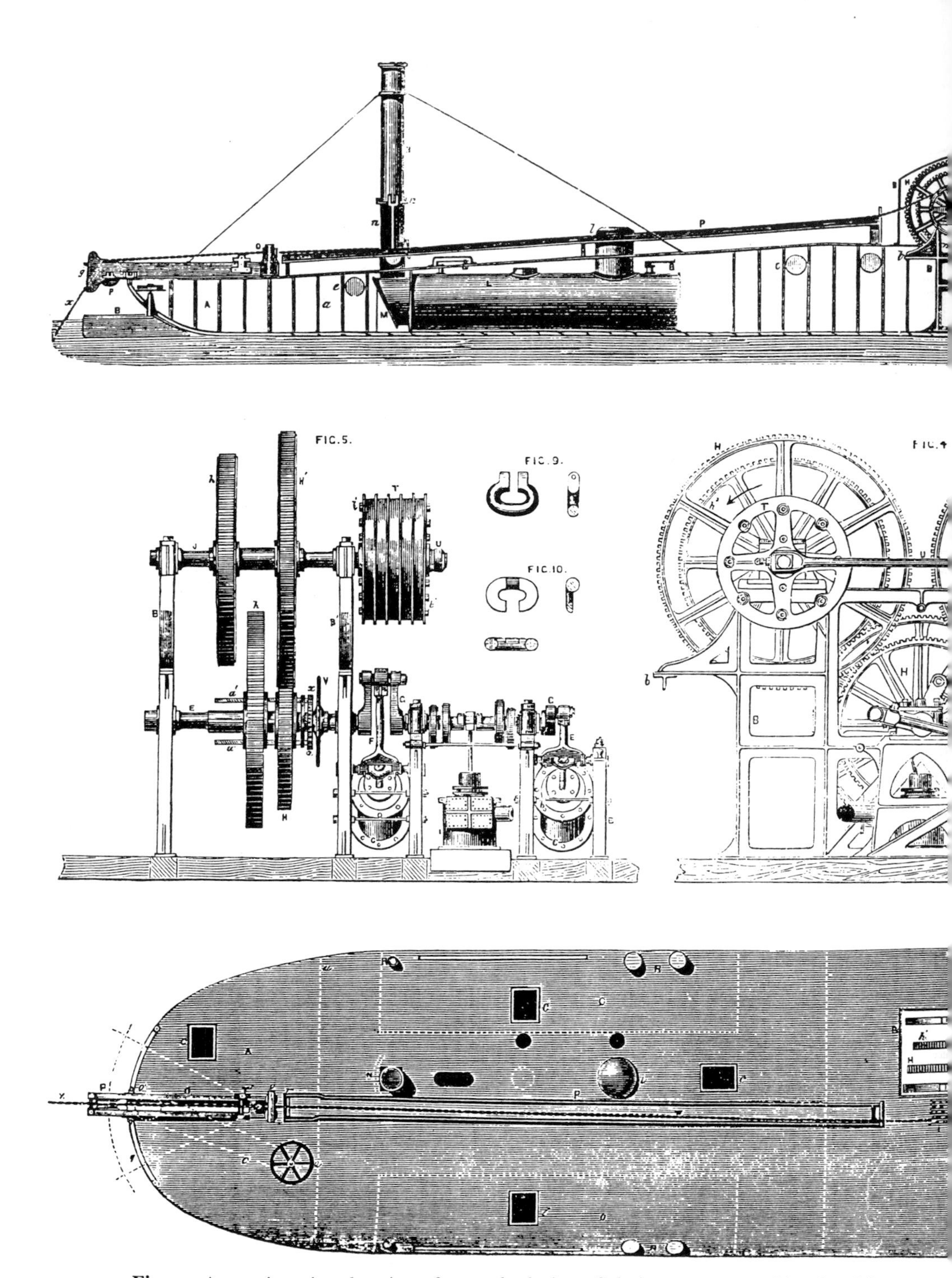

Fig 24 An engineering drawing of an early design of chain steamer, working in 1867 on the Seine. The chain can be seen coming on board at one end, passing round a series of power-driven pulleys in the middle, and then passing out on a slightly different alignment at the other.

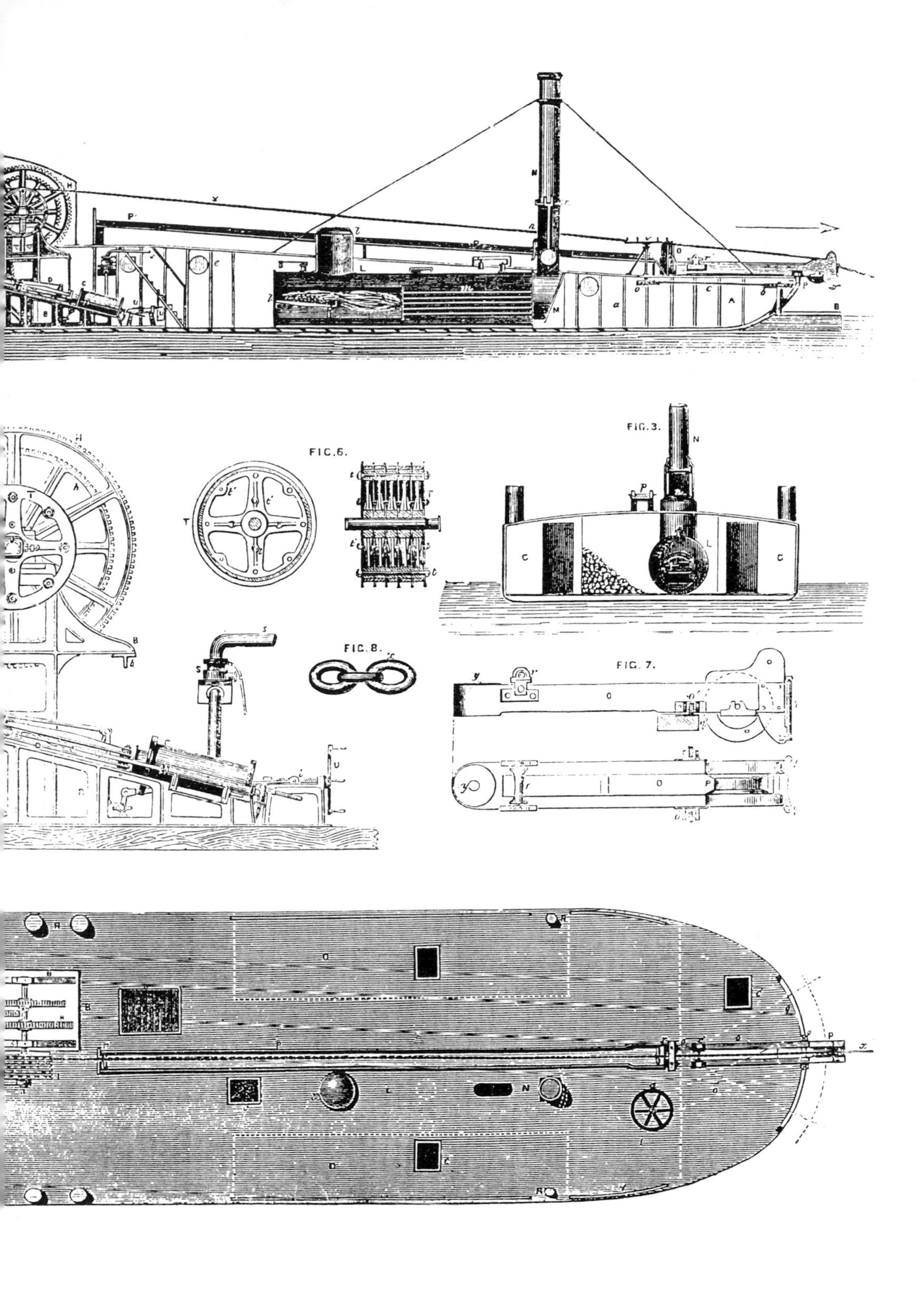
FIG. 6.
FIG. 3.
FIG. 8.
FIG. 7.

Fig 25 Chain or cable steamers could haul heavy loads, even against a current, as here on the upper Elbe in 1870.

Chain- and cable-haulage

To enable more and bigger barges to move further inland along the great rivers, channel improvements and often lock-building were required. Until they could be provided, the immediate need was for more powerful and efficient tugs. Steam-tugs had long been used on waterways: usually side paddlers or screw vessels where the water was deeper, because they were more manoeuvrable, sternwheelers where channels were shallower. Tugs could often haul three or four large barges, passenger steamboats sometimes one or two, but neither were powerful enough for more fast-running and less improved rivers. New techniques were wanted.

We have seen primitive tugs on the Volga being warped upstream first by men turning capstans and then by the horses introduced in 1810. A small horse-machine was indeed tried on the Saône in 1820, using 2.5 km of rope. But a broader future lay with steam-tugs hauling themselves along a chain laid on the river or canal bed. Such a tug had been proposed for Britain's Standedge tunnel in 1819, and worked for some years from 1824. Then in 1826 a steam chain-tug started working through Islington tunnel on London's Regent's Canal.

A submerged chain had also been tried on the Seine at Rouen in 1825. Later, in 1852–3 an experimental 2-mile chain was laid through Paris where the current ran fast under the bridges and horse-towing was difficult. It became a success, and a company was formed to lay a 71 km chain from Conflans at the mouth of the Oise to Paris. Because this section of the river was fast-running and yet regular in its flow, it suited chain-traction, and the line was a success. An extension chain in one 97 km length was then laid upwards to Montereau, a channel being cut for it in the sills of 12 shallow-lift locks, only used in times of low water. In 1867 the company was using nine towing steamers each able to haul 20–30 barges each carrying 100–250 tonnes, except when passing locks. Then tows had to be split into three or four sections.

In operation, the double-ended towing-steamer took in the chain over bow-rollers, passed it round a series of hauling drums, then back over the stern, pulleys at either end making sure that when entering a curve, the steamer was not pulled out of line by the chain, and that the latter was replaced in its original alignment. For this reason comparatively heavy chain (usually about 2.75cm in diameter) proved best; experiments with cables, otherwise often more efficient, showed that they tended to creep towards the inside of curves, so taking the tows out of the river channel.

Chain- and cable-towing now spread widely: it was used in Germany on the upper Rhine, Main to Bamberg, Neckar, Elbe (no less than 758km from near Hamburg to Melnik in Bohemia, with another 107km along its tributary the Saale), the Berlin waterways and the Oder. It was also seen in the Low Countries, and as far away as the Danube, the upper Volga and the Sheksna.

Because towing-steamers were powerful and economical, they could haul long trains of barges: thus the system was not really suited to rivers with locks. It worked best on those that were comparatively shallow and had an appreciable current. These conditions were ideally suited to upstream towing. But because the steamer could work only very slowly along the chain, an attempt to tow downstream could cause barges to overrun the steamer. In early days on the slower upper Seine, barge-trains were exchanged when two steamers met, but later steamers usually ran free down the chain, barges either being towed by a tug or descending without power.

Scattered technical improvements were made over the years, but as river channels were bettered, lock-building programmes introduced, and the power and efficiency of tugs and later of self-propelled craft increased, so chains were shortened, though some lengths survived WWI, as on a stretch of the Elbe and the upper Main, where in 1925 Negley Farson got a tow by a chain-steamer to Bamberg along with eight barges. He gives us glimpse of how it worked:

> It never stopped. When it wanted to deposit a barge, its steel arm, which guides the chain over the bow, was hauled to one side like a rudder; the captain slowed down the engines, rang a bell—and the barge cast off, using its headway to make shore and anchor. We were continually dropping and picking up barges in this fashion... Twice the *Kette-Boot*

> broke its chain. Wild racing aft to catch the last links slipping overboard; champing of the ratchet lock biting into the forward links; hammering as the sweating crew...riveted in a new link.

And at Bamberg:

> The *Kette-Boot* shed her chain, proceeded under water turbine...(3)

Such jet propulsion was quite often used. Macgregor had in the 1860s seen on the Meuse near Liège 'a fast steamer, the Seraing, propelled by water forced from its sides...'(4)

After it had been generally superseded, chain traction was retained for tunnel haulage, powered at first by steamers, later by electric power, as in Britain's Harecastle tunnel between 1914 and 1954. It is still used in France for the two tunnels on the St Quentin Canal and Pouilly on the Bourgogne. A variant was haulage by an endless moving cable, to which barges attached themselves. Such cables were installed in Britain's Braunston tunnel in 1870, and later in France's Foug tunnel on the Marne-Rhine Canal.

The River Rhône was its own law—shallow, its current rapid, its channels silting and changing. No continuous chain could be laid because of the instability of the bed, so other expedients were tried. One was a caterpillar tug, fitted on each side with endless moving chains which ran on rollers above, but below rested on the river bottom, being adjustable for depth. They did enable the tug to go upstream at 4km/h against a 11km/h current, but remained an experiment. Again, tugs were given a strong wheel with steel teeth, which enabled them to grip the river bed and so move upwards in the fashion of a rack railway. A third was to provide sections of cable about 20km long, which could be paid out by a descending tug and wound in when it returned upstream.

Channel improvement and lock-building on the larger rivers

Chain- and cable-towing was a very useful stop-gap. The real answer to greater and more efficient utilisation of large waterways, however, was to be channel improvment and lock-building. Then, with wider, deeper, less obstructed waterways whose current had been lessened by dams and locks, increasingly powerful tugs hauling bigger barges, or larger self-propelled craft, could be used. With the provision also of modern handling equipment at redesigned inland harbours, the stage would be set for expansion.

Canalisation moved upwards in scale as engineers became able to tackle, and governments willing to finance, bigger and bigger projects. In France lock-building on the Seine between Paris and Rouen was continued in the 1860s, while the Belgians were building twelve locks on the Meuse between Liège and Namur in 1862–70, and continuing with another eight from Namur to the French frontier in the 1870s.

River beds began to be bettered by dredging, easing or making cuts across bends, and by contracting the channel width, usually by building groynes projecting out into the current in order to force it towards the river's centre, which would thus be deepened. In the nineteenth century, the Rhine, Rhône, Elbe, Weser and Niemen were among rivers thus benefited. Some dredging programmes were very extensive: the Dutch state between 1868 and 1924 spent 13.5M guilders on dredging the Waal channel of the Rhine to maintain 3.5m of depth. The first steam bucket-ladder dredger was used in Sunderland harbour in 1797. By 1830 twenty or so had been built in Britain, and their use had spread to the Continent and North America. Then came steam shovels (dippers) and grabs (clamshells) and, about 1880, large-scale suction dredging, mainly by the Dutch.

The construction of very large and usually power-operated locks originated in docks, then was transferred to ship canals and rivers. The Seine led the way. Under authority of 1879, the river between Paris and Rouen was to be given 3.2m of depth, with new locks of the then very big dimensions of 152m×17m; further down, a Tancarville-Le Havre canal to by-pass the lower estuary was begun in 1880 (finished 1887), with 180m locks at each end, these latter having hydraulically-powered gate machinery, pumps, capstans and swing-bridges. By the 1890s, craft up to 1000 tonnes could navigate in normal conditions. As a result, in 1892 tonnages of 1¼ M to 3⅓ M were carried on different sections of the river.

On the Weaver in England, locks were built between 1874 and 1878 able to take 1000 tonnes at a locking. Water-powered by using two Pelton wheels to each gate, they were 70m × 13m, with 4.6m over the sills and a lift of 2.6m.

The Main followed. In early days and into the steamboat age, the river had been busy enough, in 1840 carrying 492,000 tons. Then railways came. River traffic dwindled, and with it went dredging, until in 1878 freight recorded at Frankfurt was down to 118,080 tons, and at one time there were only 31cm of water over the bar at the Rhine junction. By then a chain-towing company was at work. Nevertheless, in 1870 the authorities considered either a lateral canal or lock-building. Having decided upon the latter, of a design that would take the company's chain, work began upward from the Rhine

Fig 26 1848 at Marktbreit on the Main above Würzburg. Here, beside the old stone-pillared crane on the wharf, goods and passengers were often landed to complete their journey to Nuremberg by road.

in May 1883. In $3\frac{1}{2}$ years five large locks (each accompanied by a fish-ladder, weir and log-raft by-pass, one of the weir openings also being designed as a navigable by-pass during floods) were erected on the first 40km to Frankfurt, where a basin, quays and warehouses with rail connections were built. These locks were designed to take Rhine steamboats carrying up to 1300 tonnes. By the 1890s depth was 2.5m, and traffic to Frankfurt up to 709,000 tonnes in 1892. By 1925 thirteen locks, now able to take 1500-tonne craft, had improved the river up to Aschaffenburg, above which the chain-steamers still worked to Bamberg.

Basic to the building of big locks, vertical lifts (see p. 131) or inclined planes (see p. 136) on waterways, or to providing them with efficient loading and unloading equipment—cranes, warehouse lifts, conveyors and such—was efficient and easily available power. At first it was steam, but steam-engines were not suitable for, eg powering lock-gates or operating sets of cranes from a single source. Then from 1851 at Grimsby docks, England, came hydraulic power, usually to be provided through high-pressure power lines from a central steam-driven plant. This enabled all appliances in one dock area to be powered from one source; by the end of the century, indeed, hydraulic power could be transmitted economically for 24km or so from a central pumping station. Thus we see new harbour layouts at Bremen and Frankfurt-am-Main designed entirely for hydraulic operation. There were hydraulic cranes, capstans and rail-wagon traversers. Warehouses had hydraulic lifts, and grain elevators and conveyor belts were so powered, as were lock-gates and swingbridges. Lastly in our period came electricity. By the turn of the century engineers were planning all-electric layouts, or using electric motors to replace steam to drive hydraulic pumps. Sometimes they had to provide for their own electric generation, but more often they could now draw it from an external source.

Thus as the new century approached, a pattern of channel improvement, lock-building and mechanised port development began to show on the large rivers, to parallel the development of more powerful tugs. Some proposals, however, failed to take off. Much planning and some work was put into a Moselle canalisation with 32 locks between Metz and Koblenz, to link the coalfields on the Rhine with the iron ore of the upper Moselle. Fulfilment had to wait for our own times.

Push-towing

We must here notice another technique that in our period proved more useful in North America than in Europe, but is now being increasingly used. To push purpose-built barges with purpose-built push-tugs (US towboats) creates units which can be steered as if they were single vessels, and need no steersmen on the barges. Should the latter be immobilised, eg while waiting for cargoes, the push-tugs can be used elsewhere. Pushing seems to have been initiated on the Seine from 1822 in ITB (integrated tug-barge, see p. 391) form, in this case a paddlewheeled tug whose bow fitted into a slot in the stern of the single barge it was pushing. Later, similar units using propeller-driven tugs with rounded bows fitting barges with concave sterns were used between 1855 and 1875 on the Marne-Rhine Canal, and later on the Seine at Rouen. We find, too, in the Rouen tugs the precursors of modern tunnel-sterns, which enclose the screws to improve water flow past them and exclude air.

In Britain W. H. Bartholomew, engineer of the Aire & Calder Navigation, introduced what he thought of as a freight train on water, made up of 25-tonne compartment boats. Each had a vertical rod at its rounded head, which fitted into a slot in the rounded stern of the boat ahead. Thus the train could take a curve. Ahead of the train of some half-dozen coal-carrying boats was a bow compartment, and at the stern a push-tug, also with rounded bow. Steering could be by the tug's rudder, or by hauling in or slackening wire ropes running from the tug through fairleads on each barge to the bow compartment. The experiment was successful, and soon up to 12 bigger boats were being used: 30- and 35-tonne and later 40 tonners, while self-acting cylinders

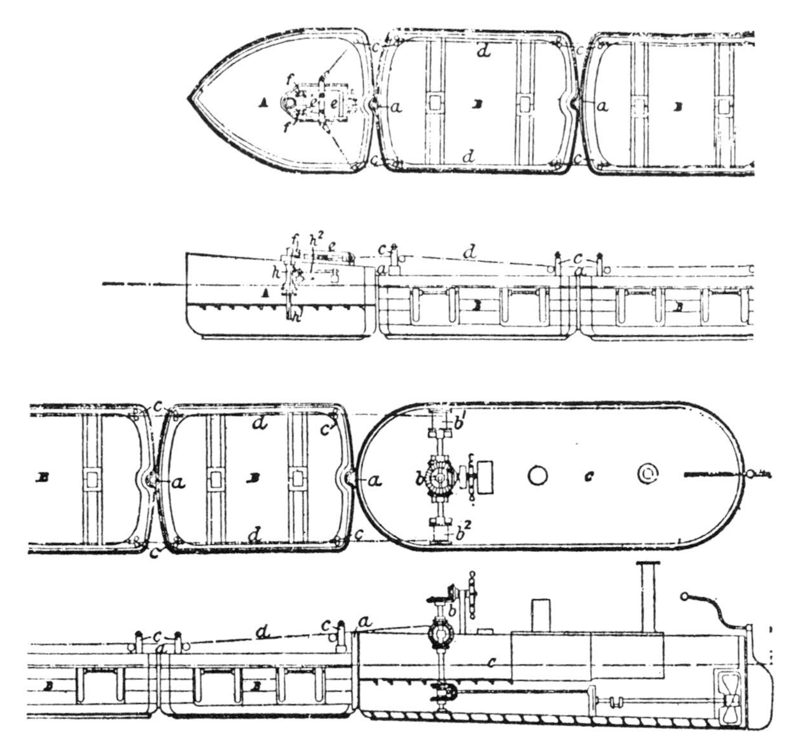

Fig 27 Part of W.H. Bartholomew's patent drawing of 1862, which shows his push-tug, compartment boats and bow section. The whole train could be steered from the tug by rudder, but also by tightening or loosening wire ropes running the length of the train.

controlling the cables now provided power steering. As the port of Goole, hydraulic hoists were built which could lift each compartment boat and tip its contents into a collier's hold. The push-towing system worked well, but it had been found that far longer trains of up to 40 boats could be worked if they were tug-hauled, spring-loaded buffers on the boats enabling them to follow-my-leader round bends. It was, however, still in partial use in 1906. Meanwhile push-towing had developed enormously in the United States, whence it was later to return again to the Old World. In 1928 the Bayerischer Lloyd company experimented on the Danube convincingly enough to order a specially built push-tug. This, the 600hp *Uhu*, began work in 1930, but her design was faulty, and it was not until after WWII* that regular push-towing appeared on the Danube. Before that, however, it began on the Congo and Kasai rivers in the Belgian Congo in 1935, using an arrow configuration of up to six barges that could be pushed by an ordinary tug, a configuration that moved post-war also to the lower Danube.

*WWI and WWII will in future be used to abbreviate World War I or World War II.

River age succeeds canal age

An important point should be made here. From the *trekvaarten* of the seventeenth century to the mid-nineteenth century, Europe had a predominantly canal age, and towards the end of the period was joined by the United States. But thenceforward a predominantly river age began on both sides of the Atlantic. Canals now needed to be built or rebuilt to dimensions that would allow craft able to navigate the big rivers to use them—craft which themselves were becoming bigger as channels improved. Those that could not be so enlarged slowly lost traffic, until in our own day they have often found a new and fruitful use carrying pleasure boats instead of commercial craft. More than one historian and politician has failed to realise that the modern river age is not a continuation of the older canal age: it is different in kind, and with far wider possibilities. Therefore it also came about in our period that railways, more efficient as freight-carriers than the old small canals, were by no means necessarily so as against rivers and river-sized canals.

The same pattern of new traction techniques, enlargement and the application of new technology was followed on canals and smaller rivers which contemporaries thought able to form part of a predominantly large-river system, or which could be made sufficiently competitive to keep down railway rates.

Improved traction

In the days before self-propelled steam or diesel barges arrived on the canals, much effort was put into experiments that sought to increase capability by replacing human or animal power either by bankside mechanical towing or by tugging services. In France, for instance, steam traction engines were used in 1867 on the canal between Dunkirk and St Omer, and in 1873 on the Canal de Bourgogne. In Britain, small steam locomotives were tried from 1888 on the Middlewich branch of the Shropshire Union Canal. Then came electric power in the 1890s. After brief experiments with barges powered from overhead trolleys and wires (successful for tunnel use, but not otherwise adopted), 3-wheeled electric tractors began work in 1895 on the St Quentin Canal's towpath. It was soon realised, however, that rail traction was more efficient. Electric locomotives powered from overhead wires, running on metre-gauge towpath rails, and exchanging barges when they met, were then introduced between 1904 and 1907. This system was to expand in France for some twenty

Fig 28 (above left) On canals various kinds of bankside traction were used including hauliers, as on the Canal du Berry early this century.

Fig 29 (above right) Another form of bankside traction: small electric locomotives powered from overhead wires.

Fig 30 (below) Used to haul canal craft, animals were often carried on board, as here on the Canal du Nivernais.

years thereafter, become fully organised in 1926, and survive until after WWII.

Britain's go-ahead Aire & Calder company supplied public steam-tugging services on its improved waterways from 1857, so successfully that it led to lock lengthening. Later in the century, compulsory tugging services were for a time to be provided on Germany's newly built canals.

Better channels and new technologies: lifts, inclines, aqueducts

The period to 1914 saw two great canal programmes, that of enlargement and standardisation in France associated with Freycinet's name (see p. 146), and that of new canal-building in Germany that followed the Franco-Prussian war. These were leaders in a wider movement to widen and deepen channels and enlarge locks or throw two into one. Quicker ways of filling and emptying locks were found, and power operation, at first hydraulic, introduced. Larger locks used more water. So we see the building of side-ponds into which some of the water used in lock-working could be stored, to be used again on the next locking—ancestors of the huge economiser basins of today—or the use of two locks side by side, interconnected so that each could use some water from the other. Occasionally, too, spectacular vertical lifts were built, to save water and because, in the days before reinforced concrete, such devices, along with inclined planes, were the only way of achieving a single large rise on a canal. The pioneer nineteenth-century lifts perhaps best exemplify the new technology.

The first modern canal lift was that at Anderton in Britain, suggested by Edward (later Sir Edward) Leader Williams when he was engineer of the Weaver Navigation, designed by Edwin Clark of Clark, Standfield & Clark, and opened in 1875 to connect the River Weaver with the Trent & Mersey Canal.

Clark had been responsible for designing and using the hydraulic presses that had raised the tubes of the Britannia and Conway bridges. He then adapted the idea to the design of a hydraulic ship lift for installation at the Thames Graving Docks*. By 1866, when Clark lectured on it, the lift had been working for about seven years, and had raised 1055 vessels of an average tonnage of 675. His lecture was given on 27 February. Since the possibility of a hydraulic lift at

*What was later called the Pontoon Dock lay on the south side of the Royal Victoria Dock at Woolwich. The ship lift was sited in the channel linking the two docks.

Anderton was mentioned in the Weaver's Act of that year, it seems likely that Leader Williams got the idea from Clark.

With a lift of 15.39 m, each of the two caissons was 22.86 m × 4.72 m × 1.52 m. These were counterbalanced, the lift being worked by removing some centimetres of water from the lower caisson. Speed was controlled by transferring hydraulic fluid from one press to another through a 1.27 cm pipe and the last 1.22 m of lift, when the descending caisson became partially immersed, by a hydraulic accumulator, powered by a steam engine. Electric was substituted for steam power in 1903, and over the next few years the hydraulic rams were taken out, each caisson being then separately operated by electric power assisted by counterweights.

The original Anderton pattern, with some improvements, was adapted by the same firm for Les Fontinettes lift on the Neuffossé Canal near St Omer to by-pass a flight of five locks dating from 1760. Much bigger than Anderton, to take 38.5 m Freycinet barges drawing 1.8 m, and with a rise of 13.1 m, it was begun in 1883 and put into service in 1888, the old locks being kept as a stand-by. In 1905 the lift passed 11,161 boats with a tonnage of 1¾M. It worked until 1967 when, as a result of the rebuilding of the Neuffossé Canal, it was replaced by a single large lock.

Another Clark, Standfield & Clark lift, this time for 400-tonne craft and with a rise of 15.4 m, was completed in 1888 at La Louvière (Houdeng-Grognies) on the Belgian Canal du Centre, then being built as a coal-carrier. Coming from the Mons-Condé Canal, the line rose by six locks, but for the remaining sharply rising 7 km to join the Brussels-Charleroi Canal the engineer, Hector Genard, decided to use vertical lifts instead of 17 more locks with their attendant problems of time taken in locking and of water having to be back-pumped. However, only La Louvière was then built, the other three planned (Houdeng-Aimeries, Bracquegnies and Thieu, in descending order from La Louvière), each with a lift of 16.93 m, being completed in 1917. These four great steel structures are still active, pending the huge new lift at Strépy-Thieu now being built (see p. 252) to replace them all.

The French and Belgian lifts, like the original Anderton, are hydraulically powered. A different system was used by Herr Gerdau of Haniel & Lueg of Düsseldorf to build the first Henrichenburg lift between 1894 and 1899 in order to give access to Dortmund from the

Fig 31 Les Fontinettes vertical lift on France's Neuffossé Canal soon after its opening in 1888. It was designed by Clark, Standfield & Clark along lines already tried out with Britain's Anderton lift in 1875. Cail et Cie supplied the metalwork.

Fig 32 Les Fontinettes has been replaced by a lock, but is preserved as an historic monument.

Fig 33 The 1899 float-principle lift at Henrichenburg. A bigger modern lift has taken its place, but the older structure is also being preserved.

Fig 34 La Louvière vertical lift on Belgium's Canal du Centre pictured in its opening year of 1888. Designed by Clark, Standfield & Clark, it was built by the Société Cockerill of Seraing, Belgium. One caisson is shown fully raised upon its piston, while a barge rests in its lowered companion.

Dortmund-Ems Canal. With a rise of some 14.5m, this had a single caisson taking 950-tonne craft. It worked on the float principle: five floats in a row beneath the caisson rose and fell in pits as water was added or withdrawn, a system that had first been patented and used in an experimental English lift built in 1796. Screws at each corner of the lift held the caisson level and regulated the motion. They, along with the pumps, were electrically powered. Effective length of the caisson was 68m, width 8.6m and depth 2.5m, the hoist-speed being 6.66m per minute. The design was good, and formed the basis for the later lift that replaced it in 1962.

Inclined planes carrying boats either floating in caissons or dry in cradles, and using various methods of keeping them level, are an alternative means of saving time and water. In what was then East Prussia is now Poland, the Oberland or Elbing (now Elblag) Canal was built between 1844 and 1860 from the Frisches Haff by Elbing and the Geerich lake to Osterode. Taking 60-tonne barges, it originally had seven locks and four inclines with rises of 20m to 24.5m, designed after study of America's Morris Canal.

The planes (still at work to carry trip-boats) have twin tracks of 3.25m gauge, on which run cradles each carried on 4-wheeled bogies. These are hauled up by a wire rope at some 1m per second, power being provided by the weight of the descending cradle and by a waterwheel. The cable continues under water from the bottom of each incline for some 90m before rising past one wheel, over a second at right angles, and back past a third (all three set in the middle of the canal) before returning to the other cradle. In the upper pound, again at some distance from the top of the plane, the cable wheels are set to the side of the canal where the waterwheel is. In 1881 a fifth incline,

Fig 35 The mid-nineteenth century inclined planes on the Elblag Canal in Poland still carry tripboats.

that nearest Elbing, and in this case powered by a water turbine, was built to replace five of the seven locks.

France also experimented with an inclined plane, which seems to have derived from a visit by French engineers to the Blackhill incline on the Monkland Canal in Scotland, opened in 1850. Steam powered, with a rise of 29.25 m and twin tracks, Blackhill carried empty barges only (downcoming loaded ones used a flight of locks) in caissons 21.35 m × 4.06 m × 0.71 m, until its disuse about 1887. The French plane was initiated by Jules Fournier, a canal-boat owner of Meaux, and opened around 1886 at nearby Beauval to connect the Marne to the Canal de l'Ourcq, there only 550 m apart, to take the latter's boats. These, 28 m × 3.6 m × 1.3 m and carrying 70–75 tonnes, then carried a considerable traffic to Paris. An iron-framed cradle carried on two 4-wheeled bogies ran on a 4 per cent slope. Propulsion derived from a water turbine utilising the head from a weir on the Marne. This actuated a cable running beside the track which transmitted its power through a pulley, friction clutches and gearing to a pinion that engaged a rack* laid up the slope. The cradle moved at 25 cm per second. Between 1888 and 1892, over 1000 craft used the plane, which continued to operate until 1922.

In England, G. C. Thomas, engineer of the Grand Junction Canal, and his brother designed Foxton inclined plane, which carried two narrow boats, each some 22 m × 2.15 m. Counterbalanced, assisted by steam power, the two caissons, connected by a wire rope, each ran transversely on eight sets of wheels carried on four pairs of rails. It was opened in 1900, and discontinued in 1910.

The Biwako Canal in Japan, with two inclined planes, had, however, American ancestry. North-east of Kyoto lies the 64 km-long Biwa lake, straddling the central mountainous neck of Japan. This had long been used for transport, but the Yodo river which ran from it to Osaka bay was only navigable at the lower end. A navigable canal connection had long been proposed, but it was not until 1885 that Kyoto's governor began building the canal for navigation and water supply. Sakuro Tanabe (1861–1944) was made engineer. He proposed that it should also be used to generate electricity, and set out to visit the United States. Among the ideas he brought back were lock drop-gates (see p. 325) and inclined planes.

The first section of the Biwako Canal, built 1885–90, some 11 km long, ran from Lake Biwa to Kyoto. It had an entrance lock, three tunnels, the longest some 2.4 km, and an inclined plane followed by a

*Except in its very early days, when a sprocket-wheel engaged a chain laid up the incline.

lock in Kyoto. This plane, with a 36m fall, was double-track, using 8-wheeled cradles to carry boats some 13.7m × 2.3m, and was powered by electricity generated by canal water. From Kyoto down to Fushimi and the Yodo river boats at first used a shallow old canal, until the second section was finished about 1894, 10.5km long with several more stone locks and another inclined plane at Fushimi. The Biwako Canal ceased to carry commercial traffic about 1914, but parts are used by pleasure craft, and the Kyoto plane's remains can still be seen.

Though often impressive, the technology of canal aqueducts built at this time was seldom exceptional. Two, one large, one small, may be considered so. Britain built one truly unique. When in the 1880s the Manchester Ship Canal was being made along the course of the River Irwell, it became necessary to replace the 3-arched masonry aqueduct at Barton that carried the Bridgewater Canal 11.6m above it by a new one which would maintain canal navigation while allowing craft to pass up the new ship canal. After having considered a high-level aqueduct connected by two vertical lifts, the engineer, Edward Leader Williams, decided upon a swing aqueduct. Opened in August 1893, it still works unchanged and efficiently. The wrought-iron swinging span, 71.6m × 5.5m, can be sealed off, like the canal pounds at each end, by watertight gates. The structure, weighing 1450 tonnes (800 tonnes of it water), and carried on a central pier, can then be swung on 64 rollers until it lies in the centre of the ship canal, and parallel to its channel. The hydraulic power used also activates a neighbouring swing road bridge.

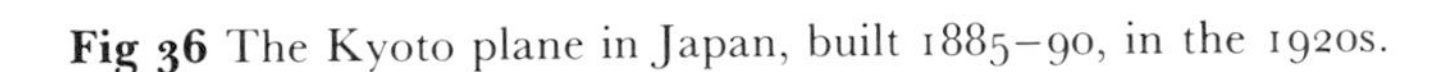

Fig 36 The Kyoto plane in Japan, built 1885–90, in the 1920s.

Fig 37 Leader Williams's Barton aqueduct, the world's only swing canal aqueduct, carries the Bridgewater Canal over the Manchester Ship Canal. Of wrought iron, 71.6m long, its centrally-pivoted span hydraulically-powered, it was opened in 1893.

A curious little aqueduct was built in Germany in the 1880s on the Ems-Jade Canal, where it crossed the Made with a clearance of 0.61 m at highest river-level. In case this might prove insufficient in a bad winter, the wrought-iron trough, 18.5 m long and 7.5 m wide, resting on masonry abutments, could be raised or lowered by rods with screw-gear, after stop-gates had shut off the canal on each side.

On his way to the second PIANC Congress at Frankfurt-am-Main in 1889, only fourteen years after Moens' cruise, L. F. Vernon-Harcourt, engineer and waterway author, found a very difrent scene:

> The Author was struck with the great industrial activity of Belgium, and the rapid development of inland navigation in Germany; the solicitude which the Belgian and German governments have exhibited for improving and extending the waterways of their countries, and the eagerness with which the local authorities second these improvements, by enlarging their ports, and providing them with the most modern mechanical appliances for facilitating trade.(5)

The British explore foreign waterways

We were introduced to W. J. C. Moens on *Ytene*. There were many like him, even if not with a steam-yacht; for the nineteenth century was a time for British exploration of waterways abroad during their time of transition. In the 1850s the *Water Lily* had been rowed as far as the Danube, and the pair-oared *Undine* with her intrepid young men from Rouen to the Rhine at Mulhouse. In the early seventies another pair-oar was rowed up the Seine and down the Loire. In the 1880s another, the *Beetle*, was taken a shorter voyage from Liège through the Ardennes to Rouen, and yet another explored the German rivers.

One is astounded by the enterprise and courage of those who, with

Fig 38 A British tourist in the Netherlands in 1881 watches a swing-bridge keeper collect his toll in the traditional wooden shoe.

only a smattering of its languages, cheerfully set off to row half across Europe, even through tunnels, and then to write about their adventures (and often illustrate them) as nineteenth-century young men did, who 'dawdled delightfully over the evening weed' or smoked 'the post-prandial pipe' or 'doffed their flannels for less conspicuous raiment' or 'after a refreshing header into the clear water, spread a humble feast under the trees, and fell to with the usual will'. One feels for the landlord who was puzzled what 'induced us to bring a boat over from England, and row from place to place, when we could afford to go by train, which, he pointed out, would be "plus vite et encore beaucoup plus convenable". And what was there to see, he asked, nothing but water and rocks, and a few trees: surely we could see these without coming so far...?' In these days of air and motorway travel, too, we can smile at Elizabeth Strutt, being rowed down the Loire in the 1830s:

> It is impossible to combine swiftness of progress with leisure for observation; and every one must be pretty well aware who travels by mail coaches, or steamboats, that he might in that way go through half the world without being any the wiser...(6)

The indefatigable John Macgregor took a canoe half across Europe and beyond it; W. M. Doughty voyaged by wherry from the Netherlands, along the North German canals, through the Mecklenburg lakes to Berlin, and up the Elbe, while the eccentric Donald Maxwell had a sailing boat built to his own design in a Swiss village miles even from a lake, and then took it down the Rhine—after which he sailed or towed a different craft from the Rhine to the Black Sea.

One is struck, too, by the social efficiency which enabled boats to be purpose-built in a month, sent unaccompanied from one Continental station to another by train, and enabled supplies to be got by post from Britain in a few days, so reliably that these services pass without comment. Our ancestors did not know how fortunate they were.

Others used public transport: we have seen Laurence Oliphant voyaging down the Volga, but nearer home the story persisted of the English lord who spent his declining years travelling up and down the Rhine on fast steamers because their restaurants were the only places where he could get a decent beefsteak.

From their books, and others like them, one has eye-witness accounts of contemporary traffic. Oliphant saw a horse-machine:

> ...seven or eight boats are lashed, one behind the other, to the immense barge which contains the horses ... On the deck of the leading barge a covered stage is erected, which serves as a stable sometimes to as many as a hundred and fifty horses. These may be observed working the huge capstan by which the boat is warped, and round which they perambulate, as in a threshing-machine. As many men as horses inhabit this floating establishment...(7)

but also

> An immense raft of pines from Perm and the Ural Mountains, on which were two or three cottages ... the wood in those regions is so cheap, that it pays to float timber from the head waters of the Kama down to the Caspian. (8)

Rafts were everywhere. Macgregor in the 1860s in France saw them being prepared near Chateau Thierry for the 150km voyage down Marne and Seine to Paris:

> ... some made of casks, lashed together with osiers, some made of planks, others of hewn logs, and others of great rough trees. The straw hut on each is for the captain's cabin, and the crew will have a stiff fortnight's work to drag, push, and steer this congeries of wood ... the labour spent merely in adjusting and securing the parts is enormous, but labour of that kind costs little here.(9)

Sailing barges too. On the upper Rhine near Lake Constance:

> Rudely-made barges ... (that) seem to turn round helplessly in the current of the deeper parts, or hoist their great square sails in the dead calm ... a very picturesque appearance, as the sail has two broad bands of dark blue cloth for its centre stripes. But the pointed lateen sail of Geneva is certainly a more graceful rig than the lug, when there are two masts, and the white sails swell towards you, goosewinged...(10)

on the Zuiderzee:

> Friesch tjalks... from Holland. High out of water were they, broad almost as long, shining with varnish or brown tar, they had great fin-like lee-boards, topsides tumbling in, green-painted houses on their after-decks, varnished or brightly-painted harness-casks, and anchor-buoys banded with red and white or blue and white like children's tops, hanging from their bows. They were mostly manned and womaned by one man, his wife, and, perhaps, a boy and girl or two; all busy ... getting in and stowing their brown sails before passing through the locks.(11)

on the Elde river, between the Elbe and the Mecklenburg lakes:

> The 'kahns', being keel-less, can only sail with a fair wind; they are far too heavy to quant [pole]; can only be towed by human labour—horses are never used in this country—and their length is so great that it is difficult to get them round corners. They sail under sprit sails, ill cut, baggy and very small...(12)

or the Danube in Serbia:

> Two barges with high sterns were dropping down river ahead, each with a large square sail which hung idle... and each one rowed laboriously by half a dozen men with huge sweeps.(13)

Beetle in the 1880s caught in a Meuse lock a glimpse of the idyllic:

> a Dutch Barge, its bulging bow and stern painted gaily in red, green and white. It was clean as a gentleman's yacht, every bucket burnished, every rope neatly coiled. Spotless white muslin curtains screened the windows of the deck-house on whose roof was a little garden of geraniums. A plump cat...lay asleep in the exact centre of the tarpaulin cover. [It was being towed by two Flemish greys] their headgear lavishly bedecked with crimson tassels, and their sturdy shoulders hidden by great blue woollen mats.(14)

Undine in the 1850s observed another, less idyllic, scene on the Canal de Bourgogne:

> In towing barges, &c., along the canal, horses are seldom, if ever, used ... when (as is frequently the case) they are one family travelling with the barge, the mother on one side of the canal and the son on the other,

> with a rope each, haul the barge along...while the father stays on board, smokes his pipe, and steers the barge.(15)

On the canal from Leeuwarden to Delfzijl, Doughty in 1890 followed:

> a trekschuit... She blew a horn for all the farmhouses, stopping when called upon to embark or lands goods or passengers.(16)

Maxwell on the Rhine in 1901 watched a floating shop supply a steam paddlewheeled tug with fruit, vegetables and bread while moging:

> The tug is put at full speed again, and the captive boat, which is built broad in the bow to stand rapid towing, flies foaming and splashing over the water. Meanwhile a spirited bargaining is going on ...(17)

and a few years later on the same river visited the dining cabin of the towed barge *Anna Jacoba*:

> The table linen was spotlessly clean, and the silver shone in the morning sun which flooded the cabin from one of the green-shuttered windows, [and the drawing-room with] a bookcase, and ... a piano, upon which the eldest daughter performed very creditably.(18)

In 1878 Robert Louis Stevenson, in *An Inland Voyage*, wrote a paragraph that, with little changed, could have been written today by anyone cruising on the Continent:

> The chimney smokes for dinner as you go along; the banks of the canal slowly unroll their scenery to contemplative eyes; the barge floats by great forests and through great cities with their public buildings and their lamps at night; and for the bargee, in his floating home, 'travelling abed', it is merely as if he were listening to another man's story or turning the leaves of a picture-book in which he had no concern. He may take his afternoon walk in some foreign country on the banks of the canal, and then come home to dinner at his own fireside.(19)

This chapter began with Moens on *Ytene*. So let it end with a canal author at work:

> Madame reminded her husband of an Englishman who had come up this canal in a steamer.
>
> 'Perhaps Mr Moens in the *Ytene*', I suggested.
>
> 'That's it' assented the husband. 'He had his wife and family with him, and servants. He came ashore at all the locks and asked the name of the villages, whether from boatmen or lock-keepers; and then he wrote, wrote them down. O he wrote enormously! I suppose it was a wager'.(20)

On river and lake and canal the young and the elderly worked side by side on old waterways and new alike. What meanwhile had governments been doing?

6
TO 1914: FRANCE TO THE AEGEAN

We have seen the multiplication of ship canals, the emergence of a new age of large canalised or improved rivers, and the beginnings of efforts to upgrade older canals to make them compatible with what was to become a European big-waterway network. Now, area by area, we shall look more closely at what happened; in particular at the small-canal expansion and standardisation associated with Freycinet's name, the extensive and successful large-canal planning and construction of Imperial Germany, and the notable improvements made after 1856 to the Danube.

France

Because of her feeling for logical state planning, France's first canal age ended late. Then came her defeat in the Franco-Prussian War of 1870–1. The peace transferred Alsace and Lorraine to Germany, and with them the use of the Rhine and Moselle as north-south routes to connect her east-west waterways. In 1872, therefore, the Canal de l'Est was begun parallel with the new frontier, from Corre on the Saône to the Marne-Rhine Canal, and then northwards as the canalised River Meuse, previously only *flottable* for log-rafts, to Givet near the border. Thence the lower Meuse, along which the Belgians were finishing a lock-building programme, would give access to that country's coalfields. Urgently needed to counter the effects of war, the

Fig 39 French waterway life in the early 1900s: (above) at Epernay, head of navigation of the river Marne, around 1907 a small steam crane unloads a barge, maybe fuel for the gasworks. A cart waits, and also a carriage—clearly someone considers the operation important enough to oversee; (centre, left) a service passenger boat has worked through the locks of the river Mayenne, a tributary of the Loire, from Angers to a lock near Château-Gontier, some 47 km higher up; (centre, right) once a familiar sight to travellers by water in France, a laundry-barge, the occupants of which were likely to exchange badinage with passers-by; (below) a barge probably loaded with wine barrels works up through a staircase pair of locks at Castelnaudary on the Canal du Midi about 1901.

northern section of this canal, 419km long, with 158 locks and a summit-level 360m above sea-level, was completed in 1880, the southern by 1882. Depth was 2m, to which the 1.7m of the Marne-Rhine was then raised.

French public opinion in the 1860s had come to consider waterways a necessary moderator of railway rates. After the war, therefore, the National Assembly appointed a commission of enquiry into railways and waterways. As regards the latter, it concluded that the system was not providing the economic benefits that it should, because, unlike railways, its components were of different gauges and dimensions. Thus, either boats had to be of the smallest size to traverse the whole network, or be confined to certain lengths and tranship cargoes travelling further. Given also slow horse-haulage and much hand loading and discharge, the whole system needed modernisation. Without it, waterways would not act efficiently as auxiliaries to, and regulators of the rates, of railways.

Charles de Freycinet, who became Minister of Public Works in 1877, took action. Legislation of 1879 authorised 8848km of new railway lines, the improvement of 4000km of rivers and 3600km of canals, and the building of 1400km of new ones, including the Marne-Saône canal already begun. Freycinet's law stated minimum dimensions to be aimed at for the principal French lines of waterway; depth 2m, lock 38.5m × 5.2m, with an available draught of 1.8m and air draught (bridge clearance) of 3.7m. Such dimensions suited the 300-tonne *péniche*, thenceforward to become France's standard canal craft.

These waterways were to be administered by the state, and one year later, in 1880, all tolls were abolished. Those not already owned were to be bought back as possibility opened: notably, on 1 July 1898, the Canal du Midi and the Canal latéral à la Garonne were reacquired from the Midi railway company.

By 1892, 4100km of French navigable waterways were of Freycinet standard, and tonnage carried since 1881 had increased by a fifth. Instead of much canal traffic being enforcedly local, *péniches* could now move from Le Havre to Alsace, from Dunkirk to Lyon, or to the Belgian coalfields round Mons and Charleroi. Freycinet's work had also engendered two notable feats of engineering.

From the Marne-Rhine Canal's east-west trunk line, a Marne-Aisne link had already been built northwards. It needed, however, another rivers link, the Canal de l'Oise à l'Aisne, begun in 1881, to give access to north-eastern France by way of the St Quentin Canal, and to Belgian coal by that of the River Sambre. For a canal only

Fig 40 Eiffel's Briare, 662 m long, the longest navigable canal aqueduct of which I know, was opened in 1896.

48 km long, it offered engineering problems enough. Having left the Canal Lateral à l'Oise, it crossed the river by a 70 m aqueduct, rose through 13 locks to the summit and the 2365 m Braye-en-Laonnois tunnel, then fell by 4 locks past a 62 m aqueduct over the Aisne to the Canal Lateral à l'Aisne. Tunnelling ran into every kind of difficulty—water, dry sand, even underground fire—and cost at least 17 lives. Because the summit lacked natural water supplies, they had to be lifted from the Aisne by a pumping station, in 1891 using three water turbines each driving two pumps, and lifting 1.5 cu m a second some 16 m to a storage reservoir. To Freycinet's impulse we also owe the world's longest navigable canal aqueduct, the huge iron structure that replaced a level-crossing over the Loire at Briare, 662 m long, 11.5 m wide and supported on 15 granite piers. Designed by Eiffel, it was opened to traffic on 16 September 1896.

Let us now glance at France in 1891. There were 12,395 km of navigable waterways, all but 858 km controlled by the state. There were 15,925 barges in use, of which 4191 took more than 300 tonnes, 3297 were 200–300 tonne, 2459 100–200 tonne, 2892 50–100 tonne, and 3085 3–50 tonne. Of them all, 8067 were decked, the rest not; 1051 built of iron, the rest wood; 13,699 had cabins, the others not. They were occupied by 19,579 men, 7917 women and 12,972 children. And 2094 boats had stables on board to house towing animals—1396 horses, 186 mules and 1604 donkeys.

Steam-barges were used on some canals, and on rivers like the Seine there were paddlewheeled, screw- and chain-tugs. Otherwise in the north and east, in this time just before electric bankside haulage began, towing was sometimes by animals stabled on board, more often by relays of horses with their own drivers, who worked stages of 16 to 32 km a day, according to lockage. These horses could be provided by the state, by contractors, professional trackers or local farmers. In the centre of France, however, the boatman and his family usually towed the smaller boats, with or without the help of a donkey stabled on board. There were still few wharf cranes, few warehouses, no system by which crews could be told of loads offering, and little rail-waterway interchange.

Another survey of 1907 showed 15,310 boats, fewer than in 1891, but with greatly increased capacity. In 1891 only some 30 per cent could carry 300 tonnes or more; now 72 per cent could. Most, especially the large ones, were owner-worked, but there were also navigation companies, the three largest owning 426, 152 and 110 craft. The number of powered pleasure craft had also greatly increased.

Though a great deal had been done under the Freycinet plan, nevertheless money made available each year was not enough to build new waterways and also modernise old ones, when all the time needs were changing. In 1903, therefore, after much discussion with local and business interests, the Baudin plan—so named after the then Minister of Public Works—was agreed. It authorised money to finish some Freycinet projects, and also two new ones, a Marseille-Rhône canal, and the Canal du Nord.

In 1835 a canal had been opened between Arles on the Rhône and Port-de-Bouc on the Golfe de Fos across from Marseille, but had proved too shallow for Rhône-Marseille traffic of the dimensions it had now reached, and was little used. Then in the 1860s Marseille, because of the opening of the Simplon railway tunnel and threat of the Loetschberg, feared loss of trade to Genoa. In 1871, therefore, a sizeable cut with one lock 160 m × 22 m, and giving 5.5 m depth, was made from Saint-Louis on the lower Rhône to the gulf de Fos; but the necessary sea passage through the Golfe was often too rough for Rhône craft, and that also was only partially successful. The authorised line ran from Marseille harbour, protected for 10 km by embankments, to a new canal tunnel, Rove, the longest and largest ever built, at 7120 m length, 22 m width and 11.25 m height above water, to Martigues. There it ran down the Etang de Berre outlet to Port-de-Bouc, whence a reconstructed canal ran back to the Rhône at Arles.

The Canal du Nord is not to be confused either with Napoleon's scheme or that of 1888 for a Canal Maritime du Nord from Boulogne to Paris. It had been actively promoted to relieve the overcrowded

St Quentin line, a substantial contributor to the 10M tonnes carried in 47,519 boats being handled in the port of Paris. The new canal was to run from the northern collieries via Arleux and Péronne to the Oise Lateral at Noyon, with a possible extension northwards to Dunkirk to give a big canal line through to the Paris area and to enable Nord coal to compete with British brought up the Seine. When war came in 1914 the Marseille-Rhône link had been made, though the Canal du Nord remained part-cut: it was not to be restarted for 45 years.

By 1913, waterway traffic had grown to 618M tonne/km, almost double the 1890 figure. The Freycinet-Baudin policies had been successful, and a hopeful sign for the future was the establishment in 1912 of the Office National de la Navigation, to provide the public with information upon inland waterways, and to seek means both to develop them and increase their use.

The Low Countries

Ship-canal development was notable in both Low Countries. In the 1870s and early 1880s, the Ghent-Terneuzen ship canal was straightened, deepened and widened, vessel tonnage then increasing from 254,094 in 1884 to 603,372 in 1897. It was deepened and improved again in the late 1890s and in 1910 to provide a modern ship channel with three sea-locks at Terneuzen and another three, used only at flood-times, at Sas van Gent. Meanwhile, the Groot Dok had been built at Ghent round the turn of the century. Canal and dock together gave Ghent a standing as a port that others found difficult to rival.

Bruges tried. Small ships could reach the city by the old canal from Ostend: in 1871, when visited by the *Amethyst*, half a dozen British schooners were there, mostly bringing salt from the Mersey. Then in 1895, the year after that to Manchester had opened, a ship canal was begun from Bruges to a new harbour at Zeebrugge (Sea-Bruges). A large anchorage was protected by a long curving pier, and in 1907 seagoing ships began to pass the sea-lock, and again visit Bruges. This lock, instead of having two pairs of mitre-gates facing in opposite directions, usual in a situation where tidal levels might be above or below the canal, had caissons rolling across the lock from wall recesses, and able to withstand pressure from either side. Electrically operated, along with swingbridges and other lock equipment, they mark an early move from hydraulic power. The canal itself was 10km long and 8m deep. At Bruges an outer turning basin was excavated,

Fig 41 A bucket excavator and suction dredger in use by the contractors Couvreux & Hersent in 1878 on the enlargement of the Ghent-Terneuzen Canal. This firm had earlier worked with de Lesseps on the Suez Canal, and these pictures suggest the modern technological background against which a decision was taken in 1880 to start work on a French sea-level canal at Panama. There also the firm were initially concerned, though they withdrew in 1882.

and two inner basins, one of which was joined by a lock to the Bruges-Ostend canal and so to other waterways. Nevertheless, Bruges found it hard to compete against the established positions of Ghent, Ostend and Antwerp. Moreover, also in 1895, a company began work on a small ship canal to Brussels, roughly following the old Willebroek Canal's line, and finished it in September 1914, just after WWI had begun. A new port at Brussels was opened after the war, in 1922.

During the 1860s, Belgium decided to develop her waterways alongside her railways. Given the importance to her of the port of Antwerp near the mouth of the Schelde, and of Ghent higher up, with its access to Terneuzen nearer the sea than Antwerp, improvements began at Ghent with the enlargement of the ship canal, and at Antwerp where extensive installations for barge traffic and transhipment were approved in 1877. The Schelde was a main traffic artery. Running from France by Courtrai to join it at Ghent was the Lys, upon which five locks were built in 1863–9. In 1863–7 another tributary, the Dendre, joining it at Termonde and running south to Ath, was canalised. From Ath the new Canal de Blaton à Ath was opened in 1868 to join the Antoing-Pommeroeul Canal, itself, linked westwards to the upper Schelde again near the French border and so to the French Canal de St Quentin, and eastwards to the Mons-Condé Canal, serving the coalfields around Mons.

The Mons-Condé was soon afterwards enlarged to 300-tonne standard, and a new canal of the same size started, the Canal du Centre (we have seen its first vertical lift, at La Louvière, opened in 1888), running eastwards from it to join the Brussels-Charleroi Canal. This coal-carrier, running up to Brussels from the coalfields round Charleroi, where it joined the Sambare (leading to France in one direction and the Meuse at Namur in the other) was also given 300-tonne standard.

Taken with works on the Meuse and Sambre, Belgium now had a much improved waterway system, better fitted to serve her industrial areas and principal ports. In 1870 her system had in fifty years expanded from 1618km to 2022km, of which 1736km were state controlled, 119km in provincial hands, 93km in communal, and 74km in those of concessionaries.

In the Netherlands the Noordzeekanaal or North Sea Ship Canal had been proposed in 1852, to replace the old Great North Holland Canal (far less useful now steamers were taking over from sailing vessels) and run wide, deep and straight for 24km from Amsterdam to Ijmuiden, the last 5km through sand-dunes. It was planned by John (later Sir John) Hawkshaw and J. Dirks of Amsterdam, authorised in

1862, and begun in March 1865 with British resident engineers and contractors. The company being unable, however, to raise enough money, in 1868 the Dutch Government guaranteed interest and redemption for shares not yet sold.

East of Amsterdam a dam, pierced by the three Orange locks, was built to cut the city and the new canal off from the fluctuating levels of the Zuiderzee. Thence the canal ran level and 7 m deep past one road and two rail swingbridges to Ijmuiden, with its two parallel locks 119 m × 18 m and 69 m × 12 m. Beyond them was a tidal stretch within embankments leading to the sea harbour, flanked by two converging moles some 1500 m long, enclosing an area of about 250 acres (101 hectares), through which ran the dredged ship channel. Each Orange and Ijmuiden lock had five pairs of mitre-gates, three iron sets facing tidewards, and two timber sets canalwards.

Building the outer harbour and moles was exceptionally difficult. After many troubles, the south mole was eventually finished, after which the celebrations were such that, the story went, the Dutch and British national anthems were played simultaneously. King Willem III on board *Stad Breda* opened the canal on 1 November 1876, and in August 1877 the north mole was completed. A by-product of construction was the drainage and reclamation of some 13,000 acres of land.

Whereas the Great North Holland Canal had regularly been closed by winter ice, it usually proved possible to keep the new one open, thanks to heavy traffic in large ships, which broke it as it formed. Deep-water quays were built at Amsterdam from 1879, followed by modern warehouses. In 1881 the company, unable to pay its way, was taken over by the state. In 1890 the canal was made toll-free and in 1896 the larger Ijmuiden lock, always too small, was supplemented by another, 225 m × 25 m, the largest yet built on an inland navigation*. Very large lock-gates had previously been hydraulically powered, but this had drawbacks in very cold weather, and so electricity was used, generated on the spot from British coal. This also provided floodlights for the moles and locks, which could now be worked round the clock, and for traffic lights. The canal was widened and deepened at the same time, its three bridges being soon afterwards replaced by ferries. Since those days the canal has been much enlarged: its largest lock is now 400 m × 50 m. With that, and a depth of 15 m, it can take 90,000 dwt vessels.

*Its dimensions were exceeded in 1914 by the new locks (330 m × 45 m) of the Kiel Canal, larger than Panama, and in length by the Davis lock (see p. 346) at the Soo, 411 m × 24.4 m.

A further step in Amsterdam's progress was a direct canal communication to the two Rhine channels. The Merwede Canal taking 2000 tonne craft, was opened in 1892 to the Lek at Vreeswijk south of Utrecht, and then on to the Waal at Gorinchem*.

Rotterdam's citizens, antagonistic to talk of a North Sea Canal to Amsterdam, determined to improve their own access to the sea, then either by the Nieuwe Maas and the Brielse Maas by way of Brielle, or by the Kanaal door Voorne from the Nieuwe Maas to Hellevoetsluis on the Haringvliet. A scheme for a New Waterway from the city to Hook of Holland dates back to 1858, and in 1859 officials of British east coast railways begun to negotiate with Rotterdam interests for the establishment of a packet-steamer line between Harwich and Rotterdam.

In 1862 the New Waterway was promoted, and 1863 authorised; work began in 1866. Meanwhile packet services began in 1863 via the Brielse Maas, though with frequent strandings. However, in 1872 the first packet, *Richard Young* of the Great Eastern Railway, sailed down the New Waterway. Much still needed to be done. In its early years it silted badly, and had to be constantly dredged. Not until 1896 was it finally brought to the state for which the 1863 Act had provided: 33km long, with 8.3m at low water and 9.8m at normal high. Before then, however, a railway had been built from Rotterdam to Hook of Holland, and in 1893 *Chelmsford* of the Great Eastern Railway opened the packet service to the Hook, rather than Rotterdam. From 1904 the Rotterdam terminal was closed and steamers called only at the Hook. By a 1917 Act, the waterway was deepened to 11m at low water and 12.5m at normal high, and commercial traffic grew steadily.

Writing in 1898, L. F. Vernon-Harcourt could say:

> Rotterdam... a small old-fashioned town when the Author first visited it rather more than thirty years ago, has been transformed into one of the principal seaports of the Continent by the considerable improvement in depth and directness of its access to the sea effected since 1863'. (1)

Basic to Rotterdam's trade was the shipping of Westphalian coal brought downriver, and the import, often in the same craft, of iron ore for the Ruhr. In 1913, for instance, nearly 5 M tonnes of coal was unloaded there from Rhine craft, and out of the nearly 9 M tonnes of imported ore handled, at least the same amount must have been transferred to barges.

* See Map 9. The ramifications of the lower Rhine and Maas within the Netherlands are tedious to describe and impossible to remember.

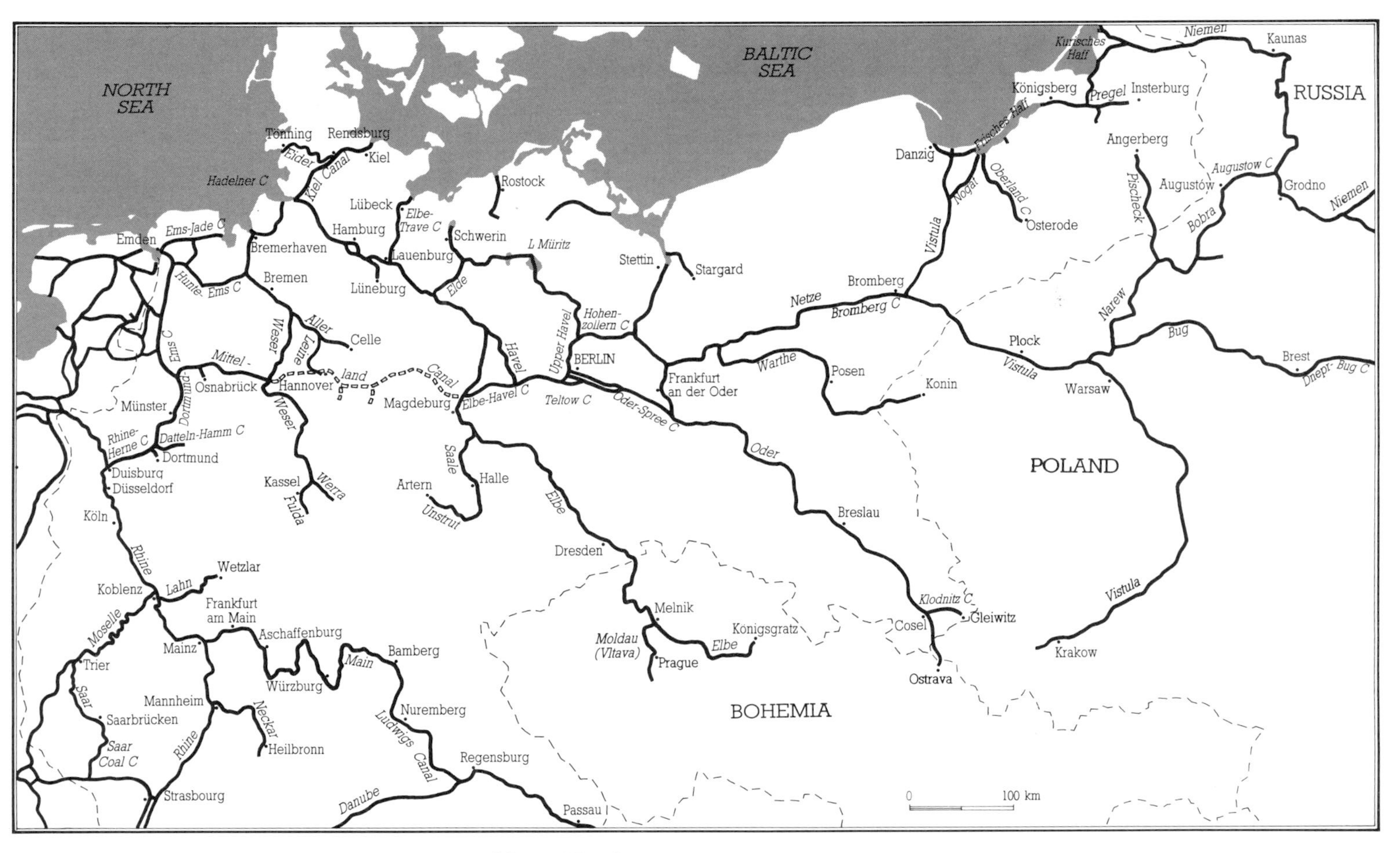

Map 7 The German canal system in 1914.

Meanwhile, though the States of Holland had in 1754 first appointed an adviser to the government on river management, no policy for dealing with flooding and the formation of ice had been evolved. It was not until 1854 that an appendix, written by two Inspectors of the *Rijkswaterstaat**, set out a policy: 'control of the summer-bed, control of the winter-bed, a general system of raising the dykes, the separation of the systems of Rhine and Meuse, and in connection with the latter digging of a new mouth of the Meuse'. This policy formed the basis of the following century's work.

Germany: 1871–87

Out of the Franco-Prussian war came the German Empire of 1871. From the North Sea and Schleswig-Holstein on the north, it was largely bounded on the west by the Netherlands, Belgium, the Rhine and Alsace-Lorraine just won from France; on the south by Switzerland, Austria-Hungary (Czechoslovakia did not exist) and, to the east, Russia. Poland was not then an independent state, and Germany ended at Memel (Klaipeda) beyond Königsberg (Kaliningrad), now in the USSR. These boundaries lasted until WWI, and it is with transport planning over this wide area—and notably with its west-east connections, centred on the imperial capital, Berlin, and having for a principal object the exchange of the products of the less developed eastern provinces for those of the more advanced west—that we shall here be concerned. Waterway development in the period came in two stages; first, a growth of river traffic, some canal-building, and many canal schemes. Then, roughly from 1887 when the Kiel Canal was begun, came extensive big-canal construction.

On the Rhine, sailing barges continued to be built, while steamers maintained a passenger service, and the power of steam paddle-or screw-tugs increased. Mannheim-Ludwigshafen maintained its position as virtual head of navigation, its river trade in 1886 at 2,443,000 tonnes being second only to Ruhrort at 2,472,000 tonnes. Other river ports like Mainz were busy improving their accommodation, and from Köln downwards small seagoing craft were being worked to England, as also to Bremen, Hamburg and Baltic ports. Upstream Colonel Gottfried von Tulla had improved the channel for smaller craft up to Strasbourg without locks by 1876, and later, helped by von

*This government body now concerns itself with all kinds of civil works, especially those of hydraulic engineering.

Tulla's plans, the Swiss Rudolf Gelpke worked upon an extended route for larger boats through to Basle, which was first reached by a barge train in 1904.

In that year, the powered Rhine fleet consisted of 1183 steamers, with tugs up to 1320hp. Whereas the average capacity of barges was still small, 182 tonnes in 1884 and 1320 tonnes in 1902, by 1906 the largest carried 2600 tonnes. In 1905 Rhine traffic at 4045M km/tonnes was over half the total for the German Empire, and in 1913, 97,000 craft passed the Rhine frontier point between Germany and the Netherlands, carrying 37M tonnes of cargo.

What of the three great tributaries of the Rhine? Cable-towing began on the Neckar in 1878, and chain-towing on the Main about the same time, in this case to be slowly replaced as large locks were built upwards from the confluence at Mainz to Frankfurt by 1886, Aschaffenburg by 1920, and Würzburg by WWII. On the unimproved Moselle steamers and small barges moved, among them in summer the 'Moselle hay-boat, a queer high-prowed float with half a rick of rich meadow-grass stacked upon it. The farmer-boatman does little more than pole the float away from the shore periodically...'(2) Plans for its canalisation, perhaps also with electricity generation, came to nothing until our own times.

To the east, the great Elbe river ran roughly south-east from Hamburg past Lauenburg (where the old Stecknitz Canal still led to Lübeck and the Baltic), past the Havel river leading towards Berlin, Magdeburg, the Saale running south to Halle, Dresden, and across the Bohemian border into Austria-Hungary (now Czechoslovakia), to Melnik, junction with the Moldau (Vltava) running to Prague. At Melnik a chain ended that in the 1880s had been laid all the way from near Hamburg.

Further east again, and also running roughly south-east, the Oder ran from Stettin (Szczecin) on the Baltic up past Hohensaaten (whence the Finow Canal ran to Berlin) to Küstrin, where the Warthe (Warta) leading to the Netze (Notéc) began the water line to the Vistula. Thence it continued by Neuhaus (Friedrich Wilhelm Canal) and the Neisse to Breslau (Wroclaw), Oppeln and the considerable river port of Cosel, where goods, mainly coal, were transferred from rail to river barges. This was the limit for most river traffic, though some continued over what is now the Czech border to Morava Ostrava. From Cosel also the 46km Klodnitz Canal, with 18 locks, ran to the Silesian coalfields. Eastwards again, in what was then East Prussia, the 43½km König Wilhelm Canal (1863–73) joined the Memel (Niemen) river to the huge Kurisches Haff lagoon.

The Oder was most important for transport: navigable with its tributaries for 1700km, it was connected with all Germany's north-eastern waterways from the Elbe to the Memel. There were three principal traffic flows: between Stettin and the Finow or Friedrich Wilhelm Canals; Polish timber-floats —16,949 in 1871 (9 logs = 1 float)—from the Warthe to the Berlin canals or Stettin; and between Silesia, the canals, and Stettin. Silesia had been developed by railways, which were by this time becoming strained. Yet, though in 1873 some 8500 barges entered and left Stettin, and carried over 500,000 tonnes in each direction, the Oder was a difficult navigation, suffering from floods in spring, low water in summer and icing in winter.

In the 1880s, the lower Oder was improved by installing chain-towing, and in 1890, Stettin harbour was enlarged. As a result, coal moved by water increased from 130,000 tonnes in 1884 to a million in 1897. The improvement of the upper river to Cosel for 400-tonne barges followed, with a 2-lock by-pass of Breslau, 2 locks thence to the Neisse, and another 12 rising 27m in all on the 24km beyond to Cosel. Between the two Breslau locks, on the Old Oder, was the winter harbour for several hundred boats. Cosel itself, with its three docks built 1895–1907 and their coal-handling facilities, became the most important waterway harbour east of Berlin.

For the decades after 1870 German canal improvement mainly took place east of the Elbe. Berlin in the 1870s attracted barges, which carried food, building materials and industrial traffic such as coal. In 1875 over 3½ M tonnes, exclusive of timber-floats, passed through the city, apart from local traffic. Between the Elbe and Berlin, the old Plaue Canal, running from the river to the Havel, was raised to 600-tonne standard and supplemented by the new Ihle Canal from the Elbe at Niegripp nearer Magdeburg (the two becoming known as the Elbe-Havel Canal). For traffic south bound up the Elbe towards Berlin, the lower Havel was in 1875–82 also canalised to the Spree at Spandau (Berlin). From the Oder, barges bound for Berlin from Stettin were being tugged to Hohensaaten and the Finow Canal—itself much improved in 1874–5—or to the Friedrich Wilhelm (later Oder-Spree) Canal at Neuhaus.

Germany: the Kiel Canal

To the north, Denmark's eighteenth-century canal across Schleswig-Holstein, German since 1864, remained almost unaltered, though now too shallow for craft of much over 100 tonnes. An Englishman

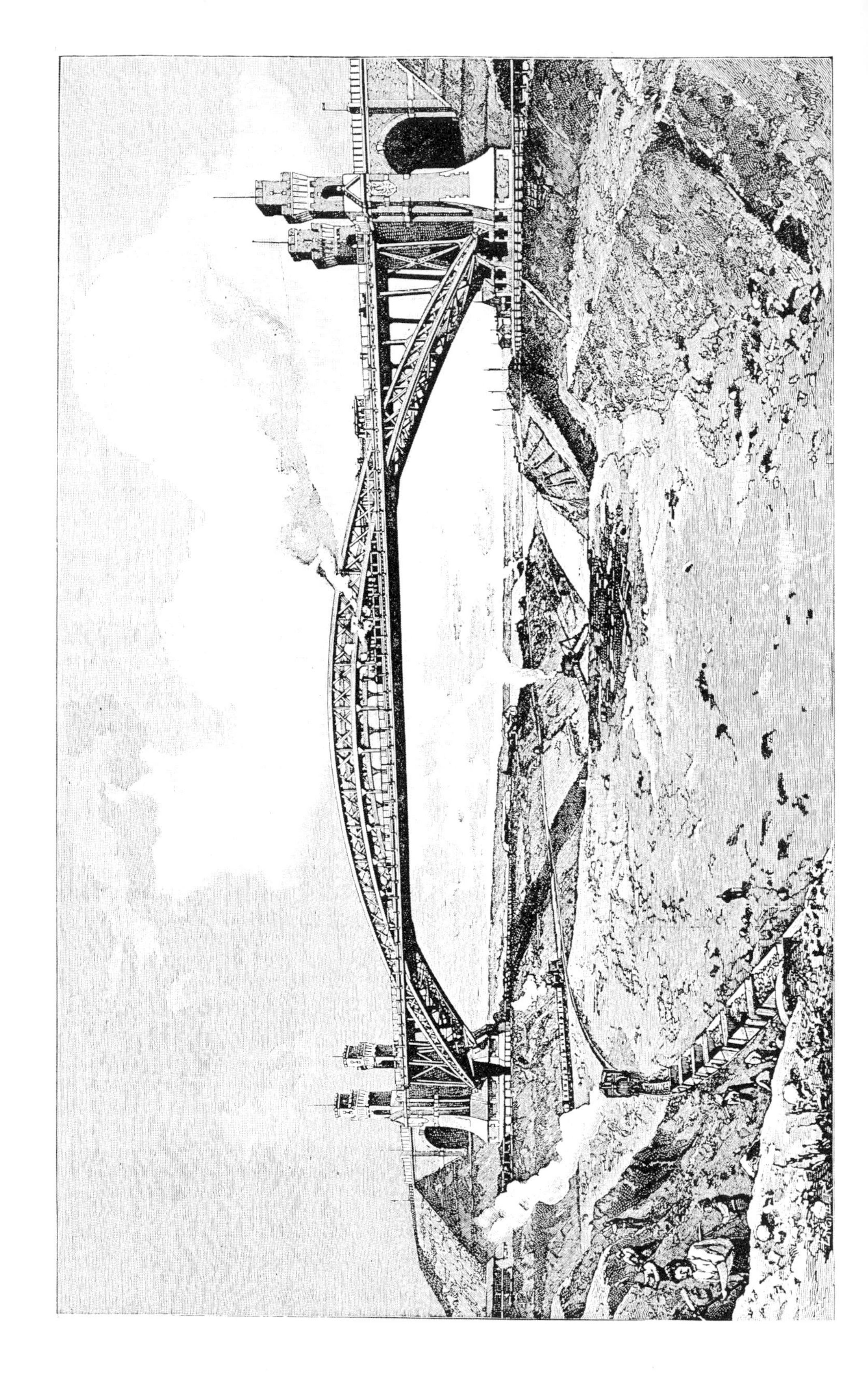

transitted it in 1887, and wrote: 'We passed a good many vessels, schooners and ketches, of about ninety tons, clumsy-looking craft with lofty square sterns, but very handy; they turned to windward in the narrow reaches of the river as smartly as a Thames barge will. Those coming from the Baltic were generally laden with timber, those from Bremen and Hamburg with coffee, sugar, and other colonial product.'(3)

The Kiel (Kaiser Wilhelm or Nord-Ostsee) Canal was built 1887–95, 99km long, within what was now German territory, to link the Elbe estuary at Brunsbüttelkoog to Kiel on the Baltic, primarily to enable German warships to move easily between the two seas. However, it also became a busy commercial route, which greatly shortened the Hamburg-Baltic distance. By 1906 it was carrying 6M tonnes of commercial traffic.

At each end of the canal were parallel hydraulically powered 150m×25m locks, their sea entrances protected by stone piers on each side*. The rest of the canal was built level, 8.5m deep, with a bottom width of 22m and provided with passing places, though future widening was provided for. At Rendsburg, a side lock gave access to the Eider. As partially at Suez and much more at Panama, cutting was done by steam excavators and dredgers. Where the line crossed marshy ground, firm banks on each side of the future canal were built with mattresses of sand.

The Kiel Canal had only been open for twelve years when, because of the increasing size of both warships and merchantmen, it was decided to enlarge it. Between 1909 and 1914, with the help of Lübecker land dredgers, the bottom width was doubled to 44m and depth increased to 11m, thus doubling its cross-section. Eleven large passing basins were provided, against eight smaller ones, and electric light throughout on both banks for night navigation. New parallel locks were 330m×45m (battleship dimensions), and they had electricitly powered sliding caisson gates, with a third to divide each lock into two compartments, the old locks being also retained. Transit took some eight hours.

* The canal was normally kept at the level of the Baltic, and therefore the locks at that end were usually kept open.

Fig 42 The Kiel Canal, nearing completion in 1893, is crossed at Grumenthal by a bridge carrying railway, road and footpath, while another of similar design is being built at Levensau. *The Engineer's* caption says: 'Many of our readers will recognise the design of the bridge as being precisely that of Sir Joseph Bazalgette, who made it about a dozen years ago for crossing the Thames near the Tower'.

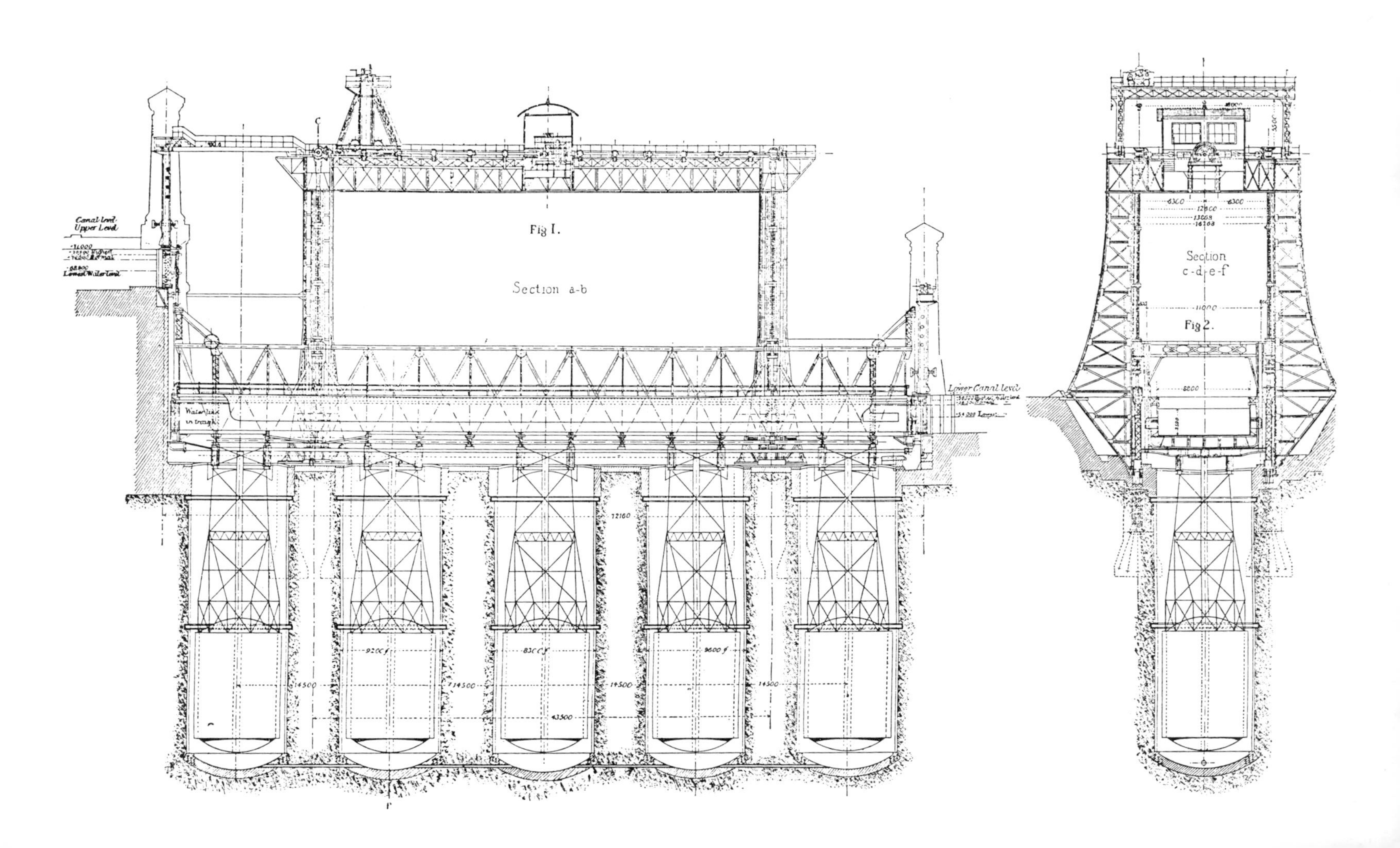

Fig I.
Section a-b
Canal level
Upper Level
Lower Canal level
Section
c-d-e-f
Fig 2.

Germany: 1887–1914

The start of work on the Kiel Canal in 1887 coincided with a worldwide surge of interest in big waterways, and with an already planned new phase of German waterway development, with three interwoven motives: to improve the nation's transport system and her access to North Sea ports; relieve her railways of traffic; and improve her military capability. The plan was based upon large-capacity (450–600 tonnes and up) barges, mechanical propulsion or traction, and the application of modern technology by using high-rise locks (10m was then thought practicable) or vertical lifts. It was intended to give the Rhine and Ruhr a water connection with a North Sea port lying entirely within Germany, (the Rhine seaports lay in the Netherlands), and to link this connection with Berlin, the Oder and Vistula, by way of Ems, Weser and Elbe.

In 1886, the first part of the plan was authorised, the Dortmund-Ems Canal, 269km long, to connect the Ruhr with Emden* and the North Sea. It was completed in 1899. As built, the canal had a depth of 2.5m and bottom width of 18m and took 600-tonne barges, though 400-tonne craft were also found useful. Some 15km from its commencement at Dortmund, it fell 14.5m by the Henrichenburg lift (see p. 133). Thence it ran to Münster, where it was lowered through two locks into what was to become the long level of the Mittelland Canal, with its junction at Bevergern. It then fell through seven more locks to the Ems at Gleesen, followed that river for a short distance, and entered the Haneken Canal. Sixteen miles long, this had originally been built as an Ems lateral canal by the Hanoverian government in 1825. It was now widened and deepened to suit Dortmund-Ems dimensions. From Gleesen downwards, both in the Haneken Canal and the river below, larger locks were provided, able to take a tug and two barges at a locking. Below Herbrum, no improvements to the Ems were needed, until the two entrance locks of the short 9.2km canal that led to the port of Emden. Locks with larger falls were electrically powered; others were worked by hand.

Notable engineering features were the Henrichenburg lift, masonry aqueducts over the Lippe (three arches of 21m span), Stever (three

*In the seventeenth and eighteenth centuries Emden's poor harbour and distance from industrial centres had kept it small.

Fig 43 A contemporary engineering drawing of the Henrichenburg vertical lift, showing its arrangement of floats. The design was that of Haniel and Lueg of Düsseldorf.

arches of 12.5m) and Ems (four arches of 12m), and several substantial embankments and cuttings. Aqueduct channels were lined horizontally and vertically with sheet lead. Water for the upper part was provided by a station pumping from the Lippe, and for the section between Münster and Bevergern from the Ems.

Once open, the canal was used not only for traffic to and from Emden, coal being exported and iron ore and timber imported, but by seagoing tug-hauled lighters from other German ports, and also by Dutch coasting vessels or barges reaching it via the smaller Haren & Rutenbroek or Ems-Vechte Canals. Traffic on the Dortmund-Ems, 470,000 tonnes in 1900, was 4 M in 1913.

The canal-building drive continued with Prussian legislation of 1905, which authorised the completion of a Rhine/Ruhr/North Sea line with the 1500-tonne Rhine-Herne Canal, 50km long from Datteln on the Dortmund-Ems Canal to the Rhine at Duisburg-Ruhrort. It had special engineering interest because built with mining subsidence in mind. Bridges could be raised hydraulically and lock-gates were usually of falling type above and sideways-moving below, to minimise leakage. It was opened, significantly as it turned out, in 1914.

The Mittelland Canal, to link the Dortmund-Ems Canal with the Elbe, and so with Berlin, the Oder and the Vistula, was also authorised by Prussia in 1905 after twenty years of opposition, notably by the railways, themselves state owned. This line built for 600/700-tonne barges struck away eastwards from Bevergern on the Dortmund-Ems with a lockless stretch 211 km long, and had by 1916 reached Hannover on its way to join the Elbe at Magdeburg. At Minden* it crossed the Weser on an 8-arched ferro-concrete aqueduct 375 m long and 24 m wide, completed in 1914. It was given links to the river, and so to the seaports of Bremen and Bremerhaven, by a pair of locks by way of the town basin and also by a single shaft-lock, built in 1912, 83 m by 10 m and with a fall of some 12.7 m, depending upon the Weser's level. The shaft-lock's construction is interesting, in that 8 tanks each side, linked in pairs to the lock chamber, and all part of the same structure, act as economiser basins, and also enable very fast operation. Pumping stations at each end of the aqueduct supply water to the Mittelland Canal itself. After completion to Minden, it was decided to enlarge the canal's capacity to 1000-tonne barges and it was rebuilt backwards to its junction with the Dortmund-Ems (also to be enlarged) by raising the water level by 40 cm. Beyond Minden it was built to 1000-tonne standard.

*The Weser and Fulda had been canalised from Minden to Kassel in 1893–7.

Eastwards of the Elbe, work was concentrated upon improving Berlin's links with the Oder, and upon the waterways of the city itself. In 1887–97, the Oder-Spree Canal had been built with 7 locks from the Spree at Spandau to the Oder at Fürstenberg to replace the older and smaller Friedrich Wilhelm Canal of 1669 on a slightly different line. Because the Oder took smaller barges, the new canal was built for 400-tonners. This modern and much shorter line now diverted a good deal of Silesian traffic from Stettin to the port of Hamburg, and was later enlarged to 500-tonne standard. The 600-tonne Teltow Canal was then built in 1901–7 as a Berlin by-pass from Köpenick on the Oder-Spree Canal past Machnow lock* to the lower Havel, and quickly attracted waterside industry.

A curiosity of the 70 km run between Zehdenick on the upper Havel and Berlin was the fleet of 118 battery-propelled steel barges each carrying some 200 tonnes of bricks, put on in 1907. They enabled boatmen without any mechanical knowledge to captain their own craft. Each battery charge enabled a run of 20–25 hours at a fully laden speed of some 4 kph. After WWI the Zehdenick brickworks declined, so the boats worked further, using additional charging points. Such barges were still operating economically in 1942.

The balance of advantage of Berlin's trade had now tilted towards (non-Prussian) Hamburg and Lübeck and away from Prussian Stettin. To help equalise it, the Berlin-Spandau Canal was partially realigned and enlarged to 600-tonne capacity. The old and small but heavily used** Finow Canal with its eleven† locks surmounting the high ridge between the Oder, Havel and Spree and taking 150–70-tonne craft now needed to be superseded. The Prussian canal law of 1905 therefore authorised the 600-tonne Hohenzollern Canal from the Havel to the Oder, the 36 m fall being taken by four high-rise locks grouped in two staircase pairs. The canal was opened in 1914.

Given these developments which placed Berlin at the centre of a quadrilateral of cities: Breslau, Stettin, Hamburg and Magdeburg, the city needed inland ports to increase accommodation and release lengths of canal and river bank for amenity use. One, the East Port, with lay-bys, quays, warehouses, cranes, grain elevator, brick-transporter and its own power station, was built between 1907 and 1913, a second and larger, the West Port, between 1923 and 1927.

*Its lower gates of the lifting type, an early example.

**In 1890, 15,451 craft passed through Eberswalde lock on this canal.

†19 through from Berlin to the Oder.

Fig 44 Towards the east before World War I: log rafts and a barge at lock 3 on the Bromberg Canal about 1908. Bromberg, then in imperial Germany, is now Bydgoszcz in Poland.

Beyond the Order, the Warthe (Warta) was already navigable for good-sized craft from Küstrin to Posen (Poznan). Between 1905–15, the eighteenth-century Bromberg Canal was rebuilt to link the Vistula* with Bromberg, and in 1891–6 the Netze (Notéc) had been improved back to the Warthe. By about 1914, therefore, Germany had provided herself with a continuous east-west large-capacity waterway line from the Rhine to the Vistula, except for the Weser-Elbe section of the Mittelland Canal, still being built. From the lower Vistula the River Nogat ran to the huge lagoon, the Frisches Haff, with its sea entrance at Pillau, whence by the River Pregel Königsberg (now Kaliningrad) could be reached. Thence there was access to the Memel (Niemen), navigable for boats up to some 150 tonnes.

Canals built to improve port access should be mentioned here. In the fourteenth century the Stecknitz Canal had linked the Elbe with Lübeck and the Baltic. A modern waterway being now needed, the 67 km Elbe-Trave Canal was begun in 1895 and completed in 1900, to take the largest 800-tonne Elbe barges and also seagoing lighters, and so link Hamburg and Lübeck. The line rose through 2 locks to its summit, then fell through 5 more to the Trave river, lock-gates being

*Navigable within German territory from the sea to above Bromberg before passing into Russian Poland, often by disposable craft of some 100 tonnes broken up when their voyages ended.

water powered using syphons and suction cylinders. Eastwards, the 33 km Königsberg tidal ship canal, completed in 1901, ran mostly between embankments along the northern side of the Frisches Haff from that city to its outport of Pillau (now Baltiysk). Previously, channel depth being only some 3.3 m, ships only reached Königsberg after part-cargoes had been discharged to lighters at Pillau. The new cut was 6.5 m deep at low tides, with a minimum width of 30 m, more on bends, and two passing places. Again, the Kaiser Fahrt, finished at the end of the nineteenth century, enabled 12,000-tonne ships to pass from Swinemunde on the Baltic along an improved Swine channel to the Stettiner Haff and so to Stettin.

Major improvements were also made to many waterway harbours, as Frankfurt-am-Main in the 1880s and again in 1906–12, Karlsruhe in 1888–91, and Dresden in 1891–3. Such improvements, which were widespread, cut down boats' turn-round times and so costs incurred, by such mechanical aids as coal chutes and tips, hydraulic, steam and electric cranes, depots served by elevated lines of rail and fixed and movable grain elevators.

The feeling for waterways that was to produce the canal law of 1905 also caused discussion on two possible large-canal Rhine-Danube links as alternatives to the old and declining Ludwigs Canal. Before 1914, plans were being discussed for a Rhine-Neckar-Danube Canal, with the development of Württemberg primarily in mind. The Neckar was to be canalised, and a canal built over a considerable watershed thence to Ulm on the Danube, whence a lateral canal would run to Kelheim above Regensburg, because of the difficulty of canalising the river. Since WWII, the Neckar has indeed been canalised, and the planned link appears on some waterway maps. An Italian engineer, Pietro Caminada, in this period suggested also a canal from Ulm to Lake Constance. Energy and resources, however, have in fact been concentrated on the alternative, a Rhine-Main-Danube canal to follow roughly the line of Ludwigs. The Bavarian diet was discussing this back in 1904, but a start had to wait until after WWI.

Other proposals dating back to the late nineteenth century were for an Elbe-Moldau (Vltava)-Danube Canal to run from Budweis for 219 km to Korneuburg near Vienna, using 15 locks and 4 inclined planes, or for a Danube-Oder-Elbe Canal originating near the present Ostrava in Czechoslovakia. The latter scheme has been kept alive, and today its prospects are perhaps brighter than ever before.

Resulting from their forward policy, German canal and river craft increased from 17,653 in 1877 to 22,564 in 1897, and carrying capacity from 1,400,000 to 3,400,000 tonnes. The cost of inland waterway transport was halved, and between 1875 and 1910, traffic increased sixfold, also increasing its percentage of the rail/waterway total from 21 to 25 per cent. Other figures show that net tonne/km increased between 1875 and 1905 from 2900 to 15,000 M, and the average distance transported from 280 to 290 km; corresponding figures for railways were 10,900 M to 44,600 M and 125 km to 151 km.

Here are comparative figures at certain harbours, in thousands of tonnes, to show the trend:

	1875	*1905*
Duisburg-Ruhrort, Hamborn	2935	19462
Berlin	3239	10114
Mannheim-Ludwigshafen	865	7117
Magdeburg	676	2008
Frankfurt-am-Main	201	1580

Here also are 1905 figures for river traffics:

	Millions of km tonnes
Rhine	4045
Mosel	6
Weser, down to Bremen	109
Elbe	2475
Oder, down to Schwedt, (except canal division)	751
Warthe	88
Weichsel and Nogat	175
Memel	112
Total:	7761

Italy

By the beginning of the 1890s waterway carrying had also revived in Italy after the initial impact of railways.

There were first the rivers—the Po running inland from the coast south of Venice past Ferrara and Piacenza to Pavia, taking large barges to its junction with the Mincio, and then 175-tonners up that to Mantua, smaller craft to near Cremona, and still smaller to Pavia, whence the Ticino taking 50-tonners ran from Lake Maggiore. The Adige left the Po near its mouth, and was navigable for some way for 150-tonners. Up the coast from the estuary of the Po, a large coastal waterway ran past Chioggia to Venice, and a smaller one, served by a myriad rivers and canals running inland, to beyond Grado.

Some 364,000 tonnes pa were carried on 129 km of small, old and interconnected canals: the Naviglio Grande itself running from Milan to Abbiategrasso and then north to Lake Maggiore, the Bereguardo southwards from Abbiategrasso, the Naviglio Pavese from Milan to Pavia, and the Naviglio Martesana, eastwards to and beside the River Adda. Then came the canals from Padua to Venice (377,000 tonnes), those converging on Brondolo by Chioggia (950,000 tonnes), and those leading from Padua towards Chioggia.

In all, inland waterway traffic, lake, river, and canal, ran out at some 6M tonnes, then about a quarter of all Italy's freight traffic. Passenger-carrying was of most importance on Lakes Maggiore and Como, on which some 1½M were carried each year, and of course in Venice, where in 1881 the long ascendancy of the gondola ended with the appearance on her canals of the French-built *Regina Margherita*, the first *vaporetto* or steamboat-omnibus.

Early in the century the engineer Pietro Caminada suggested a canal from Genoa under the Alps to Germany, to take 600-tonne barges. He proposed that for the mountainous section of the route, double slanting tunnels of some 15.25m diameter should be used, within which staircases of locks would be built. Craft would be steered by guides projecting upwards to bars on the tunnel roof. It was hopefully said of this terrifying system, that

> A model...has been constructed and found to work so perfectly that much attention has been drawn to the project, which is calculated to transport fifteen million tons from the Mediterranean to the North Sea in a single year...(4)

A trans-Alpine canal of some kind is still discussed from time to time.

The Danube and Balkans

The Treaty of Paris, 1856, established the European Commission of the Danube, to remove the sandbanks between Isaktcha at the head of the delta and the Black Sea. The Commission's jurisdiction was extended to Galatz (Galati) under the Treaty of Berlin, 1878 (which also empowered it to operate independently of territorial authorities), and Braila by the Treaty of London, 1883. This last set up the International Danube Commission to give the river above Braila international control also. In 1856 Sir Charles Hartley had been made engineer-in-chief to the European Commission, upon which the United Kingdom was represented.

The character of the Danube was that of a sediment-bearing river subject to great floods and changes. Fast and difficult down to Vienna, with rapids and sandbanks, it remained fast to Komárom in Hungary, though because of shoals below Pressburg (Bratislava) this part of the river was in 1879 said to be fully navigable only for 202 days a year. Beyond Komárom it is easily navigable to the Carpathian gorge, after which it spreads out on the flatlands towards the delta. Its three main navigable tributaries are the Drava and Tisa in Hungary, and the Sava in Yugoslavia.

In 1856 the depth of the channel inwards from the ancient Black Sea port of Sulina averaged about 2.5m in low water. In the following year the European Commission's dredging and training work began. By 1870 the Sulina channel's minimum depth was 4m. Meanwhile, in June 1880, when the largest steamer able to navigate was 1462 nrt, a new entrance to the delta 1006m long and 7.3m deep was begun. It was opened in December 1882, after which the old Kilia (northern) branch of the delta was no longer used.

Trade to Sulina, 680,000 gross tonnes in 1859, was 1,530,000 in 1883, two-thirds then being in British ships, which at high water went up to Galati or Braila to discharge merchandise and coal, and load grain. At low water they usually loaded at Sulina from lighters, each carrying 300–1000 tonnes and drawing 2.5m to 3.6m, which had been towed by steam-tug from Braila.

Depth in 1889 was 4.88m, by which time some sharp bends in the channel had been eased by cuts. Helped by two British-built suction dredgers, *Delta* and *Hartley*, introduced in 1890–1, further cuts and additional deepening went on for the remainder of the century, by about 1902 giving a minimum depth of 6.1m. Hartley himself was succeeded in 1872 as resident engineer by Charles Kühl, but remained chief engineer till 1907.

Higher up, the Danube fell 26m in 111km of winding, often narrow and sometimes rocky gorge through the Carpathians, with the so-called Iron Gates at its lower end. Before steam, the passage was often dangerous, and both freight- and passenger-traffic sparse. Then, when steamers came, three primitive cuts were made to improve transit at Djevrin, Sip and Mali Djerdap. By the 1878 treaty, Hungary (then part of the Austro-Hungarian empire) was made responsible for improving the Djerdap passage, and given power to charge tolls. She then prepared a plan to cut a 2m-deep, 60m-wide channel through its rocks and shallows, and at the Iron Gates to build a lateral Sip canal on the Serbian (now Yugoslav) side with two locks 155m × 36m to overcome the 4.4m rise there.

Work began in 1895. The channels were cut, but instead of a locked canal, Hungary built the Sip Canal, 2.5km long, stone-pitched on both sides, with a fall in level over the distance of 3.7m. After experiments with chain-towing the International Danube Commis-

Fig 45 On the Danube in the 1930s; (top) A two-funnelled tug ahead and a towing locomotive behind work a tow through the Sip Canal at the Iron Gates; (above centre) the locomotive and DDSG's *Oesterreich* towing together up the Sip Canal; (left) the passage of the Kazan gorge.

sion then supervised the laying of rails on the Serbian bank, a heavy steam locomotive being provided to help haul barges up. (Higher up the gorge, a special tug also helped tows against the current.) The job was done by 1898. Whereas 1175 craft passed the Iron Gates in 1899–1900, 2686 made the passage in 1912, with nearly tripled tonnage, though tolls levied did not meet interest charges.

Upwards to Vienna, traffic was mainly worked by the paddle-tugs and barges of the DDSG. By 1885 groynes had been built in places, but without much effect on the normal depth of 1.2m–1.5m, which fell to 0.9m at Gönyö, where a steam-raking system was used. From Gönyö north for some 112 km to Pressburg (Bratislava) the channel was notorious, barges sometimes having to tranship part of their cargoes to lighters, while their tugs made multiple trips. Above Pressburg, chain-towing was experimentally introduced on the 61 km to Vienna. Used for upstream towing only, with a maximum load of 1000 tonnes in four barges, against about 500 tonnes in two or three, it proved uneconomic, and had been abandoned by 1889. At Vienna itself, resulting from an Imperial Commission of 1866, a 16km canal, 305m wide and 3–3.66m deep had been cut to the city and back again to the Danube, mainly as a measure of flood control, but also for navigation.

Above Vienna to Passau, 653km, improvement work had by then produced 1.2m of depth, except at one set of rapids where it could be only 0.9m. Above Passau, navigation continued to Kelheim, where Ludwigs Canal came in, and then continued after a fashion for 100-tonne barges to Ulm.

Of the Danube's tributaries, no improvement work had been done on the Sava and the Drava. The latter was navigable for some 138km, and the former to Sisak, 595km. The Tisa (Theiss) was navigable for 764km to Tokay. Between 1832 and 1879, however, it had been shortened by some 483km as a result of flood-relief works across the plain.

In what was then the Austro-Hungarian empire and is now Yugoslavia a canal had been built between 1795 and 1802 from Bačko Gradište on the Tisa to the Danube near the present Hungarian border. However, the Danube stream moved west, and between 1851 and 1868 the canal was extended westwards to meet it, the former outlet then becoming disused. Bezdan lock, near the entrance to the Danube, is claimed to be the first built of concrete in Europe. It was completed in 1856. This canal has many names: the Franzen, Ferencz, King Peter I or Veliki Backa. Another, the Ferencz-Josef, King Alexander I or Mali Canal, was begun in 1855 and completed in 1872 to run from it to the Danube again at Novi Sad.

These canals had been built for irrigation and flood-prevention purposes, but in 1870 a private company* was given a 75-year navigation concession upon them. They were also empowered to canalise a branch of the Danube running from south of Baja in southern Hungary, and use it as a navigable feeder to the King Peter I Canal. The same company also altered the Tisa end to terminate at Stari Bečej, the old Bačko Gradište outlet having become disused owing to eastwards movement of the Tisa channel. The company had traffic of 396,000 tonne/km in 1876 and 970,000 in 1880, but in 1885 was paying no dividend. By 1913, however, traffic was 432,990 tonnes with an average haul of 330km. As improved for navigation, the King Peter I Canal was 123km long, with 5 locks and took 600-tonne craft; the King Alexander I was 69km long with 5 locks, including a staircase pair at Novi Sad, and took 330-tonne barges.

The Bega Canal was also built for 74km from the Tisa a little above its junction with the Danube by way of Petrovgrad (Zrenjanin) to Temesvar (Timosoara), then in Hungary, now in Romania. It seems originally to have had 4 locks, 2 additional ones being added in 1900 and 1908. The locks below Petrovgrad took 1000-tonne barges, those above, 650–800 tonnes.

The Corinth Canal

The Corinth Canal makes a fitting end to this chapter. We have seen how, in classical days, the *diolkos* carried boats across the short but mountainous isthmus, 6km wide, which separates the Ionian Sea and the Aegean, and caused ships to make a voyage of some 320km round the southern tip of mainland Greece. Nero had started to cut a canal, others had planned and surveyed it, but the wave of French enthusiasm for isthmian canals after the opening of Suez, combined with one man's vision, brought it about.

Colonel (later General) Istvan Türr was a Hungarian who in 1849 was persuaded by Italian patriots to go over to them from Austria-Hungary. In 1859 and 1860 he was one of Garibaldi's right-hand men in the conquest of Sicily. Then in 1875 this elegant figure with the splendid moustaches formed a French syndicate to seek a concession from Colombia to build a Panama Canal. They got it, and in 1879 de Lesseps bought their rights. Having made a great deal of money, Türr left the syndicate and negotiated a Corinth Canal concession with the king of Greece. DeLesseps approved, a company

*This concern, whose English name was the Francis Canal Co, had in 1873 a capital of over £1 million, and had raised large sums on the British market.

Fig 46 A 500 francs bearer certificate of the original French Corinth Canal company. Issued at 325 francs paid up, the holder subsequently provided his final instalments of: 100fr and 75fr. Interest was paid during construction, and the holder was able to cash coupons (not shown in the illustration) for 12 of these half-yearly payments before the company failed.

was floated in 1882 and French contractors began work in that year along Nero's line under a Hungarian, Bela Gerster, who had previously engineered the Franzen Canal.

Using sophisticated boring machinery and dynamite, work went ahead, until in 1889, after having raised a good deal more than its original capital, the company collapsed in the wake of de Lesseps' Panama liquidation of the same year. A Greek company was then formed, which took over the French concern without any payment for work done. Cutting began again in June 1890, and on 6 August 1893 the canal, spanned by a single road-rail bridge, was opened by the

Fig 47 The original (French) Corinth Canal company at work in 1886. Cutting is being done in a series of levels, spoil being loaded into small tipping wagons which are run by inclined plane to the lowest level, 5 ft above the sea. There they are pushed by locomotive to a heavy steam dredger and lines of barges. These last, when loaded, are taken out to sea to be emptied.

king in his royal yacht after tribute had been paid to General Türr. It was 6342 m long, 8 m deep, with an average depth of cutting of 306 m over some 4.2 km, and an extreme depth of 459 m. Width at bottom was 22 m, at surface some 25 m. Above the waterline there is a benching of some 1.5 m; then, in the main cutting, the sides rise with a batter of 1 in 3 down to 1 in $1\frac{1}{2}$.

Traffic began to move through, but the 4.8kph current, unequal depth, and difficulty of navigating at night were deterrents, and receipts were lower than had been hoped, being only enough to cover maintenance. So in 1908, the year of Türr's death, the second company was also liquidated, and a third took over the canal. Improved since then, and kept open through wars and earthquakes, the Corinth Canal still serves the shipping of the eastern Mediterranean. With its towering and almost vertical rock walls, it is perhaps the most spectacular canal ever built.

From France to the Black Sea and the Aegean, inland waterway transport was being developed, rivers were being improved and canals built. What was happening and to happen still further to the east and south?

7
EAST AND SOUTH, 1869–1945

Our period is that of the universal steamboat, navigating rivers and lakes from the Seine to Siberia, China to Africa and Australia. Long before railways it appeared in unlikely places and along difficult navigations, carrying passengers and freight, and often hauling a barge or two. Steam- and diesel-tugs worked alongside it when enough traffic offered, and round it older types of craft still earned a living in traditional ways.

Finland

Finland's many lakes and rivers, linked by some 66 km of short canals, some incorporating locks (such as the flight of three south of Kuopio on the line from Nyslott) had 6600 km of marked fairway, 1.8 m to 2.4 m deep, the most important of which connected with the Saimaa Canal (see p. 88). Steamboats were the principal means of transport along these from mid-May to mid-November of each year, when waterways were ice-free; today they have diesel-powered successors. They were supplemented by timber-barges often bound for the Saimaa Canal, huge timber-rafts, and such oddities as church boats, carrying 80 to 100 people and rowed with twenty pairs of oars.

On the partly locked Uleå river, tar-boats were in the 1900s taken down the rapids manually so that the tar could be exported from Uleåborg on the Gulf of Bothnia.

Russia

European Russia in the 1880s had 30,550 km of navigable waterway and another 61,000 km practicable for timber-rafts. Her advantage was in her wide plains; therefore in gently flowing rivers, and the ease with which canals could be cut between them. It was, however, lessened by great differences between winter and summer river levels, 12m at Rybinsk, and the formation of ever-shifting shallows and sandbanks in flatter river sections, which sometimes compelled steamer landing places to be moved. Her disadvantage was her long winter, when northern rivers were closed from October to May, southern from November to April. During these periods goods, mainly timber, grain, coal and industrial raw materials, were stored, or else carried by sledge or rail.

The Volga and its tributaries offered an enormous navigation system, 12,000km for steamboats and 22,500km for rafts. It became usable at Tver (Kalinin), and we have glanced at steamboat services worked thence past Rybinsk (Shcherbakov) to Nijni Novgorod (Gorki), then past Kazan, Samara (Kuybyshev) and Astrakhan to the Caspian Sea, itself a lake. Nevertheless, in 1881 47,272 large timber-rafts also floated down the Volga or its tributaries, and in 1887 some $8\frac{1}{4}$M tonnes, mainly of corn and petroleum products, were carried in 30,000 craft other than steamers. The entrepôt function of Astrakhan is shown by figures of 1890. From upstream 2099 laden vessels arrived carrying 298,350 tonnes; and 2801 left to go back, with 1,265,000 tonnes. Towards the Caspian 2301 sailing vessels and 2211 steamers left, and 2257 and 2201 arrived.

A major river improvement was undertaken on the Moskwa. In 1873 a Russian company co-operating with French capitalists was given a concession to canalise the 160km from the city to the Moskwa's junction with the Volga's tributary the Oka, with seven dams and locks big enough to take Oka barges. Cable-towing was first tried, but soon given up in favour of horses, and by the early 1890s the Moskwa was carrying 200,000 tonnes of merchandise a year, as well as much timber. Meanwhile, the estuarial channels of the Volga were being dredged, and on the upper section between Rybinsk and Tver, a 337km chain had been laid and ten towing steamers built.

Of the three routes that linked Volga and Neva, the Vychene-Volotski, Tikhvine and Mariinski, the last-named was most used, and so was selected for development. By 1842 it was clear that its 160-tonne barge capacity needed increasing and the number of its 49 pound- and two flash-locks lessening, especially as the locks included several staircases. Improvements were made piecemeal as pressures made themselves felt along the line, and included three new canals. That they were needed is made clear in this description of the transhipment traffic waiting at Rybinsk in the 1870s: 'Hundreds of huge barks lie there, often packed so closely together that the whole river looks like a great dockyard.'(1) By 1885 a 150-tonne steamer had been able to pass from the Caspian to the Baltic, and by 1896 capacity was up to 350-tonne barges and lock numbers down to 39. Transit time for the 1144km from Rybinsk to St Petersburg had been halved to 45 instead of 90 days, and partial steam-tugging introduced. Traffic had increased. The 1860–9 average annual tonnage of 470,600

Fig 48 The *Illustrated London News* in 1885 produces a highly-imaginative picture of the Morskoy, or St Petersburg & Cronshtadt, Canal, soon after its opening by the Emperor and Empress of Russia on 27 May.

had increased in 1880–9 to 639,600, in spite of an increase also in competing railway carryings. By 1893 the figure had reached one million, and in that decade the Mariinski route was chosen for further upgrading to 600–800-tonne standard.

Ship-canal fever reached Russia with the Morskoy Canal and that proposed between the Baltic and the Black Sea. The former was cut in 1878–85 to provide a ship channel between St Petersburg and Kronshtadt, and so save double transhipment, eg at the former, of export merchandise from smaller to larger barges, and at the latter, from large barges to ships. From the Neva in the city the line, 27.5km long and 7.4m (by 1908, 8.5m) deep, ran first in the river, then separated from it by twin embankments, for the remaining distance as a dredged channel.

Railway competition seriously affected canal traffic, except when channels were enlarged as on the Mariinski route, but the larger rivers much less. Figures given in 1894 show a tonnage* carried on the

*One assumes timber-rafts are excluded from these figures.

enormous extent of the Russian waterways (32,762,000) as rather less than the British (36,855,000), but greater than the French (23,320,000) or the German (13,700,000). But whereas the average haul in Britain was 63.57km, in France 136km and in Germany 350km, in Russia it was 1019km. Therefore the tonne-km total was over three times that of the other countries combined. Yet, whereas state expenditure on the waterways per tonne-km was 3.19 centimes in France and 0.43 in Germany, it was 0.064 in Russia.

Steamboats continued to serve the bigger rivers. In 1914 services were being run by several companies on long stretches of the Volga between Tver and Astrakhan, and also 800km up its tributary the Kama to Perm, up the Kama's tributary the Belaya to Ufa, and for 640km up the Oka from Nijni Novgorod to Ryazan. Stops were made at floating shore stations, where passengers could eat, rest or wait between steamers.

In western Siberia regular steam-tug services had begun in 1846. Traffic grew rapidly, and the number of powered vessels almost doubled every decade. Barges were large and usually wooden, and in the early 1890s there were up to 500, with an average capacity of 985 tonnes, each normally making two long trips each season. River traffic in the 1890s in western Siberia was about 328,000 tonnes, of which 262,400 was along the Tura and Tobol rivers, cargoes being mainly agricultural, along with coal and salt. Timber movements by raft were additional. Until passenger-steamboats appeared in the 1890s, people had travelled on the tugs—or the barges.

Fig 49 A Russian river steamer of Czarist days on the middle Volga near Nijni-Novgorod.

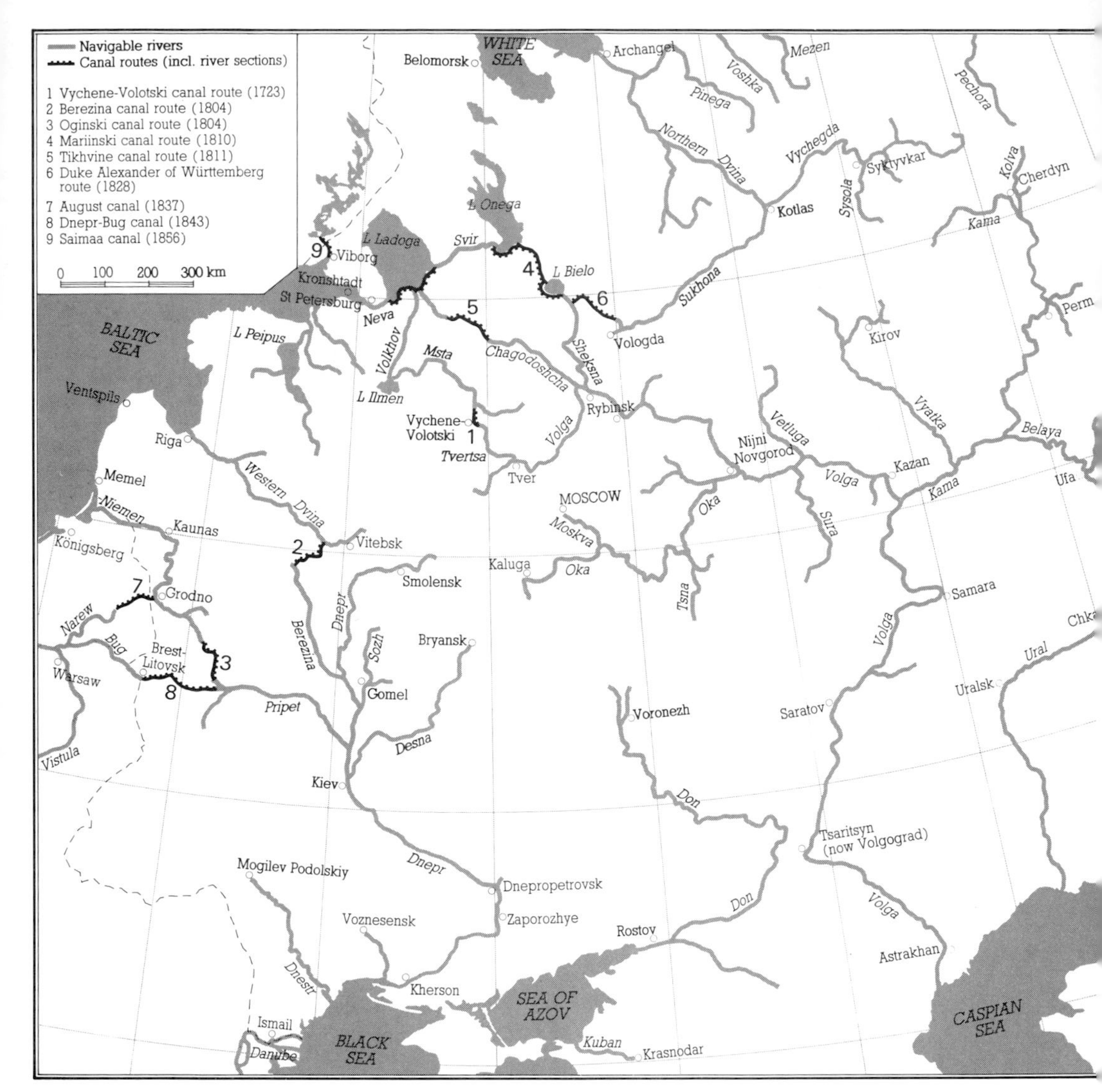

Map 8 The principal Russian waterways in 1914

In 1882 work began on a canal to link the Ob and Yenisey rivers by way of the Ket and the Kas. It was to have 33 locks and a cutting 6.6 km long. Semi-finished in the mid-1890s, it only had enough water for 80-tonne craft at the beginning of the short navigation season, and soon became disused. No waterway, however, linked western Siberia to European Russia. Siberian river traffic bound west had to be taken on the very difficult sea voyage. The situation changed when in 1885 the Perm-Tyumen railway was built across the Ural mountains.

In the last quarter of the nineteenth century Russia was as anxious to be linked with the province of Vladivostok as earlier the United States had been to California or Canada to British Columbia. The first

Fig 50 A Siberian river steamer at Omsk on the river Irtysh about 1911.

improved links were by telegraph, road and waterway, though even in 1890 it took seven weeks to get from Tomsk to Vladivostok, using steamers up to Tomsk and on the Amur. Railways followed, until French capital enabled the Trans-Siberian to be built. On 28 December 1899 it was finished (except for a train-ferry across Lake Baikal) from Moscow to Sretensk on the Shilka, a tributary of the Amur, and from Kharbarovsk on the Amur to Vladivostock. The gap of well over 1600km had to be covered—though not in the winter—by wood-burning, usually side-paddle steamers, smaller on the rapid, narrow Shilka, large on the broad Amur.

Francis Clark*, who took steamer on 1 June 1900 from Kharbarovsk to Sretensk, was an early passenger on the new all-steam route to Europe. He travelled for 20 days on three steamers and also, water being low, on a tug-hauled passenger-carrying barge, and saw numbers of Russian emigrants who were moving the other way as settlers in eastern Siberia, which in 1868 had been ceded to Russia by China. They travelled on steamers, rafts—some of which carried horses—and tug-hauled barges. Clark gives us one unexpected navigational detail—that the Amur was lit by oil lamps on poles for night navigation, these being attended by lamplighters in canoes.

The Trans-Siberian opened throughout, using the Baikal train-

*Rev Francis E. Clark, *A New Way around an Old World*, 1901, S. W. Partridge.

ferry, in 1903. The rivers indeed kept much of their traffic, but its direction of flow altered. Because waterways ran mainly north and south, railways east and west, tug and barge services now became feeders to the railway lines, while continuing to exchange goods within an area made more prosperous by the railway. External river trade disappeared. Until the building of the Tyumen-Omsk railway in 1913, for instance, Tyumen and Tura between them transhipped 370,000 tonnes a year between rail and water.

From the 1890s, passenger-steamers, also towing barges, appeared on Siberian rivers for the five months during which they were navigable. People and goods were carried for hundreds of miles to and from town and village, wharves while timber-trafts floated past, worked by long sweeps manned by crews who lived upon them in huts.

Soviet Russia put major waterway improvements in hand as part of plans to strengthen the country's economy and strategic ability. The White Sea-Baltic canal-river-lake line incorporated 227 km of canals and canalised rivers with 19 locks able to take seagoing craft. It was opened in 1933 from Belomorsk on the White Sea south to Povenets on Lake Onega, whence by the Svir river and Lake Ladoga large craft could reach the Neva, Leningrad (St Petersburg) and the Baltic, thus avoiding the journey round Norway.

A Leningrad-Moscow route for large craft was provided by rebuilding the old Mariinski waterway and the upper Volga navigation, and replacing 44 smaller locks with 9 larger between Rybinsk and Leningrad. From the rebuilt waterway the Moscow-Volga Canal, 128 km long and 5.5 m deep, built partly with labour released from the White Sea Canal, was opened in 1937. It had two purposes, to enable navigation and also generate electricity at its eleven locks 290 m × 30 m × 5.5 m. A Russian statement at the time of opening said that it was planned ultimately, to carry 3.6 M tonnes a year and 5 M passengers.

A third important waterway, the Volga-Don Canal, to link the lower river at Volgograd (Stalingrad) to Kalach on the Don, and so to the Black Sea, though approved in principle back in 1887, was begun in 1938. It was completed in 1952, 101 km long with 13 locks, (9 of them rising 88 m from the Volga to the summit reservoirs, and 4 locks falling 44 m). This, like the Moscow Canal, is remarkable for the huge and elaborate architecture of its lock structures.

We can glance now only at the Dnepr. This great river, third largest in Europe after the Volga and the Danube, would have been navigable from the Black Sea for 2075 km to Dorogobuzh, were it not for a series of rapids below Dnepropetrovsk. Efforts had been made from Peter the Great's time onwards to by-pass these and side-cuts had indeed been made, though they were only useful for very small boats. Otherwise traffic was limited to downward timber-rafts and merchandise-carrying craft that could be broken up at the end of the voyage. Success came with the Dneprostroi scheme, a predecessor of the somewhat similar works carried out later at the Danube's Iron Gates. Between 1927 and 1932 a dam for hydroelectric generation was constructed below the rapids, which backed up the river water to give a minimum depth over them of 5.3m. Navigation access was by a 3-rise staircase of locks, each 120m × 18m and taking craft drawing up to 2.85m.

Two points should be noticed about Russian inter-war development: that a waterway's improvement was often linked to work undertaken primarily for hydroelectric purposes; and that though tonnage carried by waterway increased from 18M in 1928 to 73M in 1940, the proportion carried by water nevertheless fell—figures open to some doubt show 13.3 per cent in 1928 and 7.4 in 1940—as that of roads and pipelines increased. Traffic was heaviest in timber, floated and barged, followed by petroleum, and then grain, salt and building materials. Geographically, about seven-tenths was within the Volga basin, and one-tenth in Siberia.

* * * * * *

The three-quarter century of our period saw changes in Asian waterway transport perhaps more radical than had taken place in European. Here are some glimpses of what was going on throughout a continent.

Iraq

The Tigris and Euphrates had carried trading craft since classical times, some boat types, indeed, being almost unaltered. Steamboat services began as part of British efforts to speed up the mails between India and Britain in the years before the Suez Canal. In 1836 Colonel F.R. Chesney took a paddle-wheeler down the Euphrates from Birecik to Basra and up the Tigris to Baghdad. The route was not chosen for the mail, but a Tigris service was provided during the nineteenth century by both British and Turkish companies.

British India

Tea made Assam. Though small amounts had been shipped since 1837, the boom in tea-garden investment began in 1864. Previously, a

Fig 51 (above) *Cochin* at work in British India: 64m long and drawing some 1.5m she was built about 1922 by Denny's of Dumbarton, Scotland, and sent out ckd to be assembled in Calcutta. With a crew of 23–24 who worked three shifts for continuous operation, she towed a 1000dwt flat lashed to each side, and worked until her company closed.

Fig 52 (right) Three waterway enterprises from India and Burma: (above) an irrigation and canal company; (centre) the Irrawaddy Flotilla dated from 1852–3, the limited company from 1875; (below) the Oriental company was one of several started in the 1850s to compete in the Ganges steamer trade. Having suffered severe losses in the Calcutta cyclone of October 1864, it went out of business soon afterwards.

government steamer had left Calcutta for Assam by the Brahmaputra once every six weeks. Now the India General company took the service over, to move into Assam hundreds of labourers recruited from elsewhere in India. Steamers acquired upper decks to carry more passengers, and hauled accompanying flats, while the introduction of the compound engine from 1872 much reduced coal consumption, and so cheapened services. Weekly runs between Calcutta and Assam started in the 1870s. Not long afterwards, another great trade began, in river-borne jute from the alluvial plains round Dacca and Chandpur. Local people quickly took to Western ways: the first Indian captained a flat in 1833, and a steamer soon afterwards. Slowly European officers worked themselves out, the last in the 1920s.

Meanwhile navigable irrigation canals continued to be built, most by government, but some also by irrigation companies. Among the former were the 175km Agra Canal (1868–74) from the Jumna, the extension of the Ganges Canal to Allahabad (1872–8), and the Sirhind Canal (1869–82) in the Punjab, the Sutlej-Patiala portion of which was navigable. An irrigation company with government

CERTIFICATE OF SHARE.

No. 38671

£20

THE

East India Irrigation & Canal Company.

FIRST ISSUE OF SHARES.

ORISSA UNDER CONTRACT.

No. of Certificate 7921

Transfer No. 1116

Irrawaddy Flotilla Company

LIMITED.

(Incorporated under the Companies' Acts, 1862 to 1929)

Registered Office — 95 BOTHWELL STREET, GLASGOW, C.2.

CAPITAL £2,160,000 STOCK

CERTIFICATE.

No. 19509

CERTIFICATE OF SHARE.

£10.

THE

Oriental Inland Steam Company,

LIMITED.

INCORPORATED PURSUANT TO ACT OF PARLIAMENT.

Capital £250,000. IN 25,000 SHARES OF £10 EACH.

O.I.S.Co. LIMITED.

THIS IS TO CERTIFY that Sir John Login of 31 Portman Square W is the Proprietor of the SHARE Number 19509 of the ORIENTAL INLAND STEAM COMPANY, LIMITED, subject to the Regulations of the said Company, and that up to this day there has been paid up in respect of such Share the sum part-guarantee built a remarkable, though less successful, companion in the south, the Sunkesula Canal in Madras. From an intake and dam on the Tungabhadra river a 306 km canal, partly for irrigation and navigation, partly for navigation only, was built in 1861 – 71. With 7 staircase pairs and 34 single locks each 36.6 m × 6 m, together with the 14-arched Hindry aqueduct 198.4 m long, it was a major work.

Engineers now began work upon waterways to run along the coasts. In 1881, for instance, the 146 km Orissa Coast Canal was begun to link the Hooghly with the older Orissa canals by a new line to Cuttack, needed by both passengers and goods. In time a waterway system, part artificial, part natural, existed along most of the east coast, and on the west another southwards from a little south of Mangalore. This included, 51.5 km north of Trivandrum, an Indian curiosity, a pair of navigable canal tunnels 4.27 m wide and together 1.2 km long.

Ceylon (now Sri Lanka)

Though the island's earlier native and Portuguese rulers had built some, it was the Dutch who, as in their other colonies, set about creating water transport lines. These, mainly running parallel to the coast, used short canals, some locked, to link lagoons and rivers. When the latter ran inland, they could sometimes be navigated. The usual craft were small padda-boats with thatched roofs, towed, poled or sailed. Longest was the 132 km Colombo-Puttalam waterway north of Colombo that remained in use until the early 1950s. It was part of a

Fig 53 A quiet canal in Dutch-influenced Ceylon (now Sri Lanka) in the 1930s.

continuous water line from Kalutara on the Kolu Ganga across the Kelani Ganga at Colombo and on past Negombo to Puttalam and Kalpitiya. Padda-boats were regularly using the southern length between Kalutara and Colombo in 1926 when locks were constructed at Dematagoda and Kelaniya to pass craft through the southern and northern embankments of the Colombo Flood Protection Scheme. Other systems existed in the Galle district in southern Ceylon, and round Batticaloa to the east.

The British made some additions, but mainly went for road- and railway-building, till the canals slowly became disused, especially with the coming of lorry transport from the 1920s. In 1975 a Canal Development Division was set up, however, which has done some rehabilitation work.

Burma

As on the Ganges and Brahmaputra in British India, so on the Irrawaddy and its tributary the Chindwin in Burma, inland navigation played an important part in national life. The Irrawaddy, reached from Burma's capital, Rangoon, through the artificial Twante Canal, is usable for 1600 km past Mandalay to Bhamo and, in low water when the current slows, through the gorges to Myitkyina. Navigation is, however, hindered by varying water-levels, and sometimes made hazardous by sudden storms that bring high winds, or cloudbursts that cause water surges sideways from dried-up water courses.

The local boats were high-sterned wooden craft with bamboo masts and square sails, that in flood-time, but with the monsoon wind behind them, carried rice upstream from the delta, and in low water with the north or north-east wind brought back cotton, edible oils and local produce.

During the Second Burmese War (1852–3), an Irrawaddy Flotilla was formed from four steamers of the Bengal Marine. Government then worked these steamers on the river until in 1865 the Irrawaddy Flotilla & Burmese Steam Navigation Co was founded to carry mails, troops and stores, and also provide a service from Rangoon to Mandalay, and with subsidy to Bhamo, the first steamer arriving at Bhamo in 1868.

Services were increased, and in 1875 the concern was renamed the Irrawaddy Flotilla Co. Thereafter the company became a personality, its captains an élite. Some of the craft it operated were fine triple-expansion-engined paddle-steamers 97.5 m long, drawing only 1.2 m, yet carrying over 4000 passengers:

Fig 54 The thronging boat people of Bangkok, Thailand, in the 1930s.

> With two large roofed barges attached, one on each side, they carry, besides passengers*, all kinds of merchandise, rice, dried fish, salt from Lower Burma, tea and oranges from the Shan States, sesamum and ground-nut oil, chillies, tobacco, and many sorts of vegetables from Upper Burma, at one time silk from Mandalay and lacquer ware from Pagan. They act as mobile markets, almost as travelling villages.(2)

There was also a service on the Chindwin with sternwheelers drawing 0.75 m.

Railways developed, but the company's great days ended in 1942 with Japanese invasion. Most Flotilla craft were withdrawn upriver to Mandalay and sunk; there many still lie.

Siam (now Thailand)

The Bangkok plain is a criss-cross of rivers and canals, some of the latter medieval, but including 50 or more European-type locks built from the 1890s onwards. Any visitor to the Thai capital is aware of the life of the boat people. A question-mark over Thailand, however, has for long been the long narrow Kra peninsula, as tantalising to projectors of ship canals as to shipowners.

Two British engineers from India surveyed it in 1863, and proposed

*The story is that the company, being Scottish owned, always served its passengers porridge and Keiller's Dundee marmalade.

a mixed communication, using tugs and long, flat barges for goods and passengers up the Pakchan river for 42km, then an 80km ship-railway:

> The boats of eight or ten tons for the river service should form the bodies of the carriages for the railroad service, patent slips being formed ... up which the loaded boats may be dragged on their own wheels, which could form the slip cradles, and the boats could be tacked on to the engine, and proceed to the other side without any delay.(3)

But the modest estimate was not acceptable, and nothing happened.

Then came the French in 1883, surveying for a ship canal, but rather uncomfortably accompanied by a British engineer on behalf of the king of Siam. Thenceforward to present times proposals have been made for a Kra ship canal some 150km long.

Indochina

Though Indochina has some 13,000km of waterways theoretically navigable—most would be better classed as *flottable*—intensive water transport has always been confined to the delta regions. The principal river, the Mekong, though 4000km long, is so broken by rapids that even in the best days of the steamboats, these craft could only work over intermediate lengths, the longest the 500km stretch between Savannakhet and Vientiane. Nowadays, boundaries not obtrusive in French days hinder traffic on the Mekong, where elderly motor ferries supplement country boats.

China

China in the earlier nineteenth century tried to insulate herself from change and from foreigners. She failed in both.

Out of a waterway system of some 74,000km able to take native or bigger craft, which includes the Yellow river and in the south, inland from Canton (Guangzhou), the Pearl river and its tributaries, we can only glance here at the Yangtse and the Grand Canal.

The Yangtse, 4800km long from Tibet to the sea, runs by sixteen of China's largest cities, and at Zhenjiang is crossed on the level by the Canal. Large timber-rafts, cargo-carrying junks and passenger-boats, poled, rowed, towed or sailed, had navigated the river for centuries. The lower river upwards for 965km from Shanghai to Wuhan (now a triple city of which Hankow (Hankou) is a part) is mostly immensely wide, with flood-banks, though in some places it narrows between hills, at other divides into several streams. Ships can reach Wuhan, though navigation is made difficult by mist or fog between October and April, and probable low water between December and March.

Above Wuhan to Yichang (Ichang) the Yangtse winds in great loops, then becomes narrower and shallower as the mountains approach. For 206 km between Yichang and Wanxian (Wanhsien) are three gorges that once were the only access route to the rich province of Sichuan (Szechuan), and which have been navigated at least from about 1100. They narrow to as little as 137 m, and as the Tibetan snows melt can vary 61 m or more between winter and summer levels. Then the river is out into hilly country, with navigation not much less difficult, to Chongqing (Chungking), and beyond for smaller craft.

Once the Yangtse had been opened to foreign trade in 1860, steamers were put on between Shanghai and Hankou. It then became a main British interest to develop services. In 1895 the first experimental steamer passed the gorges to reach Chongqing. On the lower river many steamer licences were issued, and passengers flocked to them; in four months 60,000 were recorded as arriving or leaving Zhenjiang. Modernisation as swift as this in many spheres of Chinese life by nationals of foreign powers all competitive for business at a time of weak central government caused the Boxer rebellion of 1900. Christopher Hibbert writes:

> Repellent enough, in the Boxers' eyes, were the foreign businessmen and overseers whose detested railways would soon be carrying 'fire-carts' and rattling, iron-wheeled wagons all over the country, desecrating burial places, disturbing the spirits of the earth, putting honest carters and porters, trackers and boatmen, muleteers and camel-men out of work. Equally obnoxious were the foreign operators of the chugging steamships on the inland waterways, the foreign mining engineers.... the foreign importers... the foreign mechanics... (4)

After the rebellion, but at a time of rapid political change, development began again. Regular summer (high-water) services through the Yangtse gorges began about 1908, and by 1922–3 ran throughout the year. The craft used were strongly built with watertight compartments, short, light-draught, usually twin-screw and with three rudders. Later, some mechanical towing stations were installed to help craft past the worst points on the river.

> It is a strange sensation hanging in the tongue of a rapid, with the ship's engines going full speed ahead, wondering whether she will ride over or not; and then, when she reaches the top, being swirled away in a backwash towards what appears immediate collision with a jagged rock; only to turn in the nick of time, avoiding the danger by inches. (5)

Yet older ways continued:

> Though steamers are now, to some degree, taking the place of the old, heavy junk traffic, the smaller local trade remains in the hands of the junkmaster, and is carried on in the time-honoured manner of China, in which time has no significance. It is still worth the junkman's while to hire forty trackers to drag his vessel over the rapids of the upper river, or to wait a week on the lower reaches for a favourable wind... The lower river junks are large and heavy... with high sterns, tall masts and a wide sail area... the upper river junks depend little on sails, but are propelled mainly by oars with trackers on shore to haul them over the rapids. Anything from sixteen to twenty rowers will stand well forward in the boat and swing rhythmically with short, little strokes. A large steering-oar projects from the bows, intended to keep the vessel's stern to the current when shooting a rapid.(6)

Though the prevailing wind blew up through the gorges to help junks, oarsmen and trackers were also needed. An eyewitness in 1951, just after our period ends, said that

> Every hundred yards or so a junk was being pulled up the river, with a long heavy rope from the prow to the shoulders of a string of men and women ... the people straining at the rope appeared at times to be holding the boat stationary while the water rushed by. Laboriously one foot after the other was put forward. On each bank the winding foot paths worn clearly on the rocks by generations of boatmen, could be seen ... the path was simply along the face of the cliff, barely 10 inches [25.4cm] or so it appeared, and at times sloping towards the river. The people were lying forward as they strained, and at places it looked as if they could not indeed have stood upright because of the overhanging rocks.(7).

Professor Middleton Smith was indeed to write of China at the end of WWI in 1919 that 'The most comfortable method of transport in China is by steamer.... It might, in fact, be excusable to say that the only highways of China worth considering are those upon which ride steamers and junks'(8) for compared to India the country had few railways. Hence the huge trade importance of the Water-served Yangtse valley and the area round Canton. In the 1930s, it was reckoned that of all the Chinese waterways, some 6400km were available to large steamers, and another 24,100km to small ones.

The Grand Canal, 1747km long from Tianjin (Tientsin) to Hangzhou (Hangchow)—in fact a series of separate canals intersected by rivers running inland—had been deteriorating. Since 1878 grain from Nanjing had left it for sea transport, and later in the century most of the route was only usable by 30-tonne boats. In 1899, for instance, the canal was reported dry for some 13.5km south of the Yangtse for about four months. Yet Lord Charles Beresford wrote in

Fig 55 The Yangtse gorges: (above) trackers hauling junks in about 1938; (below) towing path cut out of the cliffs.

that year that 'It would be impossible to over-estimate the importance of this canal to trade and commerce if opened up and rendered navigable'.(9).

Professor Middleton Smith, writing in 1919, also says:

> the most remarkable piece of engineering work in the Far East has been gradually falling into a worse and worse state of repair... (it) does... illustrate one of the national vices... the refusal to expend money upon maintenance.... It can still be repaired. It is still capable of doing good work. It can play a by no means unimportant part in the industrial development of the country. (10)

The canal was then far from disused. After stating that it 'is the only traffic road for from six to eight million people', he quotes from the *North China Daily News* of 20 April 1917 an account of the canal during the previous twelve months—floods in the summer of 1916 that prevented its use; fifty winter days of ice; drought in the spring of 1917. The paper wrote: There are hundreds of large goods boats lying like logs in mud... All passengers have to be transferred several times, each time they are duly "squeezed"* and, moreover, have to accept what accommodation they can get.'

Twenty years later, again, the Grand Canal was still open, but neglect, silting and railway competition had lessened its importance. Steamboats and tugs, in addition to native craft, did however use the southern section north of the Yangtse, and south of the river to Hangzhou. On the former, about 177km long, and 46m to 92m wide, shallow-draught tugs and motor craft towing passenger-and cargo-boats could operate. By 1936, too, three locks had been built for craft of up to 900 tonnes drawing 2.5m, and other improvements made. The latter section, lockless, some 37m wide except at bridges, and with good stone embankments with a towpath on top, enabled craft drawing 1.4m to 1.8m to work. It was heavily used by junks and native passenger- and cargo-carrying craft, and also by small tugs towing up to six barges.

Japan

From the seventeenth century, coolies had towed boats on Japanese rivers, whose channels were sometimes improved by landowners who caused obstructions to be removed and shallows dredged. Irrigation canals might also be used for navigation. In the great Kanto plain wherein Tokyo stands, for instance, the Minuma-daiyosui Canal carried rice to Edo, the then capital city with a million of population, and was later linked to the Shiba river by two canals each with three locks.

In modern times some nineteenth-century canal-building took place with the help of Dutch and other engineers; for instance, the Biwako Canal (see p. 137) and the Tone Canal, completed in 1890, between the Edo and Tone rivers. From the 1910s to the end of WWII, also, city canals were built at ports like Tokyo, Osaka and Nagoya, cargoes from ships being transferred to barges to be taken to local factories or railway transhipment points.

*Compelled to pay a bribe.

Fig 56 Australia's paddler, *Murray River Queen*, enters Lock No 1 at Blanchetown on South Australia's Murray. The Rhine-type passenger vessel *Murray Explorer* Follows.

Australia and New Zealand

And so south-east to Australia, where most inland navigation was concentrated upon the River Murray system, with its main tributaries the Darling and the Murrumbidgee, a system that serves one-seventh of the surface land area of the Australian continent. First explored by Sturt in 1830, commercial traffic commenced in 1853 with the historic voyages of William Randell in the *Mary Ann* and Francis Cadell in the *Lady Augusta*. Over the next thirty years, over 6400 km were opened up by the captains of wood-burning paddle-steamers, taking stores up to isolated settlements and bringing down wool for shipment to Europe.

At the height of river activity, some 200 steamers with attendant barges were plying the rivers, though lack of water during dry seasons meant that navigation was often only possible for six months of the year. Many schemes were put forward for locking and other measures, but rivalry between the states through which the rivers flowed meant that nothing was done until the river trade was threatened with railway competition, part from a lock and weir in 1894 at Bourke, 1200 miles above the Darling-Murray junction.

Eventually, the River Murray Waters Act passed the Commonwealth Parliament in 1915 and, delayed by WWI, the first and lowest Murray lock was opened in 1922 at Blanchetown. Thereafter, rising costs led to the abandonment of the scheme by which much of the Murray, Darling and Murrumbidgee were to be made fully navigable, and effectively limited through navigation to the Murray itself. Nevertheless numerous weirs were built, eleven of them as far as Mildura above the Darling junction incorporating locks. In flood-time the weirs can be virtually dismantled, the locks being taken out of use and boats put through a navigable pass. Later, other dams for environment control were built at the river's mouth, one of which, at Goolwa, incorporates a pleasure-boat lock.

Commercial river trade continued on a reducing scale until after WWII, ceasing in 1963 when the paddler *Marion* steamed from Berri to Mannum to become a museum piece. Since then, however, pleasure-boat traffic has increased, in houseboats, trip-boats and such luxury passenger crafts as the *Murray Explorer*, *Murray River Queen* and *Murray Princess*.

Until the 1930s New Zealand had some river trade; indeed, possible revival on the Wairoa and Waikato near Auckland is currently being studied.

Africa: the Nile

Sailed cargo-boats have probably worked on the Nile for some 5000 years, thanks mainly to the prevailing wind helping them upstream, and the current down. The building of Nile dams was undertaken primarily to control water flow in the interests of agriculture, by husbanding the floodwaters of June to September, and releasing them into a network of irrigation channels from February onwards.

First came barrages across the two mouths of the river a little north of Cairo. Planned by a French engineer, Mougel, the Damietta (535 m long) was begun in 1843, the Rosetta (465 m) in 1847. Hurriedly and

Fig 57 African water transport: an early Nile tourist steamer.

poorly built, they were finished in 1861, each with two navigation locks, but partially failed until C. C. Scott-Moncrieff, who had learned his hydraulic engineering in India, rebuilt them in 1891.

Higher up, at Aswan, John Hawkshaw had been called in to recommend improvement to the navigation through the cataracts. When water was low, only sailing cargo-boats could pass upwards, given enough men to bow-haul them, or run downwards on the current. The cargoes of tug-hauled barges had to be transhipped, usually by camel. Hawkshaw in 1869 proposed a channel widened to 30.5m at bottom, with a large lock at the lower end.

Nothing was done as minds turned to what became the Aswan-Asyut dams. Planned by Sir William Willcocks, the Aswan dam was built by British contractors and opened in 1902, 1950m long and 20m high, with 180 sluices* and three locks on a navigation channel to the west, a work that has been called 'one of the finest dam-building achievements of all time'.(11) It fed water down to the 830m-long Asyut barrage half way between Aswan and Cairo, with its navigation lock that was wide enough for the biggest tourist boat. So successful was the Aswan-Asyut scheme that in 1912 Sir Benjamin Baker raised the dam to 27m; in 1933 it was again increased to 36m, with consequential changes to the locks.

South of Egypt, Sudan government steamers worked for all or part of the year over 3570km of the Nile and its tributaries.

Africa: Central, West and East

Steamboats to supplement age-old dugout canoes had been put on Africa's lakes and rivers as soon as ingenuity could get them there. Some, indeed, like the *Lady Nyassa* in 1862, had to be carried by porters piece by piece from the Zambesi river to be assembled and launched on Lake Nyasa (Malawi). Once there, they were heavily used for missionary work, trading and passenger-carrying. Later, the colonial powers maintained numerous passenger and freight services until after WWII, in the interests of the developing economies in their care.

The most extensive system was in the Belgian Congo (Zaire) where, of some 15,300km of navigable river, about 12,000km were regularly used by about 130 mail-, passenger- and freight-carrying steamers of up to 500 tonnes, run by two transport companies and a number of industrial firms. The main artery, the Congo river, not being navigable for its whole length, the navigable sections were

*The great number of sluices enabled the highly fertile Nile silt to pass through.

Fig 58 A stern-wheeled steamer on the Rufiji river in German East Africa (now part of Tanzania) c. 1910. With boiler forward in order to distribute weight and keep an even keel, the type was common on African waterways.

linked by railways. The Belgians also maintained variations upon steamer, tug and barge services on Lakes Tanganyika, Albert, Kivu and Meru. Pioneer steamboat on the Congo river was H. M. Stanley's 13.1 m-long paddle-steamer *En Avant*, carrying 7 tonnes, in 1881. In the 1930s its successors included the sternwheeler *Capilaine Hanssens*, 1690 tonnes, the side-paddler *Kalina*, 1100 tonnes, and 550hp sternwheel tugs. The largest barges were then 76m long and held 1756 tonnes. Sometimes these were lashed alongside the steamer, but usually towed astern.

A notable French achievement on the island of Madagascar was the excavation, under Gallieni's administration at the turn of the nineteenth century, of canal links between lagoons along the eastern coast south of Tamatave, forming the Canal des Pangalanes, 460km long. The French also ran steamer, tug and barge services, using smaller craft, in Gabon, and from Brazzaville up their 5300km of navigable waterways north of the Congo. In west Africa, they ran regular services over 1782km of the River Niger, and also between Saint Louis and Podor on the Senegal, except when water-levels were too low. Most of the rivers in the area carried some sort of trading or passenger craft. The most interesting navigation was perhaps the upper Niger. Rising in Guinea, it became navigable at Kouroussa, whence steamers and barges ran to Bamako (now in Mali). There were rapids between Bamako and Koulikoro, by-passed were rapids between Bamako and Koulikoro, by-passed by the Sotuba Canal and

Fig 59 Two early sternwheelers on the Congo: (above) from a stamp issued in 1898 by Leopold II's Congo Free State, which was in 1908 transferred to Belgian sovereignty; (below) from a share certificate of 1927 of the local shipbuilding firm, Chantier Naval de N'Dolo. It is easy to see how the design of the later craft has developed from that of thirty years earlier, which in turn derived from still earlier vessels built for use on Indian rivers.

the Barrage des Aigrettes. At Koulikoro (also the railhead for the line to Dakar in Senegal) the Niger was navigable to Timbuktu and on to Ansango, by steamers, tugs and lighters. Timbuktu itself was a major trading point where the Niger reached most closely into the Sahara, the city being linked to the river at Kabara by a side channel.

In Nigeria British firms and organisations maintained a network of passenger, mail and freight services on the Niger, Benue and Cross rivers. To the east, British services on the central African lakes were about 1930 providing 13 passenger-carrying steamers, varying from 60 to 1200 tonnes, and from sternwheel to twin-screw. Another two worked on Lake Nyasa to the south. Considerable tonnages were then carried on Lake Victoria Nyanza—in 1929, 129,372 tonnes to and from Kisumu. As some lake ports were given rail transport, so the lake services developed new ones.

We have travelled in this chapter from Finland to central Africa by way of China and Australia, before World War II began changes that were to leave us with a very different water-transport world.

8
INTERLUDE BETWEEN WARS

The inland waterway world of Europe between two great wars was one, there being no Iron Curtain, though becoming increasingly distorted by the rise of the Nazis, and with them the decline of internationalism.

The first task of 1919 was to repair the damage of war, very severe in north-eastern France and Belgium, and then to maintain the existing system against the growing pressure of road transport. Out of post-war French exhilaration at having regained control of the upper Rhine, German determination to find her pre-war waterway-building momentum, and the need of the early 1920s for much more electric power, three great schemes were born. They were indeed worthy to be ranked beside the Suez and Panama Canals, but maybe greater than they, because planned to be executed over two generations: the canalisation of the upper Rhine and of the Rhône, and the building of the Rhine-Main-Danube Canal, itself linked with major improvements to the Main and Danube. The three schemes found themselves part of a general drive towards large-scale waterways taking craft of 1000 to 2000 tonnes. Though this activity had lost some momentum by 1939, nevertheless much was achieved, for which those who came after have had reason to be grateful.

The new plans were created against the background of the navigation clauses of the 1919 Treaty of Versailles. These internationalised five rivers: Rhine, Elbe, Oder, Danube and Niemen*, and for the first four established commissions, upon which the United Kingdom was represented. In November 1936 the German government denounced the navigation sections of the treaty, announcing that they would begin bilateral negotiations for states to have the same rights on German stretches of rivers as they granted German craft on their own. The treaty also provided that the Kiel Canal be demilitarised and opened on equal terms to the ships of all nations with whom Germany was at peace, a regime that lasted until the Nazis.

*The Niemen (German: Memel), rising near Minsk in Belorussia, flowed through what was then the independent state of Lithuania, Kaunas being the principal inland port, to the Baltic in the Kurisches Haff north of Königsberg in East Prussia (now Kaliningrad in the USSR).

Fig 60 A train of log-rafts from the river Niemen (in its lower reaches called the Memel by the Germans). Soon after this note was issued in 1922, the port of Memel and its hinterland became part of Lithuania as Klaipeda. It is now in the USSR.

The Rhine

During the years from 1840, much of the Rhine running through the plain above Strasbourg had been confined between artificial banks some 230m apart. This narrowing deepened the channel and improved navigation, so making Basle accessible by water in 1904. From 1906 onwards, mainly at the cost of Alsace-Lorraine (then German) groynes were also built at right angles to the bank to help force the flow towards the centre. These, however, increased the curent's speed and so erosion, which by now had lowered the river over 6m.

The natural river could not be made a satisfactory navigation, for its water-level varied too much, from flood when the alpine snows were melting to likely summer shortage. Alsace-Lorraine, and so her Rhine frontier, having been restored to France, the Versailles treaty provided for the building of a lateral canal between Basle and Strasbourg, the Grand Canal d'Alsace. It was to begin at Kembs below Basle, and include eight locks whose dams should also provide hydroelectric power. Planning began in 1922, Strasbourg was created a port in the same year, and in 1932 the Kembs power station and pair of locks (185m × 25m and 100m × 25m) were operational.

In 1925, on the lower Rhine up to the gorge, a visitor would have seen sailing barges and fishing sloops, descendants of those that had worked the river for countless years, intermixed with steam and other craft. Most Rhine traffic was carried in small ships or tug-hauled barges, four or five, aggregating 5000–6000 tonnes dwt, usually being towed together, with upward tows reduced in the gorge section at Salzig and Bingen, where barges had to be left behind and fetched by the tug on a second rùn. The largest individual craft on the river at this time measured some 120 m × 14 m, and carried about 3600 tonnes.

Serving the big craft then, as now, were the market boats; 'We... waved to the market boats which lie in midstream to provision the tugs and barges as they pass. They scuttled alongside as we took our pick of the piles of onions, turnips, butter, eggs and huge loaves of sour bread on their decks. They also sold a moderately good brand of beer.'(1)

Fig 61 A medallion to commemorate the completion in 1932 of the first section of the Grand Canal d'Alsace with the opening of the hydroelectric station and locks at Kembs. Note the paddle-wheeled tug.

The Rhône

In 1921, the French decided upon a more difficult task, the canalisation of the turbulent and fast-running Rhône, as part of a river development plan in the joint interests of navigation, electric generation and agricultural improvement, the Compagnie Nationale du Rhône* being established in 1934 to carry out the works. The Edouard-Herriot port was built just below Lyon, followed by two dams on the upper river. After that, construction began of 12 huge schemes, each to comprise a dam on the river and a diversion canal, with a power plant and navigation lock towards its downstream end. The total fall was to be 152 m. Works were interrupted by WWII; the first scheme therefore opened in 1952.

* It was made up of representatives of local organisations, concerned public bodies, and users.

France

In the 1920s the French inland waterway scene changed greatly. Since about 1870, the system had been maintained to keep down railway rates and take bulk and low-cost cargoes off the rails. But now, road transport was quickly expanding, and pressure on rail facilities decreased as every year more passengers and remunerative freights transferred to road. Waterways were therefore no longer needed to hold down railway rates—lorries were doing that. Government now saw the continued maintenance of railways that were becoming steadily less profitable, or more unprofitable, as necessary to restrain road transport rates, and the smaller waterways as increasingly superfluous.

By 1926, some 6000 km of French waterways had an annual traffic density of 4500 tonnes per kilometre, another 5000 one of only 112 tonnes, and the remainder, mainly rivers, nothing at all. The government then reclassified the network, dropping 2297 km of rivers and 163 km of canals from the active list, to leave a total of 10,598 km, made up almost equally of rivers and canals. Thereafter until WWII inland navigation was little regarded, and development money went elsewhere. Even on the Canal du Nord work was not restarted.

Action was, however, taken in that year to organise mechanical towing. Bow-hauling by *toueurs* had by now almost ended, leaving unpowered craft to be towed by animals, occasionally oxen, usually a pair of horses or mules, up to four donkeys, or maybe a horse and donkey together. Two state-supported companies were formed to provide bankside electric haulage mainly on Freycinet canals. At its greatest the system was used on some 4000 km of canal. Since only single track was laid, when two barges met, lines were exchanged and traction units reversed. As self-propelled craft increased, however, so bank towage declined, and on 1 January 1969 the last electric 'mule' was withdrawn.

The Low Countries

Most spectacular of inter-war Dutch achievement was the Afsluitsdijk, the great dam some 30 km long that seals off the sea from the Ijsselmeer*. It was put to the Dutch parliament in 1901 by Dr C. Lely (after whom the new town of Lelystad was later named), finally authorised in 1918, and completed in 1932 with two large locks,

*It was now that this name replaced Zuiderzee.

together with their protecting harbours. Thereupon the whole Zuiderzee became an inland lake, much of it to be reclaimed as new land into polders. Water channels left between these act as fresh-water reservoirs, and also give access to the ports and canals round the sea's former shores.

Amsterdam's trade only used the dyke's locks to a small extent. Most was by the North Sea Canal to Ijmuiden, or down the Merwede Canal to the Lek and so on to the Waal. This took 2000-tonne barges (and carried 21M tonnes in 1939), but in 1937 a new 4300-tonne Amsterdam-Rhine Canal was begun, an enlargement of the Merwede from Amsterdam to near Utrecht, and then a new cut to Wijk bij Duurstede on the Lek. Here the river was crossed on the level, with locks on either side (since the river could often be higher than the canal), and the canal then taken onwards to a further lock at Tiel on the Waal. It was opened in 1953.

Let us look now at some Dutch and Belgian works intended to improve water transport's ability to compete with rail and now road.

The Meuse (Maas) had always been a difficult river for navigation, flooding in winter and running shallow in summer. The central section from the French border to Liège is Belgian, and by the end of WWI had been canalised for over half a century. Thence a by-pass canal ran across the Dutch border to Maastricht, whence traffic for the Rhine could use the 600-tonne Zuid-Willemsvaart to 's Hertogenbosch and the lower Maas. As the industries of Liège grew, and the Dutch colliery area of Limburg expanded, so an improved Meuse navigation through to the Rhine was needed; yet to canalise the river itself was politically difficult, because for part of its length it formed the Dutch-Belgian frontier.

Action in the Netherlands included a rebuilt water highway from the Waal near Nijmegen to the Maas near Heumen, the Maas-Waal Canal; the canalisation of the Maas greatly to increase its carrying capacity between Nijmegen on the Waal and Roermond; the building of a new canal, the Juliana, as a Maas by-pass thence to rejoin the river near Maastricht; and a branch to link the Juliana Canal to the Zuid-Willemsvaart.

The 13.4km Maas-Waal Canal was begun in 1920 and opened in 1927 with two locks, one 260m × 16m, the other 270m × 16m, both of which could be divided by intermediate gates. For these locks, a much improved system of culverts for filling and emptying the chambers, based on that of the Panama Canal, was designed, with two main lines running the lock's length, and 26 branches connecting with the chamber. The second lock, because it needed to operate with high-water levels in either direction, was given rolling gates sliding sideways on rails into the lock structure. The 2000-tonne Juliana Canal, 35.6km long, with four locks 136m × 14m as built, was begun in 1915 (the Netherlands was neutral in WWI) and opened in 1935 to replace the Zuid-Willemsvaart line. It was an immediate success.

These improvements to the Meuse line would be likely to benefit Dutch ports, especially Rotterdam. So a great Belgian enterprise was undertaken, the Albert Canal, built between 1930 and 1940. It left the Maas a little below Liège, and ran through deep cuttings to reach the plain of the Schelde at Antwerp. With six locks, it replaced the roundabout 1856 route via Bocholt. The Liège iron and steel industry and the Belgian coalfields round Charleroi now had high capacity access to the plain of the Schelde and Antwerp.

By the mid-1930s development plans were ready for the Meuse above the Albert Canal junction. They included a new river port at Monsin island just below Liège, the replacement of several old river locks by fewer new ones (which were to include hydroelectric plants), and the reconstruction of several obstructive bridges. By 1935 the canalisation works at Liège were finished, and those above were well ahead when war came.

The Twenthe canals in the Netherlands south-east of the Ijsselmeer are industrial branches rather than link canals. The Twenthe textile and salt district near the German frontier largely depended on road transport to carry raw materials and finished products. After WWI plans were therefore made for new waterway construction, notably by a 1350-tonne canal from Zutphen on the Ijssel (which runs from the Rhine near Arnhem to the Ijsselmeer) to Enschede. Built between 1925 and 1936, with 3 locks 140m × 12m and a guard gate off the Ijssel, the main canal is 50km long. Later, it was given a similar sized branch running north for 18km to Almelo.

One interesting point is that the canal's flood-banks were set far enough back to allow for widening to 2000-tonne standard. A second was the provision of vertical lift gates on two of the three locks. These were unusual at that time, and are still so on canals. They were used mainly because soil conditions made it difficult to build culverts in the lock walls. Therefore gates were chosen which could be operated even against full water pressure. For the third lock, also unusually, the two lower gates were of the sliding type, but suspended from above instead of running on wheels below.

Work also started in 1935 upon new canals in the provinces of Groningen and Friesland, then served by a network of 250-tonne and smaller waterways, hampered by old bridges and locks. A new canal, at first 1000-tonne, later 1350-tonne with provision for enlargement to 2000-tonne, was cut from Groningen across some of the Friesland lakes, including the Sneekermeer, to Lemmer on the Ijsselmeer, whence Amsterdam was accessible. The enlarged waterway was named the Van Starkenborgh Canal in Groningen province and the Princes Margriet Canal in Friesland. A branch, the Van Harinxma Canal, was built past Leeuwarden to the seaport of Harlingen outside the Afsluitsdijk, and to the east the connecting Eemskanaal from Groningen to Delfzijl was enlarged.

Germany

The 1000-tonne Mittelland Canal, intended with its extensions to be Germany's great west-east link, had reached Hannover in 1916. After WWI work began again, and in 1938 it reached the Elbe. The line ran level from the Dortmund-Ems junction to Anderten, where two parallel locks took it to the summit. At Sülfeld it fell by another parallel pair to the level of the top of the Rothensee lift. This, finished a year later, lowered 1000-tonne craft 15m* to the Elbe, with access downstream to Hamburg and upstream to Bohemia, and also to the connecting waterways to Berlin and further east.

Though the canal itself had locks 225m × 12m taking two 1000-tonne craft and a tug, the lift has a single caisson with usable dimensions of 85m × 12m × 2.5m. Like that at Henrichenburg, it works on the float principle, the weight being taken on two vertical hollow cylinders and the tank being guided by screw spindles in each corner.

It had been intended that the lift should give access to the Elbe, but the canal itself was to be carried over the river on a steel trough aqueduct, 900m long, 30m wide and 2.75m deep, with twenty concrete side arches and a 3-arched central structure, the middle one having a clearance width of 100m and a navigable height of 6m above highest water. On the Elbe's far side at Hohenwarthe there was to be a float-operated vertical lift with twin caissons and a fall of 19m, to take the canal directly into the Ihle Canal and the route to Berlin and the Oder. Both aqueduct and lift were begun**, but Germany's post-war division has inhibited completion. Barges still have to use the Rothensee lift and cross the Elbe in order to move eastwards.

The Südflügel or south branch of the Mittelland Canal was being

*At medium water on the Elbe. The lift can work between 10.58m and 18.67m.

**Several concrete spans were built on the left bank, and the lift's four shafts, two for each independent caisson, were sunk.

Fig 62 Niederfinow lift, opened in 1934, works by counterbalance helped by electric motors.

built before WWII. This involved the improvement of the Saale from its junction with the Elbe above Magdeburg to Halle from 300-tonne capacity to 1000 tonnes, the extension of the Saale navigation upwards from Halle to Merseburg, and the construction of an Elster-Saale Canal from near Merseburg to Leipzig. The Saale scheme was completed to Halle, but the navigation upwards to Merseburg and beyond remains unimproved. The canal to Leipzig, begun in 1933 and intended for completion in 1942, has never been restarted.

Between the wars, the waterways linking the Elbe at Rothensee to the Oder at Hohensaaten were enlarged to 1000-tonne standard—the Ihle, Plaue, Havel and Hohenzollern* Canals. The Hohenzollern

*The Hohenzollern is now called the Oder-Havel Canal.

then had four power-operated locks at Niederfinow, with passing places between, electric locomotives working barges into and out of each. Pressure upon these locks so increased, however, that in 1926 work started upon the Niederfinow vertical lift to replace all four. Opened in 1934, this, like the British Anderton after reconstruction, was activated by counterweights and electric motors. Its single caisson, 85m × 12m × 2.5m (as at Rothensee) taking 1000-tonne barges, rises and falls 36m, the highest lift built to that time. At Hohensaaten the canal communicated with the 1000-tonne navigation channel of the Oder. A second 1000-tonne line linked the Ihle-Plaue Canals to the upper Oder at Eisenhüttenstadt via the Rivers Havel, Spree (or the Teltow Canal) and the Oder-Spree Canal.

Once opened, the Mittelland Canal enabled Ruhr coal for the Elbe and Berlin areas to travel direct, and not via Rotterdam and Hamburg or Stettin, with troublesome transhipments en route. Eastern Germany's grain and timber could likewise move west along this direct link between the Baltic ports and the Rhine. Nevertheless, Hamburg's access to the Ruhr and Rhine by inland waterway was still a roundabout one, by way of Elbe, Rothensee lift and Mittelland Canal, and even before the latter's completion discussions began upon a shorter route. The need for it sharpened by post-war political division, it has now been built as the Elbe Lateral Canal (see p. 253).

When C. S. Forester cruised the lower Elbe in 1929, he saw six or seven barges of up to 300 tonnes each being towed by a tug, and noted

Fig 63 A drawing made c. 1890 of a type of bridge found on the River Elde in what is now East Germany. The centre section forms a miniature lift-bridge, and enables the masts of sailing barges to pass.

that whereas on the Seine each barge was separated from the next by a long towrope and separately steered, on the Elbe there was a long line from tug to leading barge, but the barges themselves were hauled close to each other with tillers lashed, only the last barge being steered. Coming downstream, some empty barges had 'big gaff mainsails'. He also noted 'plenty of big cargo steamers of four hundred tons or slightly more, sternwheelers with two funnels side by side'.(2)

Eastwards from Hamburg to the Oder, the route was by the Elbe and Havel to the Hohenzollern Canal, but in the 1920s work began on the enlargement of the earlier canalisation of the Elde river, a tributary of the lower Elbe, to provide a route thence by the Mecklenburg lakes direct to the Hohenzollern Canal. By the time of Forester's voyage, however, the old sailing barge traffic had mostly gone, though considerable pleasure use had taken its place. Enlargement was later dropped.

Meanwhile, back on the connecting Dortmund-Ems Canal, the original Henrichenburg lift with its 950-tonne dimensions, had been replaced in 1917 by a shaft-lock with a fall of 14M, able to take 1500-tonne barges. It had ten economiser basins, five on each side*. In peacetime, however, this waterway line to Emden did not supplant such foreign ports as Rotterdam and Antwerp, though traffic, 4M tonnes in 1913, had increased to 9M by 1928, so requiring large extensions to Dortmund harbour. In 1937–9 much of the line was rebuilt, sometimes on a new alignment, and enlarged to 1000-tonne standard to conform to the Mittelland. The work continued into WWII, and was not finished till 1959.

Westwards, the Rhine-Herne Canal linking the Dortmund-Ems to the Rhine, which was carrying over 7M tonnes in 1922, had by 1931 been supplemented by the 1000-tonne Wesel-Datteln Canal to the more northerly Rhine port of Wesel. Eastwards, the lower Ems was in 1936 given a bigger link to the Weser than the Ems-Jade Canal—the 1000-tonne** Küsten (Coastal) Canal from Dörpen on the Ems above Emden past Oldenburg to the canalised Hunte river which enters the Weser below Bremen.

A start was made, too, upon a notable new river canalisation for 1350-tonne barges, that of the Neckar, Heilbronn being reached before WWII temporarily stopped work. Previously the natural river

*All major German canal locks were now being built with economiser basins or side-ponds to save water. For instance, the three parallel locks of different sizes at Handorf (Münster) on the Dortmund-Ems Canal had in 1939 ten basins between them.

**Now enlarged to Class IV (1350 tonnes).

had been navigated by the 350-tonne *Neckarschiff*, drawing 1.65m.

In the years before WWI, the equivalent of the French 'mules' was the German state tug organisation, the Reichsschleppbetrieb, which in 1930 towed 52,567 barges with cargoes of 17,333,000 tonnes, and in 1937, 73,995 of 25,959,000 tonnes, the average load per barge having risen from 583 to 611 tonnes, and the haul from 64km to 84km. This organisation had been set up under the Prussian canal law of 1905, paragraph 18 of which read:

> Only the towage provided by the government can be used on the canals from the Rhine to the Weser and the branch to Hannover, and on the branch canals of those waterways. The establishment of mechanical towage on these routes is forbidden to private individuals. The movement of ships with their own propelling force over these routes is to be permitted only under special licence...

Towing fees were added to the tolls levied for using the canals—these being levied almost entirely on loaded craft—and also to charges for loading and discharging.

Meanwhile the Kiel Canal was growing steadily busier. In 1937, with coal (in both directions) the principal traffic, the total net tonnage passing, at 22 M, was greater than Panama (19,977,000) but less than Suez (27,236,000).

The Rhine-Main-Danube Canal

Ludwigs Canal was still navigable, but little used, so the Reich and the states of Bavaria and Baden agreed in 1921 to create the Rhine-Main-Danube Canal Company to build a 1500-tonne waterway from Aschaffenburg* on the Main 87km above its junction with the Rhine for 301km to Bamberg, a canal thence for 201km to Regensburg on the Danube, and then an improved river navigation for 173km to meet the fully navigable Danube at the Austrian border at Jochenstein just below Passau. The company was to build the waterway for the state, handing over the navigation works as completed, but retaining ownership of the associated hydropower plants, revenue from the sales of electricity being used to finance the works directly or to repay loans raised on the capital market. Provision was made for additional funding by central government and Land budgets when necessary. At that time the *Mainschiff* of some 420 tonnes, horse-towed or hauled by a chain-steamer, worked upwards from Aschaffenburg to Bamberg. By WWII, however, 24 locks had been built to

*Locks had already been built below Aschaffenburg.

Rhein-Main-Donau Aktiengesellschaft
zu München

vom Deutschen Reich und Bayern verbürgte, reichsmündelsichere, mit 5% verzinsliche, hypothekarisch eingetragene, zu 102% einlösbare Teilschuldverschreibungen im

Betrage von M. 300 000 000

(Teilbetrag der Gesamtausgabe von M. 323 050 000)

eingeteilt in sechs Gruppen von je M. 50 000 000 mit je Stück 250 Teilschuldverschreibungen zu M. 20 000, Stück 500 zu M. 10 000, Stück 3000 zu M. 5000, Stück 5000 zu M. 2000 und Stück 15 000 zu M. 1000.

BUCHST. D 2000 MARK GRUPPE VI № 007257

Teilschuldverschreibung
über
Zweitausend Mark
Deutsche Reichswährung

verzinslich zu 5% in halbjährlichen Zielen am 1. Februar und 1. August jeden Jahres und einlösbar zu 102%.

Ausgegeben laut Beschluß des Aufsichtsrats vom 30. Dezember 1921 mit Genehmigung des Bayerischen Staatsministeriums für Handel, Industrie und Gewerbe vom 16. Januar 1922.

Wir verpflichten uns, dem Inhaber dieser Teilschuldverschreibung den Betrag von
Zweitausend Mark
mit fünf vom Hundert zu verzinsen und mit einem Aufgelde von zwei vom Hundert, also im ganzen M. 2040, zu zahlen. Im übrigen wird auf die umstehend abgedruckten Bedingungen der Ausgabe verwiesen.

MÜNCHEN, im Februar 1922.

RHEIN-MAIN-DONAU AKTIENGESELLSCHAFT

Der Aufsichtsrat Der Vorstand

Für das Kapital und die Verzinsung dieser Teilschuldverschreibung haben das Deutsche Reich und Bayern durch Erklärung des Reichsverkehrsministers vom 30. Dezember 1921 und des Bayerischen Staatsministers der Finanzen vom gleichen Tage die gesamtschuldnerische Bürgschaft übernommen. Vorstehende Teilschuldverschreibung nimmt zu gleichem Range mit den übrigen Teilschuldverschreibungen an der für den Betrag von M. 300 000 000 laut umstehender Bedingungen bestellten Sicherheit teil. Für diese Teilschuldverschreibung haben wir das Amt eines Grundbuchvertreters im Sinne des § 1189 des Bürgerlichen Gesetzbuches und eines Vertreters des jeweiligen Gläubigers nach Maßgabe der vorbezeichneten Bedingungen übernommen und werden nur diesen entsprechend über die Sicherungshypothek verfügen.

MÜNCHEN, im Februar 1922. Deutsche Bank Filiale München

Der Kontrollbeamte

Fig 64 Stock certificate issued by the Rhine-Main-Danube company in its first year of activity. Money raised on the capital market was to be a major source of finance for the project, repayments to be made from revenue from sales of electricity generated by the company's power plants.

Würzburg, above which the chain continued in use. The Danube's stretch between Passau and Regensburg had also been improved, with one lock, Kachlet, a little above Passau. Further work had to wait for our own times.

By 1936 traffic on Ludwigs Canal was down to 41,700 tonnes. Then came the war, during which more was carried than ever before; indeed, some German E-boats with difficulty got through it to the

Black Sea. Its last traffic passed in 1944, before it was war-damaged. Afterwards, restoration began, the Nuremberg-Kelheim section being ready by 1947. In 1951, however, work was suspended in favour of the Rhine-Main-Danube Canal, which has been built over much of the old line.

Czechoslovakia

The Elbe through Germany does not require locks, but as it reaches Czechoslovakia its gradient increases, while its water-levels show greater variations. From the turn of the century, therefore, while the territory was still part of Austria-Hungary, canalisation began on the Elbe itself and also on its tributary the Moldau (Vltava) to Prague, initially for 700-tonne barges drawing up to 1.8 m, locks being built in pairs, one of 73 m × 11 m, the other four times the size, 146 m × 22 m (though with 11 m-wide gates) for barge trains.

Between the wars Czechoslovakia carried the Elbe canalisation from the Vltava confluence upstream to Kolin, now with single locks 85 m × 12 m to take 1000-tonne barges.

Sweden

In 1916 the third flight of locks to be built at Trollhättan was opened, this time with chambers 90 m long, able to take 2000-tonne ocean-going ships drawing 4.1 m.

The Danube

After WWI an Inter-allied Danube Commission was set up to re-establish normal trade, until in 1920 the International Danube Commission was restored to control the whole river from Ulm to the sea. It was followed in 1922 by the Statute of the Danube, which (a) opened the river to the trade of every nation; (b) gave every riparian state one representative on the Commission (two for Germany, considered as Bavaria and Württemberg), plus representatives of France, the United Kingdom and Italy; (c) made the Commission the supervising (though not the money-providing) body for navigation improvements; and (d) enacted that transit traffic should be free of customs duty or exceptional taxation. Navigation improvements were made at the Iron Gates and elsewhere, until in 1936 Germany denounced the navigation sections of the peace treaties and withdrew her representatives. Thereafter the Commission started to disintegrate ahead of WWII.

Fig 65 A London-registered barge on the Danube, the *Norfolk* of the Anglo-Danubian Co.

During the inter-war period most of the traffic was carried by craft of the riparian states, the biggest fleets being the Austrian, German, Romanian and Yugoslav. British, French, Dutch and Greek flags were also to be seen, mostly concerned with the Romanian oil traffic in which foreign capital then participated. Britain, indeed, had its own Anglo-Danubia company. More goods were carried upstream—mainly oil, bauxite, cereals, ore and coal—than down, mainly coal, salt, timber and manufactures.

By 1939 the Danube below Regensburg had become busy. A service of express passenger craft, such as those of the DDSG, served the principal cities from Vienna downwards, while below the confluence with the Drava (in Yugoslavia) the river had a regular service of steam paddle-wheeled cabin-passenger craft which picked up and set down at towns down to the estuary.

Most barges, running up to about 1000 tonnes, were towed by paddlewheeled steam-tugs, low, wide in proportion to their length because of the paddles, and with two tall funnels, though diesel tugs were beginning to appear. Steam and diesel self-propelled barges could also be seen, sometimes powering one or two other barges, maybe lashed alongside. A few bargemen still took unpowered craft downstream, relying on a huge steering oar as did their contemporary lightermen on London river:

> Old wooden barges drift past, high in bow and stern, almost level with the water amidships. A man stands on a raised steering-platform built ahead of the slanting black cabin aft. With a pole he shoves a forty-foot long tiller. Clusters of men, like galley slaves, stand amidships, straining at long, flat-toed wooden sweeps. They chant as they sink the sweeps in the water. We have seen them this way in Serbia, rowing into the sunset.(3)

Danube barges were—and are—characteristic, with their high-perched stern cabins with trellised sides, often covered with creepers and bordered with bright flowers in pots and boxes. Some had cranes for use at small riverside wharves. Common then, but gone now, were floating mills, looking like barns adrift, two craft, one large, one small, moored in the current, a waterwheel between them turning steadily, served with corn and relieved of flour by small boats from the shore.

Tows might consist of eight or more large barges, though through the Djerdap passage, two barges were all that tugs could manage upstream in the worst sections, the Kazan and the Sip Canal—and perhaps four downstream in the fast-running current. Therefore tugs had to make several journeys to take a full tow upstream. The Danube is a mile or more wide above and below the gorge, but in those days it narrowed at one point to less than 100 metres. A pilot was compulsory; there were places where a one-way system was used, controlled by bankside signal stations, and others notable for shallows, sunken rocks or whirlpools, while at the bottom of the gorge, between the island of Ada Kaleh and the Iron Gates themselves, the current ran swift over a maze of rocks and rapids. The lockless and fast-running (16–18 kph) Sip Canal had been made there on the Yugoslav side, where upcoming tugs, their funnels belching black smoke, a steam locomotive attached by towline, hauled their two barges up the canal and into the gorge beyond.

Barge ports dotted both sides of the river, the huge ports of the estuary—Giurgiu, Galatz and the rest—being matched to the smaller but still large wharf of upriver cities such as Belgrade, Budapest, Bratislava, Vienna and Resenburg, frontier ports where craft passed through Customs, and the many smaller wharves between.

Off the main river, the Sava was navigable by Danube barges to Sisak below Zagreb, and further by craft measuring some 30 m × 5 m. In addition, the tributary Kupa from Sisak was navigable for 135 km to Karlovac, and the Una for 70km to Bosanski Novi by similar smaller craft. Danube barges could pass a short distance up the Drava to Osijek, smaller craft for 138 km in all. On the Tisa, with a normal depth of some 2 m but subject to bad floods, Danube barges could pass the Yugoslav frontier to Csongrád in Hungary, while passenger services ran to Szeged. By the Tisa and the King Peter I Canal, too, barges could avoid low water in the Danube, and also reach the Bega Canal for Temesvar (Timosoara). Sternwheelers were working on this canal when war began in 1939. When it ended, the old Europe that had largely survived World War I, had gone. Two Europes were to replace it, divided by a curtain less penetrable indeed than iron.

9
SINCE 1945: BROAD WATERWAYS

War divides Europe and changes the Old World

World War II caused destruction of craft and structures, and brought most waterway development to a stop. In France, 5200km of navigations had been rendered impassable in 1940, though by a year later all but 50km were again in use. Then much worse destruction began, by bombing, resistance activities, and later the retreating German armies, until 8250km were out of action. Again they were reopened, this time in improved state. In West Berlin alone, 80 waterway bridges were destroyed and 258 barges sunk, while on the Mittelland Canal the Minden aqueduct over the Weser had been successfully bombed.

Work on this canal's planned aqueduct over the Elbe and twin-caisson Hohenwarthe lift beyond had continued after war began, then stopped. It has never been resumed, for the Mittelland Canal is a symbol of the major political change that has come over Europe. There are now two German states* whose boundary cuts the canal east of Sülfeld locks, and a western enclave, West Berlin, within the eastern state. Their boundary is part of another, the iron curtain, which runs across Europe and has created two politically antagonistic blocs out of a continent that pre-war was in a real sense a unity.

The Mittelland Canal and its extensions had been planned to link the rivers of Imperial Germany from the Rhine to the Vistula; after 1918 they were to join Rhine to Oder. Today the political motives for the project have disappeared, and the considerable waterway traffic (some 4.5M tonnes pa) which moves upon it between the two Germanies has to make do with inadequate infrastructure. The GDR finds itself cut off from its two natural seaports, Hamburg (now in the FGR) on the Elbe and the former Stettin (now Szczecin in Poland) on the Oder, with only partial control of either river, and with only one major port, Rostock, and that not canal-connected. Hamburg, separated from its natural hinterland, has been given new access to the

*I shall use FGR to denote the (western) Federal Republic and GDR the (eastern) Democratic Republic.

Mittelland Canal a little west of Süfeld locks and entirely within FGR territory by way of the Elbe Lateral Canal.*

Because the GDR's transport routes have been forced into a north-south direction, whereas waterways run east and west, Berlin's situation has partly changed. While West Berlin is a focus for $8\frac{1}{2}$ M tonnes of waterborne traffic, East Berlin moves only $\frac{1}{2}$ M tonnes into or out of barges. The only post-war improvements made by the GDR are the enlargement of the Brandenburg locks and a new 34 km by-pass of the Havel to enable GDR barges to avoid West Berlin. The FGR, however, has to pay large sums to the GDR to maintain the waterway links between West Berlin and the FGR, along the GDR sections of the Elbe-Havel, Teltow and Mittelland Canals, and to maintain the Rothensee lift. Traffic is still hindered by the Elbe level-crossing, which frequently offers a navigation depth far below that of its linking canals on each side.

Post-war times found both banks of the lower Danube within the iron curtain. Only Austria and the FGR remained. At the Danube Conference of 1948 Ernest Bevin, then British foreign secretary, pressed for continued internationalism on the Danube, but Russia insisted upon limiting membership of the International Danube Commission to riparian states, on which Austria and the FGR are in a permanent minority. So non-riparian states like the United Kingdom lost their seats and the Danube changed its status: how restrictively can be understood when one realises that 76 per cent of pre-war Danube traffic was carried in non-riparian vessels.

The ending of British rule over a united Indian sub-continent was accompanied by its division on lines of religion, not economics, into two independent countries, India and Pakistan, and later by the latter's further split into Pakistan and Bangladesh along boundaries that took no account of waterway routes. Burma and Ceylon (now Sri Lanka) also became independent; the former to choose economic isolation, the latter a more outward-looking stance.

Further east, the increasing influence of Maoism in China's communist party, which took control from the nationalists in 1949, was not conducive to the application of modern technology to inland navigation, though it produced sometimes navigable irrigation canals. Since the fall of Maoism at the end of the 1970s, however, waterway development has sharply accelerated.

From the late 1950s one African country after another gained its

*Czechoslovakia does maintain port facilities at Hamburg under the Versailles treaty, and both Czechoslovakian and East German traffic uses it. In October 1983 the former country was negotiating to use the Elbe Lateral Canal in times of low water in the Elbe.

independence from Britain, France, Belgium and Portugal. Civil strife, frequent changes of political direction, financial difficulties, lack of skilled staff, the multiplication of independent states to replace four operationally decentralised but centrally influenced and financed empires, has meant some progress, some false starts and so waste of scarce resources, and fewer achievements than plans.

Standardisation and barge operation

In spite of war damage, by the end of 1950 the Netherlands' barge capacity was above that of 1938, Belgium's had reached it, France's was back to some 80 per cent, and West Germany's to some 70 per cent. In traffic carried, France and Belgium had almost reached pre-war levels, the Netherlands to about two-thirds, and West Germany to some 60 per cent, her figures being seriously affected by Hamburg's changed trading position.

In western Europe generally, the post-war period has seen large waterways accepted as a third arm of surface transport, economical of state capital because its share in total freight movement has been higher than that of expenditure upon its infrastructure. Industry has therefore tended to site itself alongside large or new waterways. Few industrial areas in Europe are not now water-served.

Water transport's steady place is shown in these figures from West Germany:

	Milliard tonne/km		*Percentage of total*	
	1974	1981	1974	1981
Waterways	51.0	50.0	26.1	24.4
Railways	69.3	62.0	35.4	30.3
Pipelines	16.9	12.6	8.6	6.1
Roads	58.5	80.2	29.9	39.2
	195.7	204.8	100.0	100.0

Yet comparative track lengths in km were:

Waterways (end-1978)	4,395
Railways (end-1978)	31,532
Pipelines (end-1979)	1,579
Roads, other than local (end 1978)	104,658

Helped by the formation of the European Economic Community, and increasingly thought of as a single network (within which in time Britain has come to be included), waterways are now used more for international trade than national. There have, however, been repercussions, and in Germany especially the strength of the railway lobby, and the ability of the railways to offer extra-low charges to water-served ports, has slowed down waterway growth. In Britain, waterways remained much undervalued into the 1980s, mainly because most freight-carrying waters had not been nationalised, and therefore no reliable statistics were available. The lack has been remedied, and we now know that waterways and coastal shipping indeed provide over 30 per cent of the tonne-km of domestic freight movements.

Europe's predominant inland waterway craft remains the versatile self-propelled barge, though there has been a steady increase in its average size as waterways are enlarged. Standardisation has been based upon the 1350-tonner, the Europa barge, navigating upon a Class IV* waterway as recommended in 1953 by the European Conference of Ministers of Transport. Radar-fitting became usual at about the same time, giving round-the-clock availability of craft, better keeping to schedules, and lower investment in new craft because old ones could be more intensively used. A new technical development is a log able to measure speed over the waterway's bed instead of through the water, so enabling that speed to be maintained which requires minimum energy.

Push-towing, with ancestry in Europe (see p. 127) and long practised in North America, began post-war with the small but successful British 'Bantam' tugs, then spread quickly wherever Old World waterways were big enough, for instance to the Congo in 1950, and later to the Seine and Rhine, with first generation push-tugs of some 2400hp. Then came the 5400hp Dutch *Jacob C. van Neck* in 1976, able to push some 11,000 tonnes, and in 1983 the German *Mannesmann V*, 5100hp, but designed to push six lighters loading up to 16,500 tonnes.

The usual push-tow is up to four, because waterway capacities will not allow more, but up to six can be seen on the German Rhine between Emmerich and Koblenz, while as I write Dutch 6-lighter

*The accepted classification is:

Class	
0	Under 300-tonne
I	300-tonne (the *péniche*), 38.5 m × 5 m × 2.5 m
II	600-tonne, 50 m × 6.6 m × 2.5 m
III	1000-tonne, 67 m × 8.2 m × 2.5 m
IV	1350-tonne, 80 m × 9.5 m × 2.5 m
V	2000-tonne, 95 m × 11.5 m × 2.7 m
VI	3000-tonne and over

Seaways constitute a seventh classification applied, eg to the Ghent-Terneuzen Canal, the channel up the Schelde to Antwerp and then by river and canal to Brussels, or those from Hook of Holland to Rotterdam or Dordrecht.

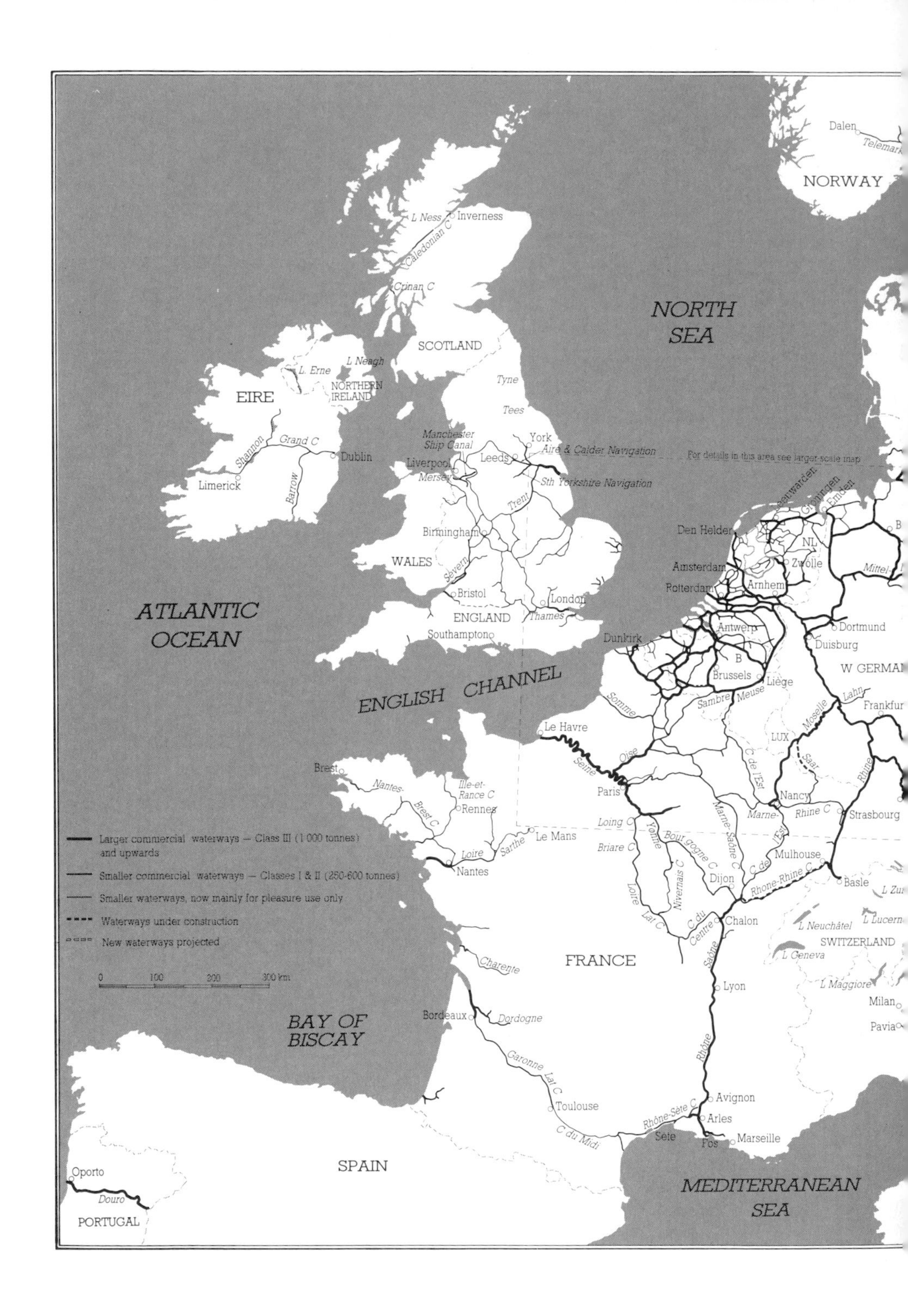

Map 9 The waterways of Western Europe.

Strömsholms C
Säffle C
Mälaren
Stockholm
Södertälje
L Vänern
Göta C
Trollhätten
Motala
Söderköping
SWEDEN
BALTIC SEA
Riga
Western Dvina
FINLAND
Nurmes
Kuopio
Joensuu
USSR
Jyväskylä
Tampere
Lappeenranta
Saimaa C
Lake Ladoga
Viborg
Lahti
Leningrad
Neva
Helsinki
Berezina C
Niemen
Kaunas
Dnepr
Berezina
Kaliningrad
Gdansk
Elblag
Elblag C
Ostroda
Dnepr-Niemen C
Rostock
Szczecin
Bydgoszcz
Oder-Havel C
Noteć
Plock
Pinsk
Pripet
Bug
Dnepr-Bug C
Brest
Vistula
Warsaw
Berlin
Warta
Poznan
Konin
Havel
Oder-Spree C
Kiev
Dnepr
POLAND
USSR
Odra
Wroclaw
Brzeg
Sandomierz
Dresden
Vistula
Koźle
Gliwice C
Krakow
Elbe
Pardubice
Ostrava
Vltava
Prague
Danube-Oder Elbe Canals
Mogilov-Podolskij
Voznesensk
CZECHOSLOVAKIA
Regensburg
Passau
Linz
Vienna
Bratislava
HUNGARY
Dnestr
Budapest
AUSTRIA
L Balaton
Sio
ROMANIA
Chilia Channel
Galati
Sulina
Braila
Szeged
Tisa
Timişoara
Sombor
Bečej
Danube-Tisa-Danube canals
Černavoda
Constanţa
Zagreb
Bucharest
Danube-Black Sea C
Kupa
Sava
Iron Gates
Giurgiu
Venice
Belgrade
Ruse
Danube
BLACK SEA
YUGOSLAVIA
BULGARIA
ITALY
TURKEY
Istanbul
ADRIATIC SEA
SEA OF MARMARA
GREECE

tows are working experimentally below Emmerich. The influence of Rotterdam has been powerful in promoting the experiment, an important factor being the removal in 1982 of the locks on the Hartel Canal, the inland waterway route that parallels the Nieuwe Waterweg sea shipping route between Rotterdam and Europoort. Full length push-tows can now move without being divided, and barge traffic can enter and leave Europoort without mixing with deep-sea ships. Older river and canal locks and curves are now being rebuilt or, as on the Belgian Albert Canal, supplemented to take 9000-tonne tows, and new ones designed with these in mind.

As a variant of push-towing one often sees a Europa barge which has been adapted to push another. More recently, new push-barges up to 3000 tonnes have been built: square-ended in front, they can push one or more lighters. The 3008dwt Rhine push-barge *Haniel Kurier 61*, delivered in 1984, pushes three Europa II lighters 76.5m × 11.4m. Such push-barges often have wheelhouses which can be hydraulically raised up or retracted, and also closed-circuit TV linking the captain to the head of the leading barge.

Barges and towed lighters may be open, covered, self-unloading with cranes and powered hatchcovers, built for special cargoes, container-carriers, the two-decker car-carriers of the Seine, or tankers for petroleum, chemicals or liquefied gas.

The difference made by changes in craft size and power can be illustrated from the experience of the Dutch firm of Europese Waterweg-Transporten (EWT). In 1939 they carried 5M tonnes in conventional craft of 400,000 tonnes capacity and with 1000 crew. Forty years later, they achieved virtually twice the performance with a fleet of half the capacity and a quarter of the personnel.

Container-carrying

The rising price of oil in the mid-seventies gave inland waterways a fillip, for studies carried out in France and Germany showed consumption levels per tonne/km to be about equal to pipelines, less than for most rail haulage and all road haulage. One result was, from about 1979 onwards, a sharp increase in the waterway carriage of containers, which had begun tentatively about 1975.

By 1983 a considerable traffic estimated at some 150,000 teus* pa, and accounting for some 20 per cent of all cargo moved on the Rhine, had built up, based on containers transferred from or to deep-sea in

*Varying sizes of container equated to Twenty Foot Equivalent Units, ie a 40ft container equals two teus.

Fig 66 A Renault car-carrier on France's river Seine at Ecluse de Mericourt, bound for Le Havre with cars for export.

ships at Antwerp and Rotterdam. A number of container-handling terminals have been opened by the principal carrying firms engaged in the trade, notably at Emmerich, Durisburg,* Düsseldorf, Köln, Mainz, Ginsheim-Gustavsburg (for Frankfurt), Mannheim-Ludwigshafen, Wörth, Karlsrühe, Strasbourg and Basle, though currently most barges do not run above Mannheim, because the containers have to be back at their home port in a week. Services are usually run on a weekly timetabled basis, with terminals operating their own road collection and delivery services. Container traffic is also developing on the Danube between Passau/Regensburg and Bulgaria, while in France there were in 1985 prospects for regular services on the Seine between Le Havre/Rouen and Paris, and on the Rhône between Marseille/Fos and Lyon.

On the Rhine, containers are not usually carried in push-tow lighters but in self-propelled barges, often fitted with a wheelhouse that can be hydraulically raised so that the captain can see over his boxes. Barge capacity ranges up to 224 teus, though the recently commissioned Dutch craft *Sayonara* carries 192 teus and can push three other barges each of 80 teus. In 1983, too, a low-profile container-carrying coaster (capacity 100 teus) stated running from Boston (England) to Duisburg. Though oil prices have now fallen back, the boost high prices gave to container-carrying has remained.

*This terminal also handles coasters carrying either containers or ro-ro trailers and sometimes also general cargo, eg the *Laila of* 2500dwt, able to carry 19 trailers or 158 teus, and and also 180 tonnes of cargo or 5125m^3 of grain.

New barge types and overcapacity

Cars from the Renault works have long been carried in special barges on the Seine from near Paris to Le Havre, but in 1983 a 3600-tonne multi-deck 650-car-carrier began to run from the Ford works near Köln to Flushing for transport to the UK.

A new development in quick-changing waterway technology was the delivery at end-1983 to a joint West German-Bulgarian company of four ro-ro* catamarans, each 114 m × 23 m, able to carry forty-nine 40 ft (12.1 m) trailers on deck. The service currently operates on the Danube between Passau on the German-Austrian border and the Bulgarian port of Vidin. It was followed in 1985 by the start of a liner ro-ro service mainly intended for trailers on the Rhine between Rotterdam, Mainz and Mannheim. Another is the 160-tonne wheeled barge able to carry a 40 ft container or other cargo, developed by a Dutch consortium in 1983, which can be hauled out of the water and towed to its destination by a tractor unit. Still another the Belgian heavy-lift barge, able to take a unit load up to 1250 tonnes and 9 m high, which can be loaded and unloaded using special trailers. Away in China it was reported in September 1982 that a year's trials had been completed of two freight-carrying hovercraft barges, which during that time had delivered 6000 tonnes of cargo over 2400 km of the river system of Heilongjiang province, where they were built. A 20–30 per cent fuel economy was claimed.

Waterway carrying generally fell back during the serious recession of the late 1970s and early 1980s due to falling industrial activity, especially in the iron and steel industries, and also because coal consumption failed to increase in spite of rising oil prices. Overcapacity therefore showed itself throughout western Europe, aided not so much by road as by heavily subsidised rail competition, notably in Germany. On the Rhine, indeed, it was estimated in 1983 that some 1.5 M tonnes (20 per cent) of the Rhine fleet was not needed.

The *tour de rôle* is a panacea for overcapacity; scrapping policies are an attempt to solve it. In the 1930s, to protect owners during the depression, affected as they were also by increasing road competition, the *tour de rôle* was introduced. This means that for most general freight contracts (as opposed to coal or oil shipments, eg to power stations, that are worked on contracts of affreightment), the contract has to be offered to the barge next in turn and available—even if it has been obtained by a shipping company owning its own craft. The *tour*

*'Roll on, Roll off' ie craft on to and off which road vehicles can be driven.

de rôle on the one hand safeguards the livelihood of the private owner; on the other it keeps in operation barges that would otherwise be scrapped, and to some extent protects inefficiency. For instance, much of the opposition to pushing six lighters in Netherlands waters has come from private owners. The shipping companies would like the *tour de rôle* abolished; private owners regard it as the Ark of the Covenant. Between them hovers the government. One way round has been found: big companies have sold some of their barges to their captains, then chartered them back and so lowered costs. In France, which operates an international *tour de rôle*, the government in 1983 also grouped owner-operators into a single marketing body, EATE*, aimed at getting additional traffics which, however, will be offered on the *tour de rôle*.

Overcapacity, combined with a steady increase in barge size, the spread of push-towing, and containerisation, also caused the introduction of policies for scrapping out-of-date barges. In Germany carriers pay a levy on freight income, and receive a premium for each craft scrapped. In Belgium by end-1982, the government had paid compensation to eliminate 405 craft averaging 470 tonnes capacity, nearly all in single ownership. Of the 8000 or so Dutch barges, some 2000 are company-owned, the rest by private owners, usually of one craft. The former are usually large modern barges, including all the push-towed lighters; the latter may be large and modern, but are more likely to be old and small, and to double also as the home of the skipper and his family. The former have to operate under labour agreements that govern manning levels, wages and working conditions, but the latter, often employing only their own families, do not. It is now Dutch policy to buy out owners of barges of less than 350 tonnes.

Some survivals

European waterways taking craft of less than 600 tonnes are tending to decline. As I write it is indeed uncommon to see a barge smaller than *péniche* (300 tonnes) size still working. One can, however, see 250-tonners, especially in the Netherlands, and now and then a still older sight, as of ancient barges piled high with recently cut reed, and towed by a diminutive tug, probably for polder construction, as I did near Westzaan in 1982. Some have, however, gone for ever: I shall always be glad that in 1960 I saw the last steam paddlewheeled tug to work on the Rhine, as later in 1969 I was to see several such

*Entreprise Artisanale de Transport par Eau.

Romanian vessels on the Danube, two in the Iron Gates in their last fierce days. Some Czech-owned sternwheelers were also still working on the Elbe in the late 1960s.

Here and there are living survivals of small-scale agriculture carried on by boat and small canal, instead of by cart or lorry and country road. In the Dutch province of Overijssel, north of Zwolle is the water-roaded village of Giethoorn; in France the 'Green Venice' area near Niort, between Poitiers and La Rochelle; and leading off the Somme near Amiens, the 300-hectare market-gardening Hortillonnages, with their 55 named canals. All three offer boat tours of an older way of life.

Short-sea traders, coasters and seagoing barges

The last two decades have seen rapid growth in the use of inland waterways by short-sea traders carrying from 250 to 1500 tonnes, though the larger craft are becoming predominant. These can be conventional ships with low air draught to enable them to pass under low lock and bridge structures* or a type of craft that can be described either as a low-profile coaster/short-sea trader or a seagoing barge. Such craft are built for the most efficient and cheapest operation—in shape by being long in relation to breadth so that they can pass locks, in fuel consumption by using diesels burning heavy fuel oil, in size under one of various 'paragraphs' which enable them to operate with minimum crews if constructed within various dimensional** limits, in design by having single hatchways with a removable bulkhead, powered steel hatch covers, box-shaped holds, water ballast tanks to help improve air draught when necessary, and retractable wheelhouses. Thus the profile can be so low that the highest points are the hatch covers.

The ships of Britain's Crescent Shipping, for instance, move as far inland as Basle on the Rhine, Paris on the Seine, Liège via the Albert Canal, and Deventer by way of the Ijsselmeer locks and the Ijssel river, and in Britain to, eg Anderton on the Weaver, Norwich (Yare), Selby (Yorkshire Ouse) and Gainsborough (Trent). Elsewhere one can find a true sea-canal-rivergoing barge working between Copenhagen and Groningen via the Kiel Canal, the north German coast, and the Eemskanaal from Delfzijl, or the low-profile container-carrying (96 teus) coasters that in 1985 worked between Vienna and

*4.4m on Class IV waterways, 6.7m on Class V and above.

**One common type in 499grt, 1500dwt; another is 999grt, 2500dwt.

Fig 67 A sea-going barge (799grt, some 2000 tonnes dwt), arrives at Delfzijl lock in the Netherlands from Copenhagen.

Trabzon on the Black Sea, with connections to Iran and Iraq.

Tug-hauled or pushed seagoing barges, some very large, are well-known in North America. In the Old World, however, the fleet of tug-hauled 200-ton barges that during WWI worked a regular service from Richborough, Kent, England, to Calais and so into the canals, until recently remained exceptional. In April, 1983, however, a Finnish barge was loaded at Amsterdam with some 13,500 tonnes of coal for Sweden, and in 1984 a design was announced of an integrated tug-barge (ITB, see p. 391) system to work between Finland's Lake Saimaa and central European waterways, while a Finnish steel company ordered one 10,440 bhp ice-class tug alternately to work two 12,800 dwt barges in carrying Polish coal and Swedish iron ore to its works at Raahe high up the Gulf of Bothnia. In 1984, also, an ITB began work between Singapore and Darwin in north Australia. It seems likely that the lower costs of ITB operation will soon provide other examples.

Barge-carrying ships

The barge-carrying vessel (BCV) has not yet fulfilled its initial promise. The BCV was an American concept (see p. 389), and most have been American owned. Europe, however, generated the Danish-built BACAT* ship, carrying LASH* and smaller barges between the Humber and Continental ports, until unofficial trade union action at Hull ended a promising service. It has also created the three German-designed BACO* liners, which run between English and Continental ports and West Africa. These successfully separate the two modes. Containers (up to 640 teus) are borne on deck and handled by a deck-mounted gantry crane, and eight barges carried in the hold, being worked floating in and out through bow-doors by tugs.

On the Danube the Interlighter company, owned by the Soviet Union, Hungary, Bulgaria and Czechoslovakia, uses large (37,850dwt) BCVs of Seabee* type. The service began in December 1978 with one BCV, *Yulius Fuchik,* working from Izmail to Karachi and Bombay. Early in 1980 a second ship, *Tibor Szamuely,* enabled a second service from Izmail to Singapore and the Mekong delta. The ships themselves are owned by the Soviet Danube Shipping Co (SDP), but chartered to Interlighter. The barges are their own, and in tows work the whole length of the navigable Danube to Regensburg. The two Finnish-built ships carry barges bigger than the 834dwt of Seabee—26 of 1070dwt—and up to 1552 teus. In late 1980 it was stated that 200 more lighters were to be added to the fleet, and in 1983 that three more BCVs were under construction.**

> By sticking carefully to very limited port calls and a mix of cargoes, backed by a long river system at one of the ends of the right trade, Interlighter would appear to have proved Jerry Goldman's† original point.(1)

In 1983 Spain put into service the BCV's latest development, *Joseph Torres,* which can carry LASH, Seabee, BACO or Interlighter barges, and also sixty 20ft containers, using both float-on and gantry-crane lifting systems.

The main lines

So much for craft. Let us now look at what has happened or is happening to the Old World's great, usually international, inland waterway routes.

*BACAT : Barge Aboard Catamaran
LASH : Lighter Aboard Ship
BACO : Barge-Container
Seabee : (see p. 390)

**One 74-lighter, 1287 teu 39,900dwt vessel, and two 513 teu 8770dwt craft, the big ship with icebreaking characteristics, being intended, it is understood, for carrying barges to and from Siberian rivers, the smaller for Danube feeder services.

†The original BCV designer of LASH ships.

Central in western Europe has been the continued improvement of the Rhine, which from Rotterdam, and by ancillary canals also from Amsterdam and Antwerp, offers access to Germany, France and Switzerland. In 1983, 127M tonnes and some 180,000 barges passed the Dutch-German frontier at Emmerich, main traffics being iron ore, sand and gravel, and fuel, liquid and solid. Its most important port, Duisburg, is a terminal for iron ore and coal for the Ruhr steelworks, and a transhipment port for cargoes moving up the river, eastwards along the canals, or to rail.

On the upper river between 1952 and 1959 three more pairs of locks (one 185m × 23m, the other 185m × 12m) with adjacent hydroelectric power stations were built on the Grand Canal d'Alsace to supplement the pair built before WWII at Kembs near the Swiss border. Meanwhile, in 1956, France and Germany had agreed that the Grand Canal should not be continued further downstream. Instead, the river itself was to be canalised, with locks and power stations located on diversion canals of varying length. Four more pairs of locks were in this way built above Strasbourg, the last opened in 1970. Since then two more pairs, their dimensions increased to 270m × 24m, have been built below Strasbourg to complete a programme that has transformed the upper river. Below, in the gorge section, the rocky river bed at the Binger Loch was blown up in 1974 to widen the channel.

A separate Dutch enterprise in the 1960s involved construction of two dams and locks on the Lower Rhine (Lek) between Arnhem and the Amsterdam-Rhine crossing at Wijk bij Duurstede, thereby raising the Rhine level at Arnhem. A principal reason was to divert some of the Rhine's flow down the River Ijssel, running north from Arnhem to the Ijsselmeer, to make that river easier to navigate and reduce the saltiness of the Ijsselmeer.

In southern France, the 310km canalisation of the Rhône to a minimum draught of 3m, begun pre-war by the Compagnie Nationale du Rhône (CNR), was resumed, the first power station and lock of 24m fall at Donzère-Mondragon being opened in 1952. By 1973, eight 195m × 12m locks had been completed with somewhat smaller rises, the biggest, Montélimar, with 18.5m. Canalisation was completed on 21 March 1980, when the twelfth and last lock, Vaugris, was opened—the only one not on a by-pass canal. Thus the formerly turbulent river below Lyon has been made available for push-tows as well as self-propelled craft, most of the traffic being to Lyon or the industrial complexes south of the city. Success of the scheme is being confirmed by regularly increasing traffic. By 1982 a fleet of barges,

Fig 68 An aspect of the Rhine: early morning in the gorge section.

Fig 69 A tanker push-barge in Pierre-Bénite lock near Lyon on France's river Rhône. Note the 12m pleasure cruiser. The vertically-moving gate of a lock such as this is counterbalanced by weights rising and falling in the side towers, electric power being used to overcome inertia.

push-tow lighters and low-profile coasters was achieving an annual traffic of more than 500 M tonne-km, with a volume approaching 5 M tonnes.

Attention has now turned to joining these greatly improved Rhine and Rhône navigations along the Saône and roughly the line of the present Canal du Rhône au Rhin to provide a large-scale waterway from the North Sea to the Mediterranean, with a link also to the Danube via the Rhine-Main-Danube Canal now building.

A modest first stage was achieved at the Rhine end in 1966, with the opening of the enlarged Huningue canal from Niffer (below Kembs) to a new inland port at Mulhouse, accessible to Rhine barges. Its enlargement to push-tow standards was authorised in 1985. To the south, the Saône has been enlarged from its confluence with the Rhône in Lyon to Auxonne, 210 km with five locks, although before traffic can develop as on the Rhône, it remains to build a 4–5 km-long canal to by-pass the historic St Laurent Bridge in Mâcon, and to complete dredging north of Chalon. Responsibility for the Rhine-Saône works has been given to the Compagnie Nationale du Rhône. The projected line from St Symphorien on the Saône to Niffer on the Canal d'Alsace is 230 km long, with 24 locks (against the present 112), and will take 1500-tonne barges or 4000-tonne push-tows.

Meanwhile, significant improvements have been made between the Rhône and the major port of Marseille/Fos which generates much of its traffic. The short canal link at Port St Louis du Rhône, with its awkward entrance, is now by-passed by the new Liaison Rhône-Fos, opened in 1983 with a lock at Barcarin accommodating 4500-tonne push-tows, and extended through the enlarged Canal de Fos à Port-de-Bouc to the eastern side of the Gulf of Fos and the Etang de Berre. The dynamism induced by canalisation of the Rhône is such that Marseille now envisages reopening the Rove tunnel, closed after a roof collapse in June 1963.

Another great post-war international route has been created by the canalisation of the Moselle. Resulting from the setting up in 1952 of the European Coal and Steel Community, work began from its confluence with the Rhine at Koblenz to Thionville in France—270 km upstream—to take 1500-tonne barges or 3200-tonne push-tows, in agreement with West Germany and Luxembourg. Previously, 300–350-tonne barges drawing 1.5 m had been towed up the unimproved river and returned downstream, loaded, under their own power. The project was completed, with 14 locks 172 m × 12 m, in 1964. To serve Lorraine industry, canalisation was then pushed further up the Moselle, to Metz with three locks in 1966, to

Nancy-Frouard with four more in 1973, and then past Toul and on to Neuves-Maisons with six locks—in all, 394km from Koblenz.

Perhaps the most interesting post-war undertaking in the Low Countries has been the Schelde-Rhine waterway, built under a Belgian-Dutch treaty signed in 1963. It joins the Antwerp canal basin and dock system directly to the Rhine by a new cut across South Beveland, at the northern end of which are the two 320m by 24m Kreekrak locks, then a channel across the Eastern Schelde and another widened cut across Tholen to the Volkerak, whence by the triple Volkerak locks there is access to the Hollandsche Diep and routes to the Rhine or Rotterdam. This 38km waterway, opened in 1975, was financed 95 per cent by Belgium, since it was mainly to enable Antwerp to compete more effectively with Rotterdam. The waterway carries heavy traffic, increasing as Antwerp's newly built dock systems show results, although shipping is adversely affected by tide-level variations and currents. Their elimination—required by the 1963 treaty—awaits completion of the revised Delta Plan, under which the Eastern Schelde is to remain a tidal sea-water estuary following a decision in 1976, dictated mainly by environmental considerations, to build a storm surge barrier instead of a dam across its mouth. Thus 'compartmentalisation' is required, with two new dams, Philipsdam and Oesterdam, to isolate the waterway and all the northern delta waters from the sea water of the Eastern Schelde. The Philipsdam will incorporate the Krammer locks, two 280m by 24m chambers and a 75m by 9m chamber for pleasure craft, all with sophisticated salt-water/fresh-water separation systems (as already at Kreekrak) to avoid salt-water penetration inland. These works are to be completed by 1987.

The route previously used by Antwerp traffic was via the South Beveland Canal, with a bottom width of only 32m (compared to 120m on the new waterway) and locks at each end, but which ended up carrying 160,000 barges a year. The canal is still busy with traffic to and from Terneuzen and Ghent across the Schelde, and it too is now being enlarged and new locks built, to accommodate large push-tows.

More important to the Netherlands is the 72km-long Amsterdam-Rhine Canal, completed in 1953. It was so successful that enlargement was begun in the 1970s, and the rebuilt waterway with second locks at Wijk and Tiel and a normally open storm gate beside the lock at Ravenswaaij, was opened in 1981. With over 100,000 movements a year of craft carrying some 80M tonnes, the Amsterdam-Rhine Canal now takes four-lighter 10,000–12,000-tonne push-tows.

Fig 70 In the 1983 election the West German Social Democratic party used a poster to attack expenditure on the Rhine-Main-Danube Canal. Their defeat ensured that construction would continue.

Work on the Rhine-Main-Danube Canal has combined an element of the spectacular with another of tension as to whether the final stages of the project would be completed or abandoned. Happily, early 1983 saw a decision to complete it. Finance comes partly from electricity generation earnings, partly from federal and state funds.

The war's end saw locks built along the Main to Würzburg, with the old chain-steamer working thence to Bamberg, and also Kachlet lock on the upper Danube. In 1959 work began on the river-canal section above Bamberg, in 1962 the Main canalisation was completed to that city, and in the 1950s work also began again on the Danube with Jochenstein lock below Passau to complement 3 earlier Austrian locks thence towards Vienna. The first canal section, 29 km to Forchheim with 3 locks, was opened in 1968, a second to Erlangen in 1970, and in September 1972 it reached a new port at Nuremberg, 70 km from Bamberg, with 7 locks having a total lift of 82 m (991 km and 42 locks from the North Sea), 55 km of the line being true canal and 15 km canalised river. The locks have multiple economiser basins to save up to 60 per cent of water used, and hydroelectric stations alongside them—common on Continental rivers but unusual on canals. The canal itself is 52 m wide at surface and 4 m deep, with locks 183 m × 12 m.

By the early 1980s, two of the five additional 230m × 24m locks on the canalised Danube between Passau and Regensburg*, Geisling and Straubing, had been built. Above Regensburg, the new 34km stretch thence to the beginning of the canal at Kelheim was opened in 1978 with two locks at Regensburg and Bad Abbach. Between Kelheim and Nuremberg channel, lock and power station construction is well advanced. The 22km Nuremberg-Roth section was due to be completed in 1985, leaving 55km built or building from Roth over the 406m summit and down to Kelheim. The locks are all to have high-capacity pumping stations, which will not only supply the summit-level, but also transfer Danube water to increase the flow of the Regnitz through Nuremberg to the Main. In 1981 cargo handled on the completed sections of the R-M-D totalled 24.3M tonnes (the 1984 estimate for the port of Nuremberg was 800,000 tonnes), while electricty generation along the line produced revenue to finance a significant portion of the capital investment. Completion is expected around 1994.

Construction of the last stages of the R-M-D was undoubtedly being delayed by western barge-owners' fears that its completion would increase competition on the Rhine from subsidised eastern bloc craft reaching it through the canal. About 1980, therefore, an addendum was joined to the Act of Mannheim with the intention of limiting operation on the Rhine to the fleets of the original signatory states and of the EEC: vessels of other countries would need special permission, at the same time ensuring that the fleets really belonged to the countries named, and were not owned by dummy companies on behalf of others. A significant step forward was taken in 1983, however, when the West German and Hungarian governments reached agreement that FGR operators would get half the cargoes passing through the R-M-D to or from Hungarian ports.

On the Austrian section of the Danube between Jochenstein and the Czechoslovakian border three more 230m × 24m locks and power dams have been built to supplement the earlier four. Two similar ones are planned above Vienna, and below two more, these with larger

*When the R-M-D was first started in 1921, the company only intended to carry out river training works on this section; the locks result from higher navigation standards and demand for hydroelectricity.

Fig 71 Rhine craft: (above) a tanker push-tow (note the control cabin raised on telescopic supports); (centre) a container carrying barge passes Bad Godesberg.

Fig 72 (below) An aerial view of Leerstetten lock on the Nuremberg-Altmühl summit section of the Rhine-Main-Danube Canal. Note the economiser basins, which can between them save over half the lock water for later re-use.

GEFO FRANKFURT
DUISBURG-RUHRORT

275m × 34m locks to enable 3000-tonne coasters to reach Vienna.

Linked to the construction of the Rhine-Main-Danube Canal further north has been the work at the Iron Gates of the Danube. Here, since the beginning of navigation, boats have struggled with fast currents, narrow channels and submerged rocks, a struggle only partially mitigated by the building of the Sip Canal. After WWII Romania and Yugoslavia agreed jointly to build a hydroelectric barrage at the Gates, flanked on each side by a staircase pair of very large locks, each 310m × 34m, with a joint lift of 34.5m, able to take nine 1350-tonne barges or a 5000-tonne ship. The resultant backing up of water through the Carpathian gorges required resiting of towns and villages, submergence of traces of Trajan's early works, the island of Ada Kaleh and the Sip Canal, and rebuilding of railways and roads at higher levels. When I passed through in the summer of 1969, the Sip Canal with its locomotive was still being used, though the bottom lock on the Romanian side was in partial use and had somewhat lessened the current. The whole enterprise was opened in 1972, and reduced the upstream transit time for full cargoes from 4½ days to 15 hours. Some 80km below the Iron Gates, near Prahovo, a new lock 310m × 37m with an 8m lift was opened in 1985 as part of a hydroelectric dam system that will later include a second lock.

Meanwhile a medley of craft moves on the Danube: push-tows (with up to 12 lighters) and self-propelled barges or, strung behind a tug, the old-fashioned kind with high stern cabins and galleries with pot-plants; cruise ships, day and overnight boats like DDSG's three paddlers, one steam and two diesel-electric; cargo-ships, hydrofoils, ro-ro catamarans, barges off Interlighter BCVs—and always the ubiquitous canoe.

The improved Danube and the coming R-M-D are producing not only serious reconsideration of the old plan of a Danube-Oder-Elbe Canal, but new studies for large-scale Danube-Adriatic and Danube-Salonika (Aegean) lines. One new Danube connection has, indeed, already been opened. In 1949 the old idea of a canal through Romania to link the Danube to the Black Sea near Constanta, so saving some

Fig 73 (above left) The Carpathian gorge in 1969, showing a new road being built above the prospective river level after the Iron Gates locks have been completed.

Fig 74 (above right) A ro-ro catámaran is in 1984 ready to leave Passau in Germany for Bulgaria.

Fig 75 (centre left) Viereth lock on the river Main.

Fig 76 (centre right) The Sip Canal locomotive in 1969, not long before its track was submerged.

Fig 77 (below) Entering the newly finished upper lock on the Jugoslav side of the Danube in 1977 after the gorge had been flooded.

Fig 78 A bridge near Basarabi on Romania's Danube-Black Sea Canal, opened in 1984.

380 km as against the route by the Russian-controlled Sulina channel, was revived, and work soon afterwards began upon the Danube-Black Sea Canal. Halted in 1953, work was resumed in 1975 upon a canal 64.2 km long, large enough to take 5000 dwt vessels or push-tows of six 3000-tonne barges with a draught of 3.8 km. The canal has two locks 310 m × 25 m, one at Cernavodă near the Danube junction, the other at Agigea near Constanta, where a new port has been built. The canal was inaugurated on 26 May 1984. This completed, the Romanians have endorsed a proposal to build a 74 km canal from the Danube to their capital, Bucharest.

In Russia the Caspian sea route via the Volga and former Mariinski line to Leningrad has been enlarged and modernised. When ice free it offers a through route from the Baltic to Iran, and recently has gained from the Iran-Iraq war. In 1984 about a million tonnes moved along the through route in 3500 dwt river/seagoing ships. The Volga itself is navigable for 7500-tonne push-tows and 5000-tonne barges that can

Fig 79 The Yangtse gorges 1984: (left) a cruise ship passes a newly-built lock at Gezhouba, and (right) a motor barge passes the gorges.

pass its 295m × 30m locks. At the northern shore of the Rybinsk reservoir the Volga-Baltic Canal, 360km long, opened in 1964 with seven Volga-sized locks, leaves the river and runs to Lake Onega, whence by the Svir river and Lake Ladoga it reaches the Neva and Leningrad.

In the Far East push-towing with European-type craft began some time ago on the Yangtse, but in 1979 the shipping administration ordered 30 hopper barges and four 6000hp Mississippi-type towboats from the Dravo Corporation of Pittsburgh.

In June 1981, when it was said that the Yangtse river was carrying 80 per cent of China's inland water transport, the first two of three parallel locks built in a power-generation dam were opened at Gezhouba near Yichang on the middle reaches of the river, one 280m × 34m × 5m, the other 120m × 18m × 3.5m. A second 280m lock is now being built. They have a normal lift of 16m and a maximum of 27m. A second dam and set of similar locks, but with 29m maximum lift, are planned some 30km higher up at Sanxia. It is intended in time to build further dams and locks to improve flood control, generate power, and lift craft some 210m through the gorges and so make transit easier, as the locks at the Iron Gates have improved the route through the Carpathians. Further upriver, the removal of some rapids has also eased the channel.

The ship canals

We have considered some of the great new inland waterway routes: what of the three isthmian canals of the Old World, the Kiel, Corinth and Suez?

The Kiel Canal carries an immense traffic; in 1984 a cargo volume of 63.7M tonnes, of which 57.3M was transit traffic, the rest being to points along its line. Of the transit traffic, 36.5M tonnes was east-west, and included a sharp rise in Polish coal exports to the west. The number of vessels (excluding pleasure craft) using the canal was 50,920, of which 56.7 per cent were over 1000grt and 41 per cent over 10,000grt. The average size of craft continues to rise: in 1984 to 2292grt against 2261grt in 1983.

As we saw earlier, the canal has been enlarged once since it was built. It is now slowly and expensively being more than doubled in bottom width and increased in surface width from 102.5m to 160m, though without deepening, to give the banks greater protection from wash and improve the ratio of the canal's cross-sectional area to the immersed cross-section of bigger ships. About one-third yet remains

Fig 80 Germany's Kiel Canal busy with traffic.

to be done; as with all old canals, time-expired high-level bridges are also coming up for replacement by new bridges or tunnels. While this work goes on, 138 vessels pass through the canal on an average day, taking perhaps 7 to 10 hours.

The Corinth Canal was blocked by the Germans in 1944, but was repaired and brought into operation again in 1949, before being again temporarily blocked by an earthquake in 1953. The new towns of Isthmia and Posidonia now stand respectively at its SE and NW extremities.

In 1888, nineteen years after the Suez Canal had been opened, nine powers signed a Convention which intended to guarantee the free use of the canal. 'It was to be open to all vessels, in time of war as of peace; its entrances were not to be blockaded; no permanent fortifications were to be erected on its banks; no belligerent warships must disembark troops or munitions in its ports or within it. If Egypt were unable to defend it, she could appeal to Turkey, or through Turkey to the signatory powers.'(2)

Under de Lesseps' original 1854 concession, the canal was to revert from the Suez Canal Company to the Egyptian government in 1968, ninety-nine years after its opening. But in 1952 Gamal Abdul Nasser drove out King Farouk, descendant of Mohammed Saïd, and established a republic. In 1956 the British occupation of Egypt, which had lasted since 1882, came to an end, and in the same year the Egyptian government took over and nationalised the canal twelve years before it was due to fall to them. Britain especially was concerned that the canal's traditional international status, as stated in the Convention, would be lost. Negotiation failed, and with France and Israel she attempted to reoccupy the canal zone. But international pressure, notably from the United States, caused troop withdrawals.

Egyptian troops now threw down de Lesseps' statue as if by doing so they could expunge his contribution to their history, and for a time Egypt maintained the canal's status. Then in June 1967 Israel invaded Egypt and occupied the canal bank, and for the first time since 1869, the canal was closed, and ships reverted to the old routes.

Eight years passed before it was reopened in June 1975 by Egypt's President Sadat, appropriately by a vessel of the P & O company. By this time ships had become used to other routes, while new craft had also become much larger. So Egypt planned an immediate programme to deepen the canal by 5m and increase its wetted cross-section to 3600sq m, so enabling it to take fully-laden 150,000-tonners, or craft up to 380,000 tonnes in ballast, as against the previous limits of 60,000 and 250,000 tonnes. Together with some straightening of curves and new by-passes built at Port Said, Timsah, Deversoir and Ballah, the work was completed in November 1980. In 1984 20,000 ships carrying 255M tonnes passed the canal.

Waterway structures

We have looked at some craft of the contemporary waterscape, and at a few of the great new routes. What of the structures that vary still more a waterscape already variegated?

The waterway scene in America is comparatively simple: most large waterways are rivers, with only a few true canals: notably the Panama ship canal, and then the Chicago Ship Channel, the Welland Canal, and sections of the New York State Barge Canal, the Intracoastals and the Tennessee-Tombigbee Waterway, all of which we shall be visiting. Almost all locks are mitre-gated, very few are built more than two abreast, and there are no lifts, inclined planes or water-slopes on commercial routes.

In Europe, however, a network of canals, large and small, links the rivers that form its arteries. On recently modernised rivers where flooding dangers exist—as they usually do—guillotine-gated locks, or those with falling or rolling gates, are usual. Modern canal locks, too, often have guillotine gates, and also economiser basins to save water. Where waterways join the sea, locks are built to work in either direction, according to relative water-levels, eg at Volkerak on the Schelde-Rhine waterway. Here also, as at Kreekrak on the same navigation, salt and fresh water can be kept separate. River or canal locks are normally built singly or in pairs, but triple locks are not uncommon. As well, in Europe and Siberia, there are inclined planes, in Europe several lifts and two water-slopes. Most are post-war.

On the other hand, North America's waterways have a splendid variety of opening bridges compared to those of Europe, many of whose older structures have been removed by war. Most European bridges are either high-level or of the lifting kind, hinged at one side, and rising vertically. Occasionally one sees a transporter, suspended as on the Kiel Canal, or running on rails, as on the Amsterdam-Rhine, or a ferry. Only a few railway bridges and a very occasional road bridge are of vertically rising type, while swingbridges only exist on small waterways. Big locks, as at Delfzijl, often have lifting bridges at each end, and slip roads, so that when one bridge is raised, traffic can be diverted round the other end of the lock, an arrangement which is also found in Canada at the Seaway's St Lambert lock at Montreal.

Pleasure craft

From large to small. The years since 1945 have seen an enormous growth of cruising, inland or partially linked with the sea. A growth, too, in hiring as well as owning pleasure craft.

In Britain, pleasure boats, first yachts and rowboats, later cruisers, had been owned and hired on her rivers, lakes and Broads back in the nineteenth century. But in the 1930s some were found on the canals, and from the early 1960s, when commercial carrying largely ended on the smaller waterways, the growth has been rapid. Today, large numbers of cruisers, many of them hire craft, move over the canal network as their owners seek holidays away from the roads. Indeed, a considerable mileage has found new life. Without pleasure cruising they would have been closed years ago; with it some have more lockages per year than ever they did in trading days.

What has happened in Britain has also been developing on the Continent. Yachting and boating are old-established on the Dutch *meers* and the lakes of Germany, Switzerland and their neighbours. Now, however, the smaller canals of France, the Netherlands, north Germany, Sweden and Ireland have lost much or all of their former small-barge business. Yet many cannot easily be abandoned, because their channels are needed for drainage. Now cruisers use them. The Hadelner Canal in 1982 still carried 350-tonne barges moving about 250,000 tonnes a year, but some 5000 pleasure craft also paid its transit and lock charges not only to move from Weser to Elbe, but to pass through a peacefully pleasant canal. Others are commercially disused. In France, the department of Nièvre in 1972 took over the

Fig 81 A pleasure cruiser passes a lift-bridge on France's Canal du Nivernais.

58 km central section of the Canal du Nivernais and has since brought significant improvements. More recently, the waterways of Brittany and Anjou have been conceded. Among officially disused waterways that have been restored for cruising are the Charente and the western section of the Canal de Nantes à Brest in the department of Finistère, while pleasure craft are now being actively encouraged on the three Parisian canals. In north Germany such small waterways as the Elisabethfehn, Nordgeorgsfehn and Haren-Rütenbroek Canals are maintained for cruisers only, while in Sweden the Strömsholms Canal was reopened in 1970. Indeed, enthusiasts and public bodies alike now work to restore old and derelict canals.

Pleasure craft also travel the bigger waterways. Twenty or so years ago only a few brave spirits like Roger Pilkington took their boats on big commercial waterways and through their huge locks. But nowadays, as one watches a couple of Europa barges exit from such locks as Herbrum on the German Ems, Delfzijl or Terneuzen in the Netherlands, one is likely to see a couple of cruisers or yachts—often

Fig 82 (above) A pleasure cruiser on north Germany's fenland Nordgeorgsfehn Canal.

Fig 83 (right) A canal structure becomes a tourist attraction.

Fig 84 (below) the DDSG's *Theodor Körner* approaches Passau on the Danube.

quite small—emerge from behind. They move everywhere, often take short sea passages, and their numbers increase yearly. As, too, one passes along the waterways, one often sees a plantation of slender masts or a gathering of white hulls, marking marinas where pleasure craft are serviced and moored. Hire-craft bases are common in Britain, the Netherlands and France (where the number of bases has increased tenfold in ten years) less so in Belgium and Germany, but each year their number grows, allowing those who do not own their craft to take to the water for a week or two.

Others use trip-boats, which multiply also. They range from the ships of the Rhine, Göta Canal, Danube, Nile, Volga, Yangtse or Murray, which take passengers for voyages of several days or a week or two, to converted barges carrying a few passengers for a week at a time under luxurious or maybe spartan conditions, to Swiss, German, French and Italian lake steamers running on regular schedules, to the round-the-harbour or down-the-river or along-the-canal boats that now ply in waterside cities of any size from Amsterdam to Wuxi on China's Grand Canal.

As in North America, so in Europe canal structures have also become an attraction to visitors. Dr Gustave Willems was farsighted when he built a viewing tower high above his inclined plane at Ronquières. Nowadays we consider normal the coach and car parks, shops, viewing platforms, information centres complete with maps, diagrams and models, and restaurants of, eg Henrichenburg vertical lift, or the lesser amenities of many a canalside site where one can sit and look at boats.

Then come the water-dwellers, all the year round or during the summer, who seem peculiarly Dutch, but are not, for they can be found from London's Little Venice or Cheyne Walk to Thailand, Indonesia and China. Their boats seldom move. For them, water is an acceptable alternative to land as a foundation for living. They lead us back to our last Old World chapter.

10
SINCE 1945: KALEIDOSCOPE IN WATER

In Chapter 9 I described in outline the quickly changing waterscape of our own days. In doing so, I was left with much of interest—waterslopes, lifts, inclinded planes; quickly changing patterns of traffic; and the building of new river navigations like Germany's Saar or Portugal's Douro, canals such as Germany's Elbe Lateral, or others little known outside Italy. In this final chapter of the Old World story, I have therefore grouped what seems of special interest by country, an arrangement likely to be easier for readers to follow than grouping by subject.

Britain

The war's end left Britain with a very few comparatively efficient though small canals and river navigations, and a network of narrow, shallow lines taking 25–60-tonne craft. Most were nationalised in 1947, though some large rivers and all estuaries were not. The newly established British Transport Commission began to modernise their bigger waterways; for the others fifteen years ensued when few believed that they had any real prospects.

However, for the smaller canals a new and prosperous future did open, thanks to those who started hire-cruiser bases or bought their own boats, and the growth of canal enthusiast groupings, notably the Inland Waterways Association. In 1963 a new body, the British Waterways Board, empowered to run canals for cruising as well as transport, took over from the British Transport Commission, now abolished, and succeeded in building up pleasure cruising into a business that government was prepared to back with money. Moreover, a combination of enthusiast work, British Waterways Board enterprise and expertise, and local authority money, has restored several disused waterways to the cruising network.

On the commercial side, two influences have combined. The few devoted professionals, mainly of the BWB, were joined in 1972 by concerned amateurs of the Inland Shipping Group, an offspring of the IWA. In combination they initiated the National Waterways Transport Association in 1975. Thus support in parliament and the media was gained, which in 1978 brought government approval of the first big waterway in Britain since 1905, the upgrading of the Sheffield and South Yorkshire Navigation from Doncaster to Rotherham from 90- to 700-tonne standard. It was opened in 1983. At the same time the creeping paralysis of many of Britain's big—though not its smaller—ports, was causing much ocean shipping to move from London, Liverpool or the Clyde to Le Havre, Antwerp, Rotterdam or Hamburg, Britain's cargoes then being sent on in short-sea ships. Thus it is that the 1500km of Britain's freight waterways carry more tonnage in ships than in barges, and are so closely linked with the coasting trade.*

France

In France, about two-thirds of inland water transport takes place on the Seine and the waterways extending north to Dunkirk and the Belgian border. These waterways have naturally been the subject of large-scale development.

On the Seine, there are now six dam and lock systems between Paris and the sea or the Tancarville-Le Havre canal, each having two or more parallel locks, the biggest chamber of each group being 185m × 24m or 12m, accommodating push-tows of up to 5000 tonnes. In 1979 the Port of Paris Authority was formed to co-ördinate short stretches of the Marne and Oise. The authority now handles over 25M tonnes a year, carried in barges up to 3000 tonnes or push-tows up to 5000 tonnes. Apart from over 250 river wharves, there are major terminals at Gennevilliers and Limay on the Seine and Bonneuil** on the lower Marne. Some traffic is within the port, but most is to and from the seaport of Rouen (100km up the estuary) or Le Havre, plus some coaster traffic.

The old canals that together provided a route from Dunkirk port past the Canal du Nord and St Quentin Canal junctions to the Escaut and on by that river to Valenciennes and the Belgian border, 190km in all, have been rebuilt to Class IV-plus standard, allowing for example 3600-tonne push-tows to work from Dunkirk (which now has much improved barge access to its docks) to the steelworks at Denain on the

*For a fuller account of post-war British experience, see the final chapter of my *British Canals*, 7th ed, 1984 (David & Charles).

**Originally built in World War I to handle military supplies.

Escaut. The route has eight locks on what is now called the Liaison Dunkerque-Escaut, one of them replacing the old Fontinettes lift (see p. 133), and six on the upgraded River Escaut down to the Belgian border.

The Escaut is now reached from the Seine by way of the Oise and the Oise Lateral Canal, the latter now Class IV, thanks to the building in 1968 of large new chambers beside each of the four locks which formerly were hindrances to traffic. Thence there is access via the St Quentin Canal of improved Freycinet standard, with its 35 twin-chamber locks and two long tunnels. A higher-capacity alternative route is the Canal du Nord, joining the Oise Lateral at Noyon to the Dunkirk-Escaut line a little short of the Escaut. This, built to Class II with 19 locks, was opened in 1965. To avoid limiting the canal's capacity, craft controlled by traffic-lights enter Ruyaulcourt tunnel (4350m) simultaneously at each end, and pass in a central widened section. The canal takes 650-tonne barges, or mini-push-tows of 800 tonnes, comprising one motor-barge pushing a dumb *péniche*. Waiting for authorisation is a much-needed high-capacity Seine-Nord waterway to by-pass the St Quentin canal. A third major projected watershed canal link, after Rhône-Rhine (see p. 227) and Seine-Nord, is Seine-Est, for which there are two possible routes from the Seine basin to the Moselle.

Fig 85 A barge is carried down the transverse incline at Arzviller-Saint-Louis on the Marne-Rhine Canal, while another waits for passage.

In the meantime, traffic between the Seine and eastern France is handled by the busy Freycinet standard Marne-Rhine, some rebuilding of which took place in the 1960s. The aqueduct at Liverdun was widened to allow two-way traffic (only to be demolished 15 years later when completion of the Moselle canalisation project made this section of the canal redundant), the navigable depth east of Nancy was increased from 1.8 to 2.2 m, the six locks in a winding canal section at Réchicourt were replaced by a single new 16 m lift lock, and above all an inclined plane was opened in 1969 at Arzviller-Saint-Louis, to by-pass 17 locks in the steep Zorn valley not far from Strasbourg. A single caisson is carried down a transverse slope at 41°, with a difference in level of 44.5 m. The caisson (the design allows for a second if necessary) is 41.5 × 5.5 × 3.2 m, its 32 wheels running on four sets of rails. Propulsion is by counterweight assisted by electric motors.

A link with the past on this canal was broken when in March 1980 the electric-mule-towage service through Foug tunnel (867 m) was suspended, barges being expected to pass through under their own power, though the electric-powered chain-tug through Mauvages (4877 m) still operates.

In the south-west, it was decided to bring both the Canal Lateral à la Garonne and the old Canal du Midi, with their 150-tonne locks, up to *péniche* standard. On the lateral canal, the French in 1974 opened at Montech near Montauban the world's first water-slope. An invention of the late Professor Jean Aubert, chief engineer of the Ponts et Chaussées, it consists of a concrete water channel 6 m wide and 539 m long with side walls 4.35 m high, at an approximately 3 per cent gradient, to by-pass five locks. Water is held back at the upper end by a drop-gate. A *péniche* entering at the bottom of the channel is pushed up, together with the wedge of water on which it floats, by a metal shield which is lowered behind it and then propelled by two diesel rubber-tyred traction units, one on each side. At the top the gate falls as the water-levels equalise. Passage takes about six minutes for the rise of 13.3 m. The idea has advantages and disadvantages; in favour is that it uses no water, and that the load borne by the traction units' wheels is only that of the shield, whereas when using either big inclined planes or lifts for large craft the load-bearing structures have to be huge. Therefore it is practicable for big push-tows or barges. On the other hand, a straight course is required, and the duration of the operating cycle would tend to restrict capacity of the waterway if a considerable height were to be overcome. Design of the projected Seine-Nord waterway currently features water-slopes on either side of the summit-level.

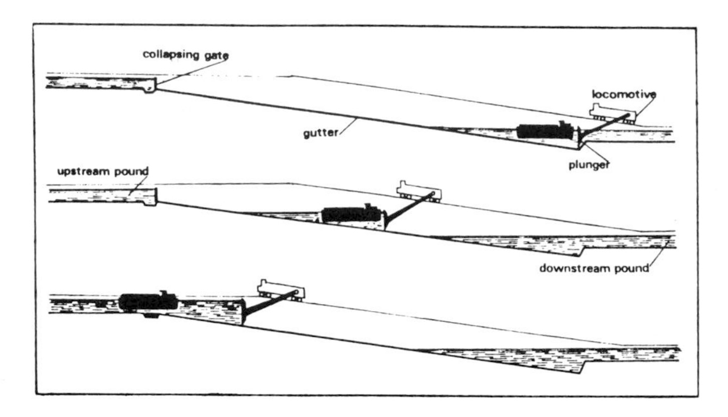

Fig 86 Diagram to illustrate the working of a water-slope.

Meanwhile, enlargement of the Midi has been going on, and in 1983, a second 'small-scale' water-slope was opened to by-pass the 7-rise staircase of locks at Fonsérannes near Béziers. The new slope, unlike that at Montech, can take not only *péniches* but also up to six pleasure craft. Upgrading has still, however, a long way to go—in 1983 to embrace the 124km section between Baziège, 26km from Toulouse and Argens, 57km short of Béziers. When completed, with all locks lengthened to accommodate 38.5m craft, it is estimated that commercial traffic between Bordeaux and Sète will be some 1¼M tonnes pa. In the meantime, there has been a phenomenal increase in pleasure traffic since the Englishman Michael Streat set up the canal's first hire base in 1969, so much so that some summer shifts of lock-keepers are being worked.

The French waterways possibly present more varied facets than those of any other country. About a quarter of the 8000km network is open to high-capacity vessels, especially push-tows, which account for a substantial proportion of the annual traffic of around 80M tonnes. The 'Freycinet' waterways make up about three-fifths of the network mileage. Their performance varies widely, and there are differing views as to their long-term prospects, but many are considered important feeders to the high-capacity routes in the main valleys; some indeed are to be improved to 'mini-push-tow' standards. That leaves, at the other end of the scale, about 1300km of smaller waterways whose future, like that of Britain's small canals, now depends on pleasure cruising, in private boats, hired cruisers or hotel barges. The policy is to downgrade these waterways to local status, with the *départements* taking over responsibility for their operation and maintenance, though with state subsidies.

Fig 87 A barge and tug begin their journey on the Montech water-slope in south-west France, opened in 1973.

The Low Countries

One axis of Belgium's waterways is the Escaut (Schelde) from its great port of Antwerp past Ghent and Tournai to the French border and its links with the Seine and the Liaison Dunkerque-Escaut. The other is the Meuse (Maas), reaching Belgium from the Netherlands below Liège, and running up to Namur and Dinant to enter France at Givet, with its tributary the Sambre leaving it at Namur to pass Charleroi also on its way to France.

Since WWII, and reinforced by the 1953 conference of transport ministers (see p. 215) the Belgians have been improving the Escaut upwards from Ghent to the French border to 1350-tonne standard but with locks able to take larger push-tows. It is at last done, the final obstacle, the old Freycinet lock at Rodignies, having been eliminated in 1984. Previously, a special difficulty had arisen at Tournai, whose citizens objected to enlargement through their historic city or any interference with their medieval Pont des Trous. After some twenty years of discussion, it was agreed that training works should be built on either side of the central piers to prevent push-tows hitting one or the other, and a one-way traffic system installed to control barge movements through the town.

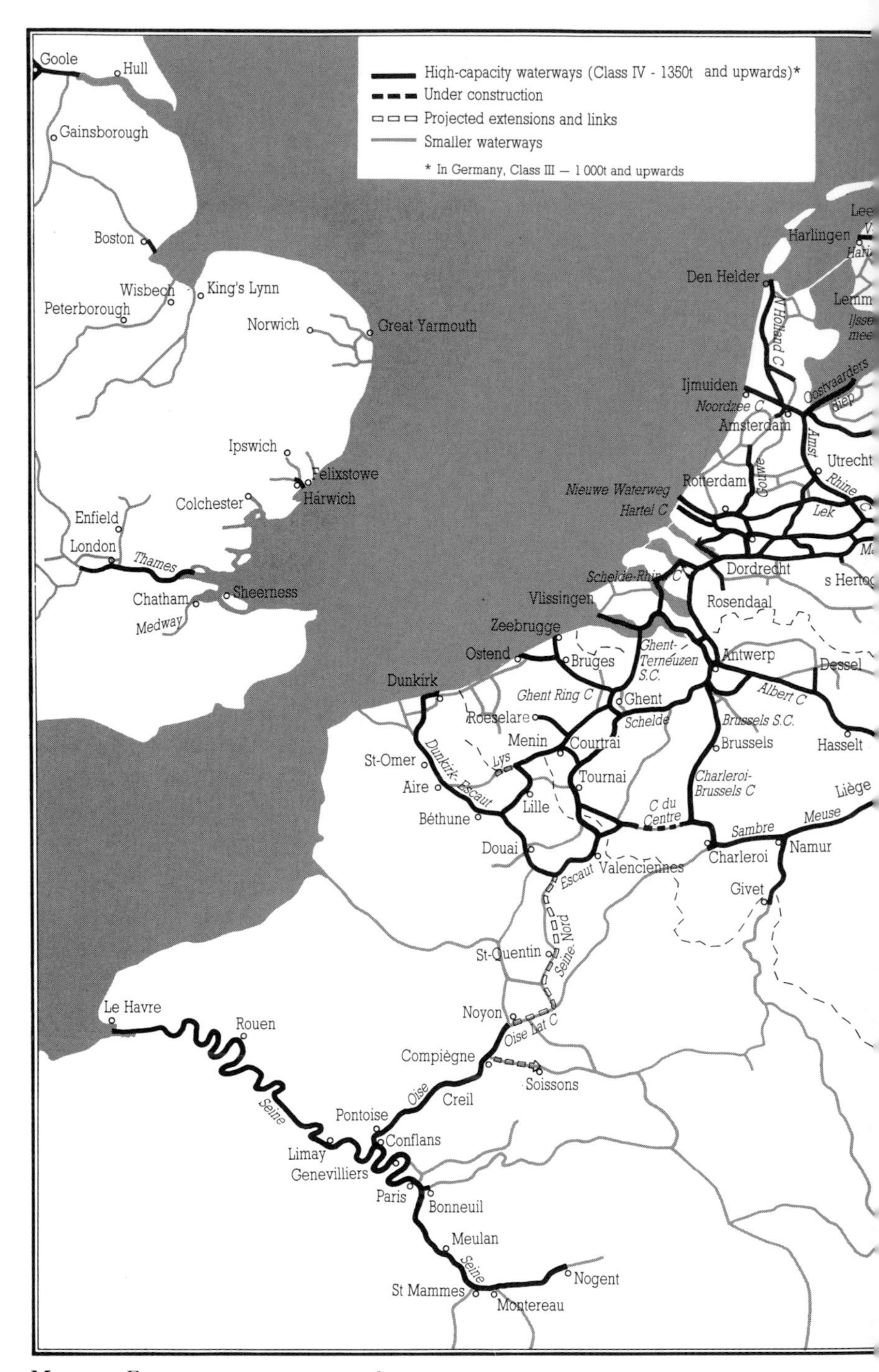

Map 10 European waterways today

Brunsbüttelkoog
Kiel Canal
Lübeck
Elbe-Trave C
Hamburg
Elbe
Lauenburg
Bremerhaven
Delfzijl
Emden
Eems C
Lüneburg
Oldenburg
Bremen
Hunte
Winschoten
Weser
Dörpen
Elbe Lateral C
Elbe
Dortmund-Ems C
Canal
Mittelland
Hannover
to Berlin
Almelo
Magdeburg
Bevergern
Minden
Osnabrück
Hildesheim
Salzgitter
Enschede
Münster
Wesel-Datteln C
Datteln-Hamm C
Hamm
Rhine-Herne C
Dortmund
Herne
Duisburg
Düsseldorf
Köln
Bonn
Rhine
Neuwied
Koblenz
Frankfurt
Kaub
Main
Aschaffenburg
Bamberg
Main
Binger Loch
Mainz
Ginsheim-Gustavsburg
Würzburg
Moselle
Erlangen
Trier
Rhine-
Nürnberg
Mannheim
Ludwigshafen
Heidelberg
Neckar
Main-
Saar
summit level 406 m
Saarbrücken
Danube
Regensburg
Heilbronn
Canal
Wörth
Karls
Kelheim
Danube
Lauterbourg
Stuttgart
Rhine
Strasbourg
0
50
100 km
to Basle

The inland port of Ghent is itself linked by large canal with Ostend and Zeebrugge on the north coast, and also with the Schelde estuary through Dutch territory at Terneuzen by the Ghent-Terneuzen ship canal, which after WWII still limited ships to 10,000dwt which, moreover, could not pass the West sluis sea-lock (added in 1910) at low tide. Modernisation began in 1953 on a basis agreed between Belgium and the Netherlands, but largely financed by Belgium. The result has been a new 35km canal, partly on the old line, partly on a new, that can take 80,000dwt ships. The main features are a new Westsluis (opened in 1969) at Terneuzen, 290m × 40m, with an intermediate gate and a low-tide draught of 10.8m, a lock, the Binnenvaart Ostsluis (opened in 1978), 277m × 24m, able to take push-tows, and the smaller Zeevaart Middensluis, 140m × 18m. Only three bridges now cross the canal: the rest have been replaced by ferries or a tunnel. At Ghent, the deep-water docks were renovated to provide 13km of quay of the same depth as the canal. Seaborne trade has grown from 2.8M tonnes in 1968 to 23.9M tonnes in 1983.

The city has also been given a 2000-tonne circular canal, opened in 1969, 23km long with two locks, which by-passes the old city route with its six lift-bridges and three locks, and links the docks and ship canal with the upper and lower Schelde, and the canal to Bruges and onwards to Ostend or, by the Baudouin Canal, to Zeebrugge. At the last-named, the harbour is being extended by jetties and a new very large lock giving access to a new inner port on the east bank of the Baudouin Canal.

Below Ghent, work began in 1967 upon upgrading the 30km ship canal to Brussels. Depth is being increased from 6.5m to 9.5m, and a cut made to avoid the lower Rupel and connect the canal directly with the Schelde through a new lock at Hingene. Completion of this enlarged line to enable 10,000dwt ships to reach Willebroek and 9000-tonne push-tows the docks of Brussels, is planned for 1990.

Only a central portion of the Meuse (Maas) is Belgian. The river enters the country from France at Givet to run past Dinant to Namur, where it meets the Sambre, then continues to Liège. This port, currently with some 200 vessel movements a day, is linked to the lower Meuse and Rhine by the Dutch Juliana Canal, but directly back to Antwerp and the Schelde by the steel- and grain-carrying Albert Canal, which is entered below Liège through a deep cutting. This canal is being upgraded for most of its length for 4-barge (9000-tonne) push-tows. At the Antwerp end, a new cut with the same profile was originally planned from Oelegem to Zandvliet, to avoid the congested and narrow sections in the older Antwerp docks. This ambitious project having been dropped under pressure from local residents and environmentalists, a more modest scheme is now under way, to bring the old line at least to 2 barge (4500-tonne) push-tow standard.

From Namur the 1350-tonne Sambre leads upwards to Charleroi, whence it continues at 300-tonne standard into France. From that town centred in Belgium's coal-producing area, the Charleroi-Brussels Canal leads north. Following the 1953 conference, the Belgians decided to upgrade this from 300-tonne to 1350-tonne standard, and from 1957 began major reconstruction under Professor Gustave. Willems, a spectacular symbol of which is Willems' Ronquières inclined plane. A new deep cutting reaching 50m depth on the summit-level replaced the former 1000m Godarville tunnel, and the inclined plane by-passed 17 locks on the older waterway, which had themselves replaced 30 on the original canal of 1832.

Each independently operated caisson of the plane carries a 1350-tonne barge lengthwise down a 5 per cent slope, overcoming a height of 68m in 22 minutes, at a speed of 1.2m/s. The incline itself, 1432m long, is overlooked by an observation tower 125m high, and headed by a concrete-pillared aqueduct 300m long and 60m wide which also serves as a waiting bay, and a 4.2km embankment with a maximum height of 25m. Each caisson (usable size 85.5×11.6m) weighs 5700 tonnes and is carried on 236 steel wheels running on rails. These are not flanged, being kept in place by horizontal guidewheels. Counterweights of 5200 tonnes run on rails in troughs beneath the centres of the caissons. Counterweights and caissons are linked through electric-powered pulleys in the headworks. When a caisson reaches either end of the incline, a locking mechanism comes into play, after which a gantry simultaneously raises the gates sealing the ends of the caisson and the pound. The only serious fault that developed under working conditions lay in the original mounting of the rails on chairs. After eight years these had to be replaced by longitudinal concrete supports for the rails to prevent deformation.

The Canal du Centre with its four old 400-tonne lifts and two 300-tonne locks runs westwards from the Charleroi-Brussels Canal to Nimy near Mons, where it joins the 1350-tonne Nimy-Blaton-Péronnes Canal, leading to the Belgian Escaut near Tournai, with a new branch, the Pommeroeul-Condé Canal (opened in 1983) giving access to the French Escaut, the Seine and the Dunkerque-Escaut waterway.

The enlarged routes up the Schelde from Ghent towards France, from Namur to Charleroi and Brussels, and from Antoing or Condé to

Mons, would form a single 1350-tonne system were it not for the old Canal du Centre. It was therefore decided to upgrade this canal also to 1350-tonne standard by by-passing all four of the old lifts with a single new one at Strépy-Thieu. With a vertical rise of 73 m, this will be far higher than any other lift so far built. It is to have two independently operated caissons balanced by counterweights as at Scharnebeck on Germany's Elbe Lateral Canal. The useful length of the caissons will be 112 m × 12 m, and each will with its water weigh 8400 tonnes. At the top of the lift craft will cross a steel aqueduct 600 m long to join the huge embankment of the upper level canal. Work began on the lift in 1982, and is already well advanced both there and on the remainder of the canal. Completion is scheduled for 1988, and La Louviére lift will then be preserved as an ancient monument.

Because the Netherlands is a country which believes strongly in the importance of inland water transport, the current picture is one of constant improvements on the main carrying routes, to eliminate traffic bottlenecks or keep pace with increasing craft size. Because of the past importance of the small canals, however, the problem of what to do with them is especially difficult. In recent decades many hundreds of kilometres have been closed, like the Apeldoorn Canal, sliced through by the Zwolle-Arnhem motorway, and the greater part of the canal network in Overijssel and Drente provinces. The tide is turning, however, with the growing interest in canal cruising as such (as opposed to the traditional recreational activity of sailing or pottering about on the *meers*). The Opsterlandse Compagnonsvaart is thus preserved as part of a ring of canals now promoted as the 'turf route', and the province of Groningen is indeed restoring one of its disused canals.

Germany

Apart from the Rhine and Moselle, notable work has been done on one river, and is proceeding on another.

The Neckar had been part-canalised upwards from the Rhine to Heilbronn before WWII; afterwards it was completed at 1350-tonne standard past Stuttgart to Plochingen, 200 km and 27 locks, in 1968. Later, the canalisation of the Saar upwards from the Moselle at Konz above Trier for 90 m to Saarbrücken was agreed, to provide that industrial area with a high-capacity link with the Rhine basin, far more useful than the 300-tonne Canal des Houillères de la Sarre, built in 1862–6, which runs south in France to join the Canal de la Marne au

Rhin. Authorised in 1973, with 6 locks 190m × 12m (a little longer than the Moselle's 172m) work began in 1975, and the line is intended to open about 1988. Radial upper lock-gates are being provided as easier to maintain and open against a head of floodwater than the Moselle's vertically-falling type, but some of the river's sharp curves will remain a difficulty. Normal practice allows a minimum radius on curves of ten times the maximum length of craft. On the Saar this will be a 180m-long push-tow, but curves such as that at Mettlach will be only some 300m. Nevertheless, by one-way working and careful lock siting, navigation without splitting the tows will be possible.

One entirely new canal has been built, the 1350-tonne Elbe Lateral (Elbe Seitenkanal), 115km long and almost straight from the Elbe above Hamburg to the Mittelland Canal near Wolfsburg, entirely within FGR territory. Two linked motives for building it were the poor state of the Elbe itself—for an average of 219 days a year there is insufficient water to float a fully-laden 1000-tonne barge—and the fact that, by keeping the Elbe in that state, the East Germans were hindering Hamburg's efforts to compete with such ports as Rotterdam. Construction began in 1968, and the canal was officially opened in June 1976, only to be closed again after a serious burst. Reopened, traffic is now building up in spite of strong railway competition. The canal has a lift at Scharnebeck near Lüneburg and an economiser lock at Uelzen with the exceptional rise of 23m. The lift, biggest so far opened, has two counterweighted caissons each working independently, as at

Fig 88 Scharnebeck lift near Lüneburg, opened 1976, works by power-assisted counterbalance.

Anderton and Niederfinow. Each measures 100m × 12m × 3.5m, weighs 5700 tonnes when filled, and raises barges 38m in three minutes.

In 1985 Bremen, having complained of its rival Hamburg's improved access to the Mittelland Canal by way of the Elbe Lateral, gained federal agreement to the deepening of the Weser up to the canal junction at Minden to take 1350-tonne barges.

The Mittelland Canal itself is also being enlarged to 1350-tonne standard by increasing its surface width from 33m to 53m, and its depth from 3m to 4m. By 1980 the enlargement had reached Minden from the Dortmund-Ems, and enlargement of the harbour locks giving access to the Weser was under way (the shaft-lock being just too small for Europa barges), with preliminary work in hand for doubling the aqueduct. In 1979 12M tonnes passed Minden, even before the enlargement was finished. Similar enlargement of the Dortmund-Ems Canal, also used in part by Mittelland traffic, was completed in the 1960s, involving the building of new cuts to ease bends and giving Handorf, near Münster, three busy parallel locks.

The new Henrichenburg vertical lift on the Dortmund-Ems was begun in 1958, and opened in 1962. It was built like its predecessor on the float principle, but with two instead of five floats beneath its single 90m × 12m × 3m caisson taking 1500-tonne craft. The maximum lift is 14.05m, and the hoisting speed 9m per minute. Such has been traffic pressure, however, that a 4500-tonne supplementary lock is being considered.

In 1982, 222M tonnes were carried on West German waterways (the cross-border figures at Emmerich being 124M tonnes), in 1983, 224M tonnes, and in 1984, 236.5M tonnes.

Switzerland

The port of Basle on the upper Rhine has grown in importance: from a turnover of 1.1M tonnes in 1930 to 3.5M in 1950 and 8M in 1981. Because 30 per cent of Swiss oil imports are handled by two terminals, Birsfelden and Au, on the 14km navigable length of the Rhine above Birsfelden locks as far as Rheinfelden, its end, the Swiss in 1980 opened a second parallel lock at Birsfelden, 190m long (10m longer than the first), to take push-tows made up of two Europa II lighters and a push-tug.

A link between the Rhine and Lake Constance (the Bodensee) has long been discussed. But the distance, some 120km, would require the building of a dozen locks and two tunnels at Rheinau and the Rhine falls near Schaffhausen. There is also a clash between those who favour industrial expansion and others who want only recreational use of the upper river and the lake.

On the Bodensee, as on other partially or wholly Swiss lakes, timetabled passenger services are the main activity, though commercial traffic is also seen. On Lake Lucerne, for instance, barges of up to 800 tonnes carry stone, gravel and sand.

Italy

Post-war Italy has some 2100 km of commercial waterways, of which about 350 km take a 1350-tonne self-propelled barge, or a push-tug with one large and one small lighter. The central waterway is the Po downwards from its confluence with the Ticino below Pavia. Southbound traffic turns off at Pontelagoscuro on to the Ferrara line for Porto Garibaldi near Ravenna, while northbound continues to Volta Grimana for the Brondolo coastal waterway to Chioggia and Venice.

Three new 1350-tonne canals are being built. In 1965 a 63 km waterway with eight locks and an aqueduct over the Adda was begun from the Po at Cremona to Milan, where an inland port was envisaged. The two entrance locks, a basin at Cremona, and 15 km of

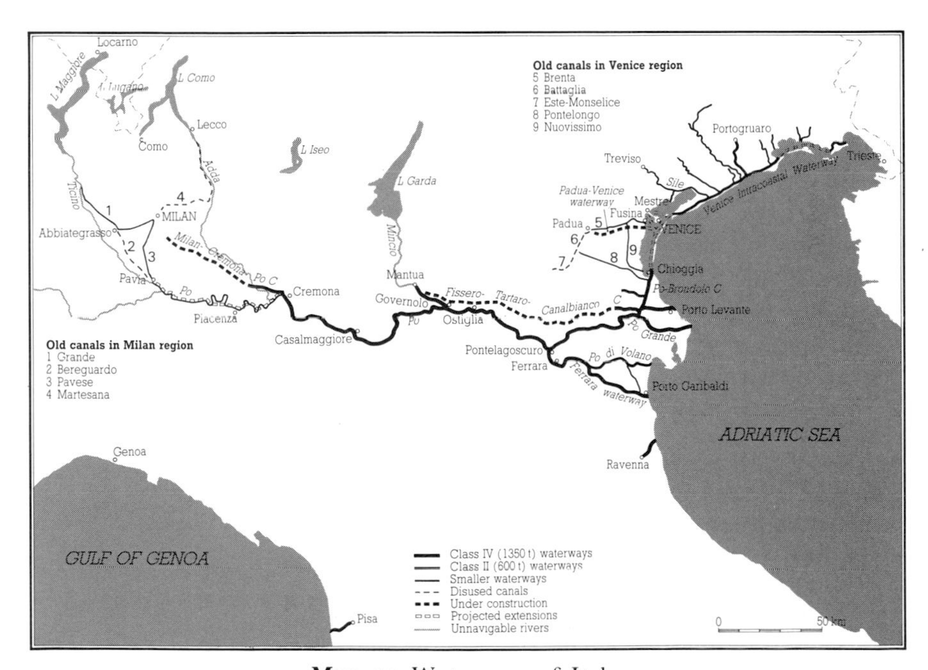

Map 11 Waterways of Italy

canal are built or building. Further progress probably depends upon more support, or less opposition, from interested parties at Milan than has so far been forthcoming.

A second and much more ambitious project is well under way, the Fissero-Tartaro-Canalbianco Canal. This, effectively by-passing the lower Po, will be 140km long (of which 60km at the eastern end had been built by 1982). It leaves the River Mincio (leading from the Po for 20km to Mantua) near its junction with that river, and runs eastwards to join the River Tartaro (currently a 300-tonne navigation); and then to Arqua and on to Adria for access to the lower Po or the coastal waterway towards Venice.

A new Brenta canal between the Venice lagoon and Padua is also being built. There has been a navigation between Padua and Venice since about 1200. Post-war, the 31.6km line left Padua by the Piovego Canal as far as the River Brenta, and then followed the Naviglio Brenta with five locks able to take 300-tonne craft, till it emerged into the Malamocco-Marghera Canal which edges the lagoon. Post-war, it was decided to build an entirely new 1350-tonne canal, 27.4km long, from a new port at Padua across the Brenta by level-crossing, and then to the lagoon south of Fusina, with one lock near Venice, the Conca Romana. By 1982 it was half built.

Fig 89 Carrapatelo lock on Portugal's Douro under construction. A shaft-lock, its rise of 34.5m is the greatest in Europe.

Portugal and Spain

The wine harvest of northern Portugal was once carried by high-pooped river boats, the *barcos rabelos*, carrying up to 60 tonnes for some 97 km down the Douro to the warehouses of Oporto. Then railway and motor lorry took over. In the 1960s, however, work began on five hydroelectric dams and locks, the latter to enable barges of 2000 tonnes or more to carry iron ore from the Moncorvo mines near the Spanish border some 180 km to Oporto for shipment. A port is also being planned just inside the Spanish frontier to facilitate more mineral exports.

Two dams and locks were completed by the early 1970s, at Carrapatelo and Régua, the shaft-lock at the former, at 34.5 m, being the deepest in Europe, with Valeira lock at 32.5 m, not far behind. Then economic and political conditions delayed further work; in 1980 it was restarted, and should be completed in 1986.

Yugoslavia

After WWII Yugoslav canals were hardly used. Then an ambitious plan was drawn up for a Danube-Tisa-Danube scheme, work on which began in 1947. This network of canals, partly flood and irrigation channels and partly for navigation, is based on the Bezdan-Bečej-Banatska Palanka line (the former King Peter I Canal), Banatska Palanka being above the entrance to the Djerdap gorge. The main line is 278 km long, the channel varying from 20 to more than 100 m wide, and 2.5 to 5 m deep, allowing big barges to by-pass the Danube loop that passes through Belgrade, especially in times of low or floodwater in the main river. By 1964 the canals were carrying some 400,000 tonnes, in which year a new port on the line was opened at Sombor.

Current investment in improved facilities for building and repairing river vessels up to 5000 dwt, probably a result of Danube development, shows a policy of expanding river carrying.

Czechoslovakia

Since WWII, Czechoslovakia has been enlarging her Elbe locks by providing the larger pre-war ones on the upper river, 85 m × 12 m, with a second chamber 190 m × 12 m for push-tows, and rebuilding those below. New locks have unusual hydraulically powered falling upper gates, which are lowered gradually as the lock fills to speed up operation without causing much turbulence. In winter, they can be lowered to enable water to flow through the lock and delay ice formation. By 1977, also, the upper limit of navigation had been extended with two locks to Chvaletice, with four more planned onwards to near Pardubice. A new traffic now began, in lignite from Lovosice towards the German border for 150 km to Chvaletice. Apart from the usual bulk cargoes, too, in 1982 there was a considerable traffic upwards from Hamburg to a terminal inside the border at Děčin-Loubi, which included containers. Higher up at Melnik, where the Vltava joins the Elbe, a container quay is being built. Thence the Vltava is being deepened for some 30 km to take 1500-tonne instead of 1300-tonne barges to a new river terminal for Prague at Radotin. That also will handle containers.

Meanwhile the old dream of a 500 km connection between the upper Elbe (probably at Pardubice), the upper Oder and, via the Morava, Bratislava on the Danube, is still alive. In 1983 the Economic Commission for Europe produced a study which concluded that a waterway to take 3600-tonne push-tows would be feasible as mainly utilising already navigable rivers, advantageous as relieving overloaded road and rail routes, and likely to carry some 80 M tonnes a year. It suggested that the Danube-Oder link should be built first.

Poland

Poland's great inland waterways are the Odra (Oder), with its branch the Gliwice Canal, and the Vista (Vistula). About a quarter of her barges move on the Vistula, the balance on the Oder or the line through Bydgoszcz (Bromburg) connecting both.

From the delta and Szczecin (Stettin), the Odra is navigable for 761 km, the last 20 km being in Czechoslovakia, but its effective head is at Koźle (Cosel). Following it upwards where it forms the Poland-GDR boundary, we pass on the GDR side at Hohensaaten the Oder-Havel Canal entrance, later on the right bank the confluence with the navigable Warta, then Frankfurt, followed by the Oder-Spree Canal entrance at Eisenhüttenstadt. Soon afterwards, the boundary continues along the Nysa (Neisse), the Odra turning south-east past Glogow and Wroclaw. From Brzeg the gradient steepens to Koźle and the Gliwice Canal*.

The latter, 41 km long, has six pairs of powered locks. Traffic,

*Originally the Klodnitz Canal, 46 km long with 18 locks to take c90-tonne barges, built in 1788–1806, it was replaced by the present shortened and enlarged canal, then called the Adolf Hitler Canal, in 1930–8. It was given its present name after WWII.

mainly coal and ores, is usually carried and worked through the locks by two 800-tonne barges and a push-tug. Gliwice terminal port with its two basins and large coal-storage capacity handles 37 per cent of Poland's inland port tonnage. At Koźle the Odra takes 170-tonne barges upstream for 45km to Raciborz, but most coal tonnages for Czechoslovakia are transferred there to rail. Below Koźle, craft like those on the Gliwice Canal work down through 13 similar sized locks to Wroclaw and two more to Redzina. Below, others are planned. Tonnage on the Odra and Gliwice Canal in 1980 was 11.6M tonnes. Very varying Odra navigation depths have been mitigated since WWII by the building of four reservoirs which together have given the river 19 more navigation days.

The Odra-Vistula link is not now heavily used: 294km long with 22 locks able to take 550-tonne barges, it runs by the Rivers Warta and Noteć to the Bydgoszcz Canal*, and thence by the Brda river to the Vistula. About 1.5M tonnes pa pass, mainly coal going east from Silesia, ore and agricultural products back. From its junction with the Noteć, the Warta navigation continues to Poznan and (by linking canals) Konin.

The natural Vistula is being canalised upwards from Sandomierz (south-west of Lublin). Sixteen locks able to take 3500-tonne tows are planned, of which three have been built (two of them on the by-pass Kanal Krakowski), the fourth being the uppermost at Dwory. Currently, push-tugs take two barges up to 600 tonnes on the Vistula, traffic having increased steadily from less than 2M tonnes in 1965 to over 9M in 1979, km/tonnage being largely building materials, stone and chemical products.

Currently, the Vistula downwards from Krakow to Bydgoszcz takes 400-tonne barges, thence to Plock 650 tonnes, thence to the Baltic 1000 tonnes. In addition to these locks, the Vistula has another at Wroclawek below Warsaw (115m × 12m × 3m), and a large one at Przegalina near Gdansk, 290m × 12m × 3.5m.

The Germans intended to extend the coal and ore-carrying Gliwice Canal to the Vistula and new markets, but instead in late-1980 the Poles began upon a new Kanal Slaski (Silesian Canal), which will initially run from above the Vistula lock at Dwory for 15km to a new coal-exporting port at Tychy, to which coal from three collieries can be taken by conveyors. Extensions could be either to the Odra south-west of Rybnik, or to the hoped-for Danube-Odra-Elbe Canal.

*Earlier called the Bromberg Canal, it was built 25km long in 1772–4 with 9 locks to take 170-tonne barges. It was rebuilt 1905–15 with 9 new locks.

Fig 90 Log-rafts are worked through a lock on Norway's Halden Canal, a waterway used only for log transport.

The old international water route from Magdeburg on the Elbe through to the Odra and then by Bydgoszcz, the lower Vistula, the Nogat, Zalew Wislany (Frisches Haff) and Pregola to Kaliningrad (Königsberg) is now little used, though Polish craft use the Nogat.

Finland

The original timber-carrying 250-tonne Saimaa Canal (see p. 88) had been highly successful. Work on complete rebuilding for much larger craft began in 1927, and was about one-third complete when WWII broke out. After it, former Finnish territory, including Viipuri (Viborg) itself, was ceded to Russia. However, agreement to continue the canal was reached in 1963, the USSR then leasing its section to Finland for fifty years. The new line was opened in 1968. Taking 1600-tonne seagoing craft, it is 56 km long with 8 locks to overcome a 76 m difference of level. The canal, mainly serving the timber and paper businesses and the growing industrial area round Lake Saimaa, has become increasingly busy, carrying well over 1 M tonnes a year. Its open water season is being increased from some 211 days pa to nearly ten months.

Some 350,000 tonnes of timber-rafts move downwards to the Gulf of Finland. There is also a heavy traffic in worked timber up the canal, carried either in lighters pushed by small Soviet push-tugs, or in motor barges either from Sweden or the area of Russia's Lakes Ladoga and Onega. Timber and paper is also carried from Lake Saimaa by self-propelled craft, many West German, to central European waterway ports.

In Finland generally since the war, timber has been increasingly moved in rafts, and not as individual logs. For these, timber-driving locks have been designed. Such are 160–75 m long, with a sector gate at each end. A current created by slightly opening the downstream gate draws the rafts in, whereupon a similar current from the upstream gate drives them out.

Russia

We can here glance only at a few changes in Russian waterways.

The Dnepr works were partially destroyed in WWII. Rebuilt to 3000-tonne standard and extended, there are now 6 locks between Kiev and the Black Sea. Two extension plans are reported: one, to continue the navigation works upwards along the Pripet river towards the old Pripet-Bug Canal as part of a Class IV Dnepr-Vistula link, the section between Pinsk and Brest near the Polish frontier, with 10 locks, being currently used by 600-tonne barges. Another scheme is to rebuild the old Dnepr-Niemen line by joining the Yaselda, a tributary of the Pripet, to the Shchava, a tributary of the Niemen that runs to Kaliningrad (Königsberg), to provide a second Black Sea-Baltic line.

The 1100km Kara Kum canal across Turkmenia in central Asia exemplifies a navigable irrigation canal. Built for the latter purpose, it was reported in 1984 to have some 150 craft using it to deliver and collect from communities that have grown up along its line.

Since 1945 Russia has seen important craft changes. The wooden barges usual before 1917 continued to be used between the wars; only afterwards did steel craft become usual. Again, the tugging of dumb barges gave way to self-propelled craft, able to navigate semi-rough water on the inland artificial lakes, and also to a considerable extension of push-towing, so that figures could show 38 per cent of freight as pushed in 1958—much higher than in western Europe at that time. Craft also became much bigger—in the 1960s 10,000-tonne tankers were working on the Volga waterways, as were pushed tonnages of 12,000. Finally, large passenger-craft including hydrofoils appeared, mainly for tourists, on the Moscow-Astrakhan and Moscow-Rostov routes, alongside old paddlewheeled steamers that still run local passenger services, eg taking 10 days from Moscow to Astrakhan. Specialised craft have also been appearing, such as very shallow-draught barges to come alongside river banks where no wharves exist, or seagoing refrigerated barges for fruit and vegetables that can also carry containers on deck.

In recent years Russia has made great efforts to develop several Siberian rivers for hydroelectric power and navigation. The greatest of these is the Angara-Yenisey river system which runs north for some 3100km from Lake Baikal to the Kara Sea. Given a poor railway network, much industrial development and a consequent growth in river freight, mainly timber and building materials, during the short summer season, the Russians plan to provide a deep waterway. Therefore each dam and hydroelectric plant must be by-passed for navigation.

At the Krasnoyarsk dam on the Yenisey, with a difference in water levels of 101 m, a double inclined plane has been built, one from below up to the crest of the dam, and another slantwise backwards from the crest to the upper level. Thus a caisson (seemingly taking craft up to 2000 tonnes) rises on the 1190m-long lower incline at a 1:10 gradient on rails racked for power and braking. At the top it runs on to a turntable, which is then swivelled through 142° before the caisson descends the second 310m incline to the upper level. This provides for water-level fluctuations of up to 13m. The caisson, with an inner width of 12m and depth of 9m, runs on 78 two-wheeled bogies, electricity powering the hydraulic drive. Time of operation from water-level to water-level is 45 minutes.

At Boguchany hydroelectric plant on the Angara river, an inclined plane carries timber-rafts and small boats over the dam on an electrically propelled cradle. This is built in three separate linked and loaded sections, each 22m long and 24m broad. This flexibility enables the cradle to pass from the lower slope on to the flat top of the dam and then to the reverse slope leading to the upper water-level.

Work has also been done on building very high locks, the biggest, at Ust-Kamenogorsk on the Irtysh river, being a shaft-lock with a 42m lift.

In 1984 Siberian rivers running to the sea were carrying some 25M tonnes pa. In 1981 Russia ordered seven river icebreakers: diesel-electric, with a shaft horsepower of 5170 (3800 kW) and a draught of only 2.5 m, they were intended to work on Siberian rivers at temperatures down to −50°C, thus offering the possibility of keeping them open for much longer than the present four to six months a year. As we have seen, the Russians are also experimenting with an icebreaking BCV.

India, Pakistan, Bangladesh

During WWI, British India's waterways had worked to good purpose in the allied cause, and at its end a network of active steamer routes operated by the India General and Rivers (the Joint) Companies radiated from Calcutta to Chittagong; to Goalundo and thence up the Ganges, not now to Allahabad, but to Buxar and Barhaj on the Gogra; or up the Brahmaputra as far as Sadiya. Britain had hoped to grant independence to a single state, but two communities which could coinhere when ruled from outside found they could not coexist. So in 1947 British India became India and Pakistan, the latter comprising two widely separated portions, one based on Sind and the Punjab, the other, East Pakistan, on East Bengal and part of Assam. Among other consequences of partition, the headwaters of the Indus fell to India, the bulk of its channel to West Pakistan, while the great port of Calcutta's access to Goalundo and the Ganges in one direction, and to the Brahmaputra in the other, now lay through East Pakistan. Then, in 1971, after war between the two states, East Pakistan became a third independent country, Bangladesh.

Some twenty years after independence the Joint Companies had disappeared. War had put replacement and repair programmes behind time, partition had disrupted their widespread and integrated passenger and freight services, state subsidies to railways had increased financial strain, and a prolonged Pakistan strike had made matters worse.

Country boats move on the Indus; otherwise present-day Pakistan has little modern inland water transport.

Bangladesh is criss-crossed by water, its three great rivers, Padma (Ganga below Goalundo), Brahmaputra (Jamuna) and Meghra, their tributaries and the canals linking them, bearing some 2000 powered craft taking goods and passengers over some 5300 km of waters navigable in the high-water season, 3700 km in the low. Among important river ports are Chandpur, Narayanganj (for Dacca), Barisal and Khulna (for Daulatpur), seaports Chittagong and Chalna. Canals to shorten voyages include the large-sized Ghasiakhali Canal, 5.6 km long, operation in 1974. It is 128 m wide and 7.3 m deep, and cuts the distance between Narayanganj and Khulna by 15 per cent.

The former operator of powered craft, the East Pakistan Inland Water Transport Authority, was after the war replaced by a Bangladesh IWTA, and in the middle 1970s traffic began to move again, mainly coal out of Calcutta and jute returning, Bangladesh having agreed to allow Indian craft to move between (Indian) Calcutta and (Indian) Assam through Bangladesh waters. Other bulk cargoes are oil, rice, fertiliser, cement and (usually by raft) timber. Before steamboats, alongside them and their successors, and still predominant today, are the estimated 500,000 country boats carrying 30–75 tonnes, with or without passengers, that are sailed, towed or rowed along the rivers, lakes and canals of Bangladesh and India. They proliferate especially on Bangladesh's deltaic plains, where seasonal floods make them virtually the only means of transport. Indeed, for many farmers, a boat replaces a farm cart.

Most of India's older irrigation canals, their locks weired, are no longer navigable, though some still have locks, but few boats. Parts of the coastal canals are busy, many of the rest disused or derelict, though with perhaps some potential to become a tourist attraction. Country and powered boats often work together. For example, the former may bring jute to baling centres, whence the bales may be taken by tug-hauled barges to ports for shipment.

India's Central Inland Water Transport Corporation (CIWTC) was set up in 1967. In 1979–80 it was in a bad way, after a 4½ months' strike, trouble in Assam, and heavy interest payments on craft, 80 per cent of which were described as 'unsuitable for operation'. However, the government has made money available to begin buying new craft over a period of years and progress had been made by 1982–3, when the CIWTC carried 44 per cent more traffic than in 1981–2, and in 1983–4 again increased the figures. A main aim of the Corporation has been to reopen the Ganga route, previously accessible by way of the Sunderbans in Bangladesh, by opening entirely within Indian territory the first National Waterway, 1581 km from Calcutta to Allahabad. This has involved the building at Farakka, some 400 km from Calcutta, of a Ganga dam and 180.7 m × 25.1 m navigation lock, primarily to provide navigation depth, but secondarily for water supply and drainage. From above the dam the 40 km Farakka Canal with a very slight downwards gradient runs into the Bhagirathi river near Jangipur. Previously navigable only in floodwater, this will provide a good channel down to a terminal on the Hooghly's Garden Reach at Calcutta.

It is now Indian government policy to develop inland water transport, principally by creating a series of National Waterways, and setting up an IWT Authority to develop them and their related terminals. Next is likely to be the Brahmaputra river system, while other candidates are the Narmada river in western India, the backwaters and canals of Kerala, the estuarial regions of the Krishna and Godavari rivers in Andhra Pradesh and the Buckingham Canal in Tamil Nadu. State governments are also being encouraged by loans and grants to develop inland water transport, while new craft construction is now helped by subsidies to entrepreneurs that effectively reduce the rate of interest paid on capital investment.

Burma

The courses of the 2100km Irrawaddy river and its main tributary the Chindwin north from the delta, in relation to Burma's geography, give inland water transport great importance. In 1976–7,38 per cent of tonne/km moved by motorised transport were carried by inland water, with another 8 per cent by coastal ships.

The rivers have unusual passenger-carrying importance. Burma's Inland Waterways Transport Corporation (IWTC) in 1977–8 carried over 11M passengers and performed 338M passenger km, operating 58 scheduled passenger/cargo services over some 6400 route km.

In 1977–8 the IWTC were using a splendid range of 164 passenger/cargo craft, 12 being still steam, the rest diesel-driven, many of them sternwheelers, with an average passenger capacity of 279. Of the total, 43 smaller and 10 larger craft were built in 1955–6 when the fleet was partially renewed; 38 later, and the balance at dates as far back as the fleet's veteran of 1914. In the same year the IWTC carried some 2M tonnes of cargo, about half being oil, nearly a quarter rice and cement, and most of the rest general cargo, much of it conveyed on passenger-ships. A push-tug had been introduced for oil barges; otherwise most cargoes moved in tug-hauled barges or self-propelled craft. By 1983 fleets again needed renewal. This time the Burmese decided that, instead of renewing older vessels, they would replace the machinery only by 360° steerable propulsion units, 36 single-screw and 18 twin-screw.

Yet older ways persist, as with the manned bamboo rafts that float slowly downsteam to the Andaman Sea, beneath which are slung teak logs or large clay cooking pots.

China

The Grand Canal is currently little used for through traffic, but much for local; sampans, scows and barges carrying farm produce, reeds, bricks and similar products. Improvement began in 1958, on a stretch 690km long in Kiangsu (Jiangsu) province running north from the Yangtse. This length is being enlarged to 1000-tonne standard, with four new mechanised locks replacing many smaller ones, and improvements also to bridges and inland ports. In 1980 this stretch carried 16M tonnes. The rest of the canal is mostly navigable for 30–100-tonne craft except for two disused sections, one near the Yellow river and the other near Tientsin (Tianjin). A project exists now to rebuild the canal north from the improved section to the Yellow river, as a means not only of navigation but of moving Yangtse water to the north China plain.

Just as India's nineteenth-century irrigation canals sometimes had a secondary navigation purpose, so in China in our own times. Purely irrigation channels have been built, like the much-publicised Red Flag Canal in Henan, but some also that carry small craft. One such is on the Shaoshan system that takes water from the Liu river, carries 10–20-tonne boats, and includes the Chutsin aqueduct over 500m long and 26m high.

In 1981 it was said that China had a current navigable network of 107,800km (the definition of navigability is not stated), upon which 330M tonnes (57.1 billion tonne/km) of freight were carried. There are a hundred inland ports that handle 1 M tonnes each, and over 300 that handle 100,000 tonnes, the two biggest being Nanjing near Shanghai, and Wuhan on the middle section of the Yangtse, at its junction with the Han river. Yet most of the navigable network carries only numberless small craft (the old ones wooden, the newer sometimes concrete) carrying a few tonnes, and poled, sailed, rowed, self-propelled or tugged in long lines. Were it not for them, the waterway scene in, say, Shanghai, with its self-propelled barges, push-tows, tugs and their barge-trains, is not very different from one in Europe.

But innovation is welcomed. Experiments are being made with hovercraft barges (see p. 220), and in 1983 two Japanese-built 11,500dwt LASH vessels were bought.

The Nile

The Mamoudieh Canal from Alexandria to the Nile, and the old Sweet Water thence to Ismailia, remain open, though little used because they are small.

On the Nile above Cairo, there are 70 m × 15 m locks at the dams at Asyut, Nag Hammadi and Isna, and a 5-lock set 80 m × 9.5 m to pass the old Aswan dam. Thence boats used to continue to Wadi Halfa below the second cataract, but since the Aswan High Dam, four miles above, was begun in 1959 by Russian engineers and completed in 1971 without a lock, navigation stops there, to begin again upon Lake Nasser above it and continue to Akasha in the Sudan. Traffic, apart from a growing fleet of hotel boats, consists of push-tows, tugs and barges, self-propelled craft and maybe 12,000 traditional Nile sailing craft averaging some 50 tonnes.

Given the High Dam, and cataracts above, Sudanese inland navigation is self-contained. It is based upon the port of Kosti on the White Nile, whence there is navigation for 269 km downwards to the Djebel Auilya barrage. This has a single lock where push-tows have to be broken. Thence local navigation continues to the sixth cataract at Omdurman near Khartoum. These routes and others less important carry annually some 200,000 passengers and 100,000 tonnes of freight in mostly elderly craft.

West, Central and East Africa

The colonial powers administered their African territories during WWII and for some fifteen years afterwards. In spite of war exhaustion, Britain, France and Belgium did much to re-equip and modernise African water transport systems upon which traffic was increasing along with economic development. The Belgians, for instance, gave the Congo and Kasai rivers new diesel-engined push-tugs able to handle up to 4000 tonnes, to work alongside their older craft. They also provided fixed-formation tows, ie barges meant to be worked as a unit. In such a tow, maybe of three in line ahead, the centre barge is straight at both ends, while the front of the leading barge and stern of the last are shaped upwards to eliminate eddy-making points, so reducing water resistance and fuel consumption.

Then came the 'winds of change'. With very few exceptions, independence from empires whose long experience included ability to run, and by example to teach others how to run, efficient passenger and freight transport services and organise a slow but steady

Fig 91 *Yapei Queen*, operated by the Lake Volta Transport Co, carries palletised bagged rice from northern Ghana.

Fig 92 Fertiliser bags being manhandled from lighters at Basse on the Gambia river.

development of natural resources, has led to deterioration, tonnages being well down on those of twenty-five years ago. Proliferation of state boundaries has made international river voyages more difficult, though lake transport has been less affected. Scarcity of finance and insufficiently experienced staff have sometimes affected the maintenance of existing craft, the ability to buy new ones, and work on river channels. African river navigation is naturally difficult, because widely varying water-levels only permit boats or tows to pass for parts of each year. Plans to mitigate the problem by building reservoirs, and providing hydroelectric dams with navigation locks, have not got far because most countries have given financial priority to new road construction. A 70km dredged channel for use when roads are flooded was, however, begun from Kalabo to Mongu in western Zambia in 1985 with Dutch help. It should be operational by 1986. Elsewhere opportunities for development in the interests of the local people offer themselves from east to west.

Fig 93 Dordrecht.

Envoi

At the foot of Wijnstraat, in the lovely old Dutch town of Dordrecht, a gate gives on to the waterside. Sit there beside the Bellevue Hotel, binoculars in hand, and look out to the point where, under radar and television guidance, traffic from the Beneden Merwede that leads back past the Waal to the Rhine or the Maas, turns right up the Noord on its way to Rotterdam, or left for the Oude Maas, the Hollandsch Diep and the route to the Schelde and Antwerp. More than 22 craft pass every hour of the year: coasters, push-tows, barges large or small, tankers, container carriers and many others, as well as an occassional cruiser or trip-boat. Then turn to your left, and walk beside the cut that runs in a semi-circle through the old city, past cruisers, yachts, houseboats—hundreds of them, many converted from small old trading craft, some newly built as replicas of the old, some just floating houses—from a window of which some Dutch *mevrouw* will peer at you through her potted plants.

From this extraordinary range of waterway life, let us turn to the New World, in some ways so like the old, in others so very different.

PART II

THE NEW WORLD

11

EARLY DAYS

Across the Atlantic lies a New World of inland navigation. Long before history began to be written, Indian canoes had used North American waters. When explorers and settlers arrived with the late sixteenth century, their ships worked up the rivers: the St Lawrence, the Connecticut, Hudson, Delaware, Potomac, James and others on the east coast, and in the Gulf of Mexico the Mississippi.

Canals for Canoes

Exploration and colonisation went on together, their needs causing the native North American canoe to be scaled up for cargo-carrying without alteration to its superb technology. In Canada, such canoes were paddled and portaged by Frenchmen along Champlain's 1615 route up the Ottawa and Massawa rivers to Lake Nipissing and thence to Georgian Bay and Lake Huron; the length of Lake Superior in 1622; by La Salle and Marquette in 1673 from the Lakes across to the Mississippi and down its great length to the sea, and from 1669–70 past the many rapids of the St Lawrence to Lake Ontario, Lake Erie and beyond. In so doing they drew a thin French circle round British colonists slowly penetrating inland from the eastern seaboard.

As colonisation extended, trade routes lengthened to supply the growing fur trade and the needs of the colonists for supplies and transport for their crops. In Canada especially, canoes* and *bateaux*** were used. The island of Michilimackinac on Lake Huron now became a base for fur-traders penetrating far westwards, to Grand Portage and the fur-trading stations north and west of Lake Superior. On the more navigable sections of other rivers, local inland waterway craft appeared, such as Durham† boats, and on the Great Lakes keeled

*Canoes might be birchbark *canots de maître*, some 30ft to 45ft long with a crew of 14 carrying some 3 tons of cargo, or, especially for the fur trade, the smaller *canots du nord*, some 25ft long, and with 8 men for crew. Intermediately there was the *batard*, with a crew of ten.

** A word generally applied to strong flat-bottomed timber boats pointed at either end, fitted for sailing, paddling or poling, carrying some 3 to 4½ tons, and usually with a five-man crew.

†These, carrying 30 tonnes or more, and mainly used on the upper St Lawrence, were long, shallow, nearly flat-bottomed, with rounded bow and a keel or centre-board. They could be sailed, rowed or poled.

and decked sailing ships, the earliest above Niagara being La Salle's trading ship *Griffon*, launched on Lake Erie in 1679 and lost on the return voyage from Green Bay, Lake Michigan. Timber-rafts, too, were everywhere floated down rivers.

Man-made navigation works began to appear. Maybe the earliest canals in North America were the tidewater cut made about 1637 from Plymouth Bay to Green Harbor, Marshfield, in Plymouth County, Massachusetts, and that built by the Jesuit Mission to the Indians near Midland, Ontario, soon after 1639. It seems to have been used to carry stone and timber to build the mission, and later, using a lock with vertically rising gates, to enable freight canoes to be loaded inside the stockade.

To improve military and civil communications up the St Lawrence at a time when Britain was trying to control those Americans who were in rebellion, Lt (later Captain) Twiss of the Royal Engineers, some Cornish miners being among his work-force, was put in charge of improvements between Lake St Louis, where the Ottawa river joined it, upwards over the 11 m. Soulanges section to Lake St Francis. It included rapids at the Cascades (Split Rock), the Cedars and, worst of the three, Côteau du Lac. At this last Lt Twiss cut what was probably the first true locked canal in North America, 900 ft long and 7 ft wide, with three locks. It was begun in 1779 to avoid 'a most tedious and laborious passage'.(1) By 1783 three short rock-cuts had been made at the remaining rapids, with part-timber, part-masonry locks able to take *bateaux* 6 ft wide and drawing 2 ft. Traffic was mainly military, merchants' craft being charged tolls.

Above Lake St Francis, *bateaux* had to be emptied at rapids and their cargoes carried past, the craft themselves being then poled, towed or dragged on log rollers. Beyond Lake Ontario the line to the west ran by portages over the 25 m strip of land that divided it from Lake Erie: one Canadian and one American portage on each side of the Niagara river that formed the frontier. Westwards again the North West Company in 1798 built another canal, 1000 yd long, with one wooden lock* 38 ft × 8 ft 9 in, with 9 ft rise, on the St Mary's River at Sault Ste Marie, the transit point between Lakes Huron and Superior, and above it a towpath along which oxen towed *bateaux* past the rapids.

To Ohio and Mississippi

Though France and Spain then held the Mississippi line, a drive

*The lock was destroyed by American troops during the War of 1812. A replica stands near the original site.

beyond the Allegheny mountains to the Ohio river and the west was a main motive for improved transport in the American colonies. We can perhaps date the beginnings of such Virginian and Pennsylvanian trade from the treaties with the Indians of Lancaster (1744) and Logstown (1748). A year later the Ohio Company was founded and a trail established up the Potomac valley, which was to become a route for river, canal, railroad and highway alike. The natural river was used up to Alexandria near Washington, whence waggons carried goods round the Great and little Falls, and then the river again to above Cumberland. Thence a trail led to the Monongahela which meets the Allegheny at Pittsburgh to become the Ohio.

In the 1760s and early 1770s, several proposals had been made to improve the Potomac's navigation, one by George Washington. Then came the Revolutionary War, and its creation in 1783 of a new nation, the United States. In a letter of 1784, Washington picked up and enlarged his vision of westward expansion centred on the Ohio:

> Extend the inland navigation of the eastern waters; communicate them as near as possible with those that run westward; open these to the Ohio; open also such as extend from the Ohio towards Lake Erie...(2)

The magnet was the Ohio.

Especially after 1783, emigrants moved west over the mountains to favoured embarkation points: Pittsburgh, Brownsville (Redstone) on the Monongahela, Olean on the Allegheny, Wheeling on the Ohio, and Kelly's Station on the Kanawha. They then used flatboats to descend the rivers until they found a place to live. So fast had been the movement that in 1803 Ohio was admitted as a state of the Union.

In 1803 the Louisiana Purchase made south and west, and notably the great Mississippi, American instead of Spanish or disputably French, and soon afterwards the end of the 1812–14 war removed the danger of British blockade of its mouth. The products of the interior now floated down the Mississippi and its tributaries on their way to the eastern states, the West Indies and Europe, on rafts and flatboats—clumsy wooden craft with a sweep each side and a steering oar, carrying 30–40 tons. These were built for the single journey and, were they successful in completing it, broken up for timber or used as a temporary house. Journeys might be from a few dozen to 2000 miles, perhaps from high up the Ohio to New Orleans, in which case the crew might have to walk back.

As trade grew, so did the size (up to 300 tons) of the flatboats intended for long voyages. Now also came keelboats, 40–80 ft long, 7–10 ft wide, sharp at both ends, drawing some 2 ft, carrying 30 tons or less, and usually with a covered centre for goods or passengers. Later, from about 1800, barges appeared, bigger than keelboats, carrying perhaps 100 tons with one or two masts and, usually, square sails. Much less trade made a precarious passage back upriver by barge and keelboat, sailed, poled, rowed, warped (kedged) or cordelled (bow-hauled) from banks that had no towpaths. A keelboat voyage of 1,350 miles upwards from New Orleans to Louisville might take three to four months.

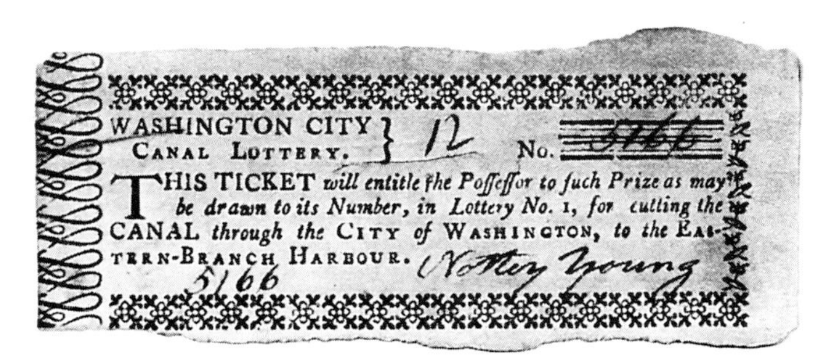

WASHINGTON CITY CANAL LOTTERY. } 12 No. 5166

THIS TICKET *will entitle the Poſſeſſor to ſuch Prize as may be drawn to its Number, in Lottery No. 1, for cutting the* CANAL *through the* CITY *of* WASHINGTON, *to the* EASTERN-BRANCH HARBOUR.

5166 Notley Young

Fig 94 A company was organised in the 1790s to build a canal from the Potomac river across the city to the Anacostia river (the Potomac's eastern branch). Maryland authorised a lottery to provide funds, but receipts were negligible. In 1809 another company was chartered. Work began in 1810, and in 1812 Maryland revived the lottery method of finance. The 2¼-mile Washington City Canal was opened in November 1815. By way of the Potomac it was linked to the Chesapeake & Ohio Canal.

Meanwhile, Washington had not forgotten his wish to use the Potomac as a route to the Ohio. In 1785 the Potowmack Company was set up under his presidency to improve the navigation. Though it faced pioneering difficulties, lacking skilled engineers and a labour force, it began to cut canals round the five main falls of the river, and by 1792 had completed three lockless by-passes at Seneca, Houses and Shenandoah falls. Two others, Great and Little Falls, needed locks. In 1795 Little Falls had been avoided by a cut and wooden (later stone) locks, and in 1802 a ¾m locked canal, partly cut through rock, was completed at Great Falls, so opening 220 miles of river from tidewater to some sort of navigation. Its five masonry locks, each 90ft × 14ft and in all rising 77ft, were an astonishing achievement for their time. The falls by-passed, the company began to clear the river bed, and also to build cuts and locks round five rapids on the Potomac's main tributary, the Shenandoah. By 1808 this too had been done, making that river navigable upwards for some 200m*.

*Throughout the New World chapters, the word 'miles' is abbreviated to 'm'.

The Potowmack Company had not, however, created a satisfactory navigation. Traffic was mostly rafts or shallow-draught expendable boats, broken up after the down trip to Georgetown (Washington) carrying furs, flour or agricultural produce. The developing country needed more. In 1815, a separate New Shenandoah Company took over that river's improvement, and in 1816 a parallel canal rather than further improvement of the Potomac was first suggested. It was to become the Chesapeake & Ohio Canal.

South of the Potomac, the James River ran from Chesapeake Bay past Richmond, Virginia, to the foot of the Allegheny mountains, whence it was temptingly near western river valleys that ran down to the Kanawha river, leading to the Ohio. In the way, however, were the Kanawha Great Falls. George Washington, a leader here also, saw these as the great difficulty in a route via the James.

In 1785, thanks to his enthusiasm, the James River Company was incorporated to build a navigation to the highest practicable point on the river at Buchanan. Between 1786 and 1793 it cut a 7m canal round the falls of Richmond: locks on the Westham-Broad Rocks stretch, opened in December 1789, were probably the United States' first. Upwards, the company did what it could to clear the river and build dams (weirs) and sluices (flash- or half-locks). By 1816 navigation was possible for some 200m to Buchanan, which however included a 4m stretch through Blue Ridge between Lynchburg and Buchanan, where the river fell 200ft by the Balcony Falls. The concern was for a time profitable, but navigation was poor and complaints many.

Fuel for Philadelphia

North of the Potomac, the Susquehanna River led north-westwards from Chesapeake Bay in Maryland into Pennsylvania. Northwards again, across Delaware's narrow neck of land, the Delaware River ran up to Philadelphia, where it was joined by the Schuylkill, itself running roughly parallel to the Susquehanna in the direction of collieries round Pottsville. From early times one-way flatboats had descended the Susquehanna from Pennsylvania and gone on to Baltimore in Maryland. Except, however, for a 1½m cut round the Conewago falls, opened in 1797, little had been done on the river compared to the efforts further north. Pennsylvanians thought this trade should also benefit their own state, and worked out a solution, a canal through the Delaware peninsula.

Such a cut had been envisaged back in 1654, and so necessary was it that a traveller in 1697 describes how:

> Sloopes...of 30 tunns are carryed over land in this place on certaine sleds drawn by Oxen, & launched again into the water on ye other side.(3)

The first serious promoter of what was to become the Chesapeake & Delaware Canal was Thomas Gilpin who, after having seen English canals in 1768, put proposals to a Philadelphia merchants' meeting in 1769. Surveys were made, but the War of Independence intervened, and after it, though a company was incorporated in 1802, interest moved towards an alternative solution, a canal between the Susquehanna and the Schuylkill.

In 1792 work therefore began upon a canal from Philadelphia to Norristown, an improved Schuylkill thence to Reading, and a canal again from Reading to the Susquehanna. The United States not having yet generated any professional civil engineers, Pennsylvanians asked a British contact to find one who could take charge of canal- and road-building. He approached William Jessop, then Britain's leading engineer, who recommended William Weston, probably the son or nephew of the veteran canal engineer Samuel Weston. William had been employed on the Oxford Canal and the building of Gainsborough bridge on the Trent. In 1792, aged 39, he agreed to work in the United States for five years at £800 pa British money. He must have been stunned—the sum is equivalent to some £32,000 pa today.

Arrived in January 1793, perhaps bringing with him America's first engineers' level, he started work designing locks for the Schuylkill & Susquehanna canal, already begun, but recommending against making the Schuylkill navigable and in favour of a canal through to Reading. By 1794, however, money had run out, with 15m of navigation and some locks built, and $440,000 spent.*

By the end of the 1812–14 war, attention had centred upon Philadelphia's need for coal. One obvious answer was to bring supplies down an improved Schuylkill river. A company formed in 1815 had by 1825 built a small dimension navigation taking 25-ton boats from Philadelphia for 108m, rising 588ft by 92 locks upwards to Port Carbon just above Pottsville, 62m of it canal, the rest slackwater (canalised river) navigation. The line included a 450ft tunnel near Auburn, the first on an American waterway. The Schuylkill Navigation was so successful that it was enlarged in 1832 for 80-ton craft, and again in 1845–7 for 170-tonners. Before then, in 1836, the downward tonnage was 570,094, of which 432,045 was anthracite.

*It was restarted in 1821 as the Union Canal, and completed in 1828. Built too small, it was never more than a moderate success.

A second coal-carrying route to Philadelphia was down the Lehigh river from collieries near Mauch Chunk to Easton, and then by the Delaware river. From 1791 efforts had been made to make the Lehigh navigable, using training walls and flashes of water. Sudden demand for soft coal, however, caused by the British coastal blockade during the 1812 war, encouraged young Josiah White to learn how best to use anthracite, and then to transport all he could to Philadelphia in one-trip wooden arks broken up at journey's end.

Having formed what soon became the Lehigh Coal & Navigation Co, White and his partner Erskine Hazard were in 1818 authorised to make the Lehigh from Mauch Chunk a one-way navigation. They did so using White's 'bear trap' lock (see p. 324) and arks some 25 ft × 18 ft, roughly built of logs held together with iron straps, fitted with boxes to hold the coal, and worked singly or in trains. The first boats moved in 1820, and by 1823 the line was fully open, carrying some 40,000 tons of anthracite a year to Philadelphia. In 1827 a 9 m gravity railroad was built to carry coal from the company's mine down to Mauch Chunk.

Canal and Lock Companies

The earliest true canals of any length in what is now the United States were built at opposite ends of the country: in Louisiana, South Carolina and Massachusetts.

Maybe we can see in Debrueil's Canal, 25 ft wide, cut 1736–40 under French rule from the Mississippi westwards to Bayou Barataria, which gave on to the Gulf of Mexico, the beginning of the Gulf Intracoastal Waterway. Later, at the small city of New Orleans, no longer French but Spanish, beside the lower Mississippi, the governor, Carondelet, in 1795 opened the canal that bears his name, 6½ m long and only 15 ft wide, as an extension of Bayou St John, to carry trade coming across Lake Pontchartrain thence to a basin beside the city ramparts. In 1803 Louisiana passed first to France and then to the United States, and in 1805 the Carondelet, now called the Old Basin Canal, was dredged, widened and improved. In 1807 Congress enacted that the city should provide land for an extension to the Mississippi. It was never built, but Canal Street is its memorial.

As early as 1786 a company was chartered to built a canal from the Santee river, by itself and its branches reasonably navigable into the interior, to the port of Charleston at the mouth of the less useful Cooper river. Money was raised privately, and in 1793 building began, with Col J. C. Senf in charge. It was finished in 1801, 22 m long, with eight locks (two a staircase pair) from tidewater on the Cooper to the summit-level, and then falling by four locks (two a pair) to the Santee. With lock dimensions of 60ft × 10ft, the canal took 22-ton boats whose main cargo was cotton from the interior. It seems to have been intermittently profitable. The canal had one odd feature: a single charge per boat, loaded or empty. As a result, boats were built that on the return journey could be carried one upon another, so paying a single toll.

The Merrimack River ran from New Hampshire's White Mountains to the sea at Newburyport in northern Massachusetts. Some seaport merchants formed a company, the Proprietors of Locks and Canals on Merrimack River, to build a canal round its only serious obstruction, the 30ft-high Pawtucket falls, just above the point where its tributary, the Concord, left it, Completed in 1796, it made easier the carriage of agricultural produce down the river to the sea.

Almost simultaneously the Middlesex Canal Company was formed in 1793 to build a canal from Boston to Middlesex village on the Merrimack above the Pawtucket Canal, by which, the promoters hoped, country products would benefit Boston's trade, while groceries and other goods would move inland.

The Middlesex Canal was to become somewhat of a prototype for American canal development, at a time when relevant engineering knowledge hardly existed. Loammi Baldwin, trained as a cabinet maker, had interested himself in science and engineering. He is probably the only man in history who, in his letter of acceptance of the appointment as Middlesex Company engineer, could write that he 'had no Experience at all. It is true I have studied the theory for many years and have been at Considerable pains to get possessed of the principles of Canaling, but I have never seen one foot of Canal which has been completed in a proper manner.'(4) Baldwin did, however, go south to Philadelphia and bring back William Weston, who re-surveyed and laid out the line before returning, and later acted as consultant.

Work began in 1795. As built, the canal was 27¼m long and 3½ft deep, with 8 aqueducts and 20 locks some 80ft × 10–11ft (some of them staircase pairs) including two tide-locks on the Charles, and guard-locks on each side of a level-crossing of the Concord river, where a floating towpath was installed which could be opened up to let logs pass down the river. The line began on the Merrimack with a basin and three rising locks to the summit-level, and ended in the Charles river at Boston, with a branch canal from the far side of the river (across which boats were hauled by cable) into the old city.

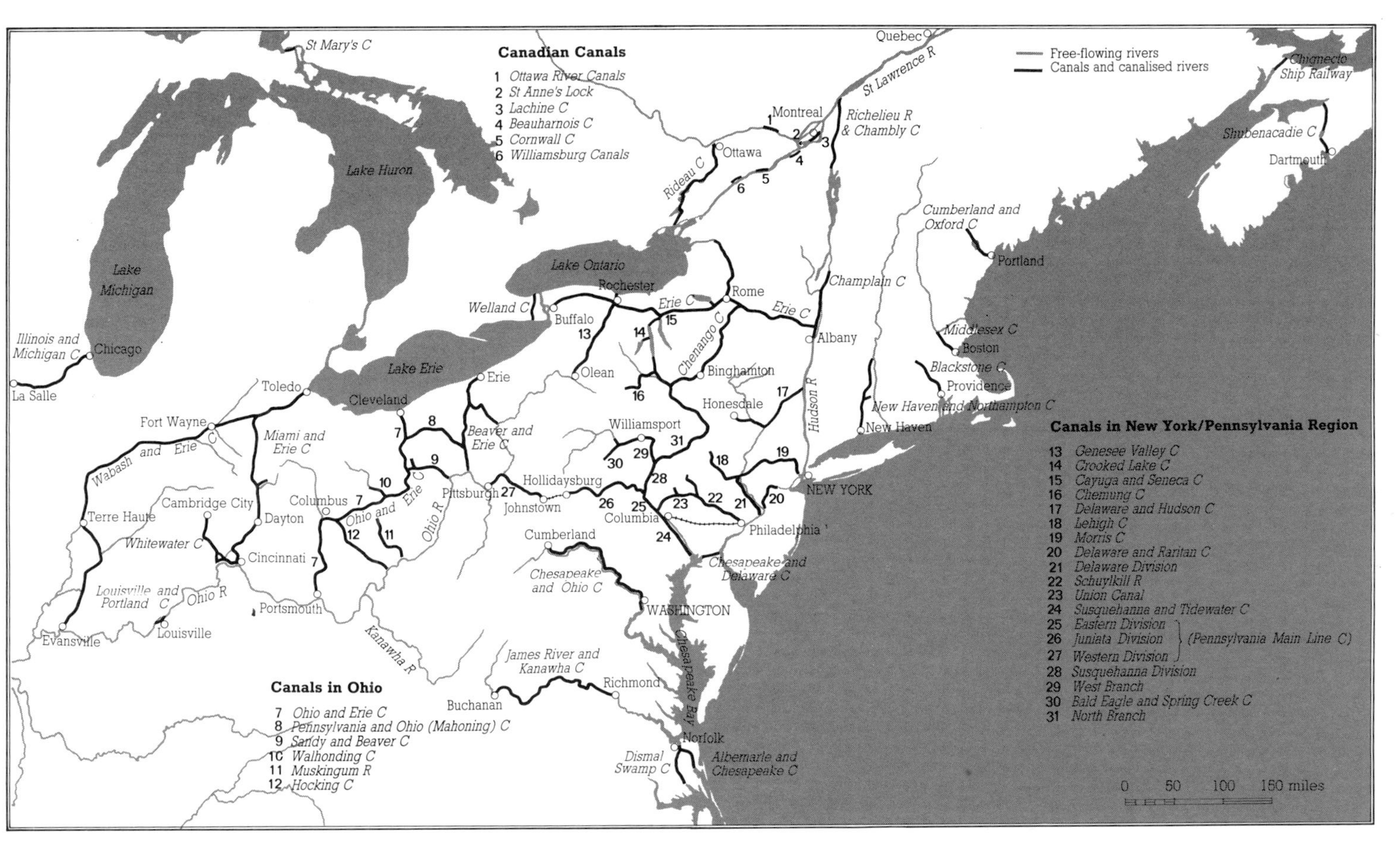

Map 12 The canal age in North America

Baldwin learned as he went along. His aqueducts had wooden troughs, and either wooden piers on stone foundations or, as at Shawsheen, the biggest, some 188ft long and 30ft above the river, stone piers. Some of his locks were of stone, others of timber to save expense. He had read of hydraulic lime that would set under water, but knew of no American source. He found a supply in St Eustatius in the Dutch West Indies; a sloop was sent to fetch some, which he used successfully on the first locks to be built, those nearest the Merrimack. Later, it was found locally.

On the last day of 1803 the canal was finished, and in 1808 Albert Gallatin was to call it 'the greatest work of the kind which has been completed in the United States'.(5) In the course of becoming so it had given novices an opportunity to learn engineering so that others in turn could learn from them.

At first, timber-rafts downwards were the main traffic, along with miscellaneous freight and passengers. A steamboat was tried in 1812, and another in 1871, but were found to be no advantage. But by 1813 a series of further cuts round falls on the Merrimack, with 25 locks, had been built with the Middlesex company's encouragement. These enabled craft to pass upwards to Concord, New Hampshire, and brought business to the canal. Traffic increased in both directions, tolls were reduced and improvements made, a state of things that lasted until the Boston & Lowell railroad opened in 1835. The canal had, however, scarcely been profitable: the first dividend, paid in 1819, was for $15.00 per share upon which $592 had been called. The highest was to be $30.00

Meanwhile, the manufacturing town of Lowell was growing up just below the Pawtucket falls at the junction of the Merrimack and Concord rivers. Its creation was due to the move there from Boston in 1822 of a group of textile manufacturers, who bought the Pawtucket company for their land and water rights and then reorganised it to provide water for America's first concentration of power looms. The company flourished. It still does, surely one of America's oldest, while Lowell's former power and navigation canal system, now supervised by the Lowell Historic Canal District Commission, is a great attraction to visitors. Within a few years, the original Pawtucket Canal locks may well be restored for navigation.

Further east, the Connecticut River was made navigable high into Vermont by the building of cuts around each set of rapids, mostly by separate little companies or partnerships. To get round falls at South Hadley, Massachusetts, Benjamin Prescott, employed by the local river company, in 1792 began a dam and cut some 2½m long, much

No. 173 DUPLICATE RECEIPT. Montague, 7 Sept 1826
John Parsons Esq To the Proprietors of the upper Locks and Canals on Connecticut River, in the county of Hampshire, Dr.

	DOLLS.	CTS.
Toll for tons of merchandize, at 91 cts. 7 m.		
Toll for 16 feet of boards, at 80 cts. per M.	12	80
Toll for other lumber equal to ft. of boards, at do.		
Toll for tonnage on boat at 17 cts. per ton,	2	
Received payment for said Proprietors, Wm Russell	14	80

I agree to the quantity as above estimated,

No. Hartland, Vt. 7 July 1835
Mr John Person, to the Company for rendering Connecticut River navigable by Water Quechee Falls, Dr.
To toll of tons of merchandize,
To toll on boat
To toll of 56 tons of lumber, 14.00
Received payment for the Company, $14.00
S Noyes Toll-Gatherer.

15 Oct DUPLICATE.
Wm Waddell
1836 To Mills Olcott & James Harris, Dr.
To Canalage and Assistance in passing by White River Falls
For Tons of Merchandize
" 250 " of Lumber 142 50
" " Boat besides Loading
Received Payment, Nathl Verbeck

NO. BELLOWS FALLS, Vt 23 May 1827
John H Carter To the Company for rendering Connecticut River navigable by Bellows Falls, Dr.
To toll of tons of merchandize,
To toll of 82 tons of lumber, $40.80
To toll of boat,
To toll of 97 sticks, 67.90
Received payment for the Company, $108.70
R Fleming

Fig 95 The upper Connecticut river was made navigable from the early 1790s by cuts round the various falls. These toll tickets from Vermont show some of the ownerships. Two cuts are well-known: that at Hadley Falls, Massachusetts, because at first it had an inclined plane, opened 1795, which was replaced by five locks in 1805; and Bellows Falls, opened in 1802 with eight locks.

of it through rock, to take craft or rafts 40ft × 20ft. Instead of locks he installed an inclined plane that must have derived from that at Ketley in England, opened in 1788. With 53ft vertical lift, giving a 13½° slope on its length of 230ft, boats were drawn up on a cradle powered by twin waterwheels, one on either side of the canal. Locks were provided at top and bottom to enable boats to be floated over the cradle before the water was drawn off. The whole by-pass was opened in April 1795. There was so much local objection, however, to the height of the dam that it soon had to be lowered, five locks then replacing the plane before the route was reopened to traffic in 1805.

Higher up the Connecticut, the best-known by-pass is that at Bellows Falls. A British-owned company was chartered in 1791 to make the cut, which was opened in 1802 with a dam and nine locks rising 52ft. In the twenties and thirties river traffic was quite considerable in the navigation season, which usually ran from late April until towards the end of November. Timber-rafts formed far the largest traffic on the upper Vermont stretches, as at Water Queechee falls at Hartland, with only an occasional boat carrying a few tons of merchandise. Lower down, as at Bellows Falls, lumber-rafts were supplemented by some sawn timber and a little more merchandise. Towards the state border, lumber disappeared from the toll tickets, and was replaced by sawn boards. As for the traders on the river, a dozen names recur again and again.

In 1792 two Inland Lock Navigation companies were incorporated in New York State, encouraged by Elkanah Watson and General Philip Schuyler of Albany. Both were practical men of affairs, but Watson had seen Dutch canals and studied American possibilities. Their main support came from land-owners looking to land values to be improved by navigation works. One, the Northern, was to open a line from the Hudson to Lake Champlain; the second, the Western, from the Hudson to Lake Seneca and Lake Ontario.

The Northern company did little before it failed, (English readers will note with interest that young Marc Brunel did some of the surveys), but the Western, with cash help from the state, had some success. William Weston brought a professional eye to the work in 1795, and in 1797 seems to have taken charge for two years, with, among others, Benjamin Wright, later engineer of the Erie Canal, under him. On the Mohawk two by-pass canals and sets of locks were built, and a third to link that river to Wood Creek. By 1798 Durham boats carrying 16 tons could use the river. However, to by-pass the Mohawk falls between Schenectady and Albany, or to reach the Oswego river and Lake Ontario, was beyond the company's resources. It survived, however, and maintained its works until superseded by the Erie Canal.

A national waterway policy takes shape

After the peace of 1783 between Britain and the loosely knit American Confederation, the commercial interests of its states began to diverge, so that the possibility of internal trade barriers became clear to men like Alexander Hamilton and George Washington, the latter concerned as he was in 1783 with the extension of navigation westwards from the Hudson along the Mohawk, and in 1784–5 with the push to the Ohio from the James and the Potomac.

In 1787, therefore, the Congress of the Confederation, in drawing up an Ordinance for the settlement of the country north of the Ohio under national, not state, auspices, agreed Article IV:

> The navigable waters leading into the Mississippi and St Lawrence, and the carrying places between the same, shall be common highways and forever free, as well to the inhabitants of said territory as to citizens of the United States, and those of other States that may be admitted to the confederacy, without any tax, impost, or duty therefor.

An act of 1789 under the new American Constitution adopted Article IV unaltered, and so provided the legal basis for the USA's free waterway policy. The substance of Article IV was soon extended by putting similar wording into the constitutions of newly admitted states, and by the Louisiana Purchase of 1803, which established the free use of the Mississippi.

Men like Hamilton, John Jay, Thomas Jefferson and James Madison now began to emphasise that the development of national transport routes, river, canal, road, should be primarily by the federal government rather than by states, so that everything done should yield the widest benefit. Two contrasting drives followed: for national development and for local improvement.

As in England, so in America something of a turnpike road boom preceded that for waterways. From the 1970s to perhaps 1825, road-building in the north-east and east, and westwards as far as Ohio, had been rapid, mainly by private turnpike companies, sometimes helped by public money, sometimes as entirely public enterprises. The turnpike decline that followed was not, however, so much due to canal and river development as to the inherent costliness of operating heavy horse teams and waggons on long- or medium-distance haulage routes.

The eighteenth century had seen some navigation construction, and many promotions: 74 inland navigation corporations were chartered between 1783 and 1800. Early in the new century, two events stand out in the development of national inland transport, the authorisation by Congress in 1806 of surveys for the National Road, and the report on roads and canals that Secretary of the Treasury Gallatin in 1808 presented to the Senate.

The former, greatest of the turnpikes, began at Cumberland, terminus of a road from Baltimore and also of the Potomac navigation, and by 1817 had crossed the Alleghenies to Wheeling, whence it was soon linked by the Pennsylvania turnpike to Pittsburgh. The second had arisen out of the incorporation of the Chesapeake & Delaware Canal company. Work had begun in 1802, but stopped in 1804 for lack of money, whereupon the company had applied for federal aid. This move almost coincided with the first use in 1805 of the small 20m-long Dismal Swamp Canal further south, which ran across the waste that divided the North Carolina areas round Albemarle Sound from their best port at Norfolk, Virginia, to the north.

Probably influenced by these, Gallatin, among his proposals, included the development of a sheltered passage along the Atlantic coast with four canals across necks of land, which included the Chesapeake & Delaware and Dismal Swamp Canals, and two others that we know as the Cape Cod and the Delaware & Raritan. This proposal was later given immediacy in the public mind by Britain's many raids on the Atlantic seaboard during the War of 1812. It was to presage the Inland, or Atlantic Intracoastal, Waterway. After Gallatin's report, debate began upon whether the federal government was constitutionally empowered to aid such projects, with the result that New York state went ahead on its own in 1817 with the Erie Canal, to which we shall return. Meanwhile the Chesapeake & Delaware lay dormant.

Happenings in Nova Scotia

Away in Nova Scotia, where transport depended upon lakes, rivers and the sea, a tiny canal mania occurred at places where links between one and the other could be improved. It began at Yarmouth with the opening in 1812 of a staircase pair of wooden locks each 133ft long to enable ships to move between Yarmouth lakes and the sea—the earliest large locks in Canada. It went on with surveys in 1823 for a Baie Verte canal across the Chignecto peninsula from the Bay of Fundy to Northumberland Strait, and another, the St Peter's Canal, to link the Bras d'Or Lakes of Cape Breton Island to the sea. Two years later Francis Hall, who had done both surveys (his experience had been on Scotland's Edinburgh & Glasgow Union Canal, a branch of the Forth & Clyde) did a third, for the Shubenacadie Canal.

No canal was ever built at Baie Verte (later the Chignecto Ship Railway (see p. 352) was to be sited there), and that at St Peter's had to wait until 1869. But in 1826 the first sod of the 54m Shubenacadie Canal was turned by Lord Dalhousie, Governor of British North America, to run by lakes and rivers across Nova Scotia from Dartmouth Cove on Halifax harbour to Minas Basin on the Bay of Fundy, and so promote trade between Halifax and the Fundy settlements. Rather improbably, it had Samuel Cunard as a principal shareholder. The canal-lakes-river line was to include 17 locks 87ft × 22ft, 6in, designed along the lines of those on the Forth & Clyde. Work stopped in 1831 in spite of a British government loan of £20,000, then restarted in 1854 on a redesigned line that included smaller locks and two inclined planes copied from those on America's Morris Canal. It was eventually opened to traffic by the steamer *Avery* in 1861.

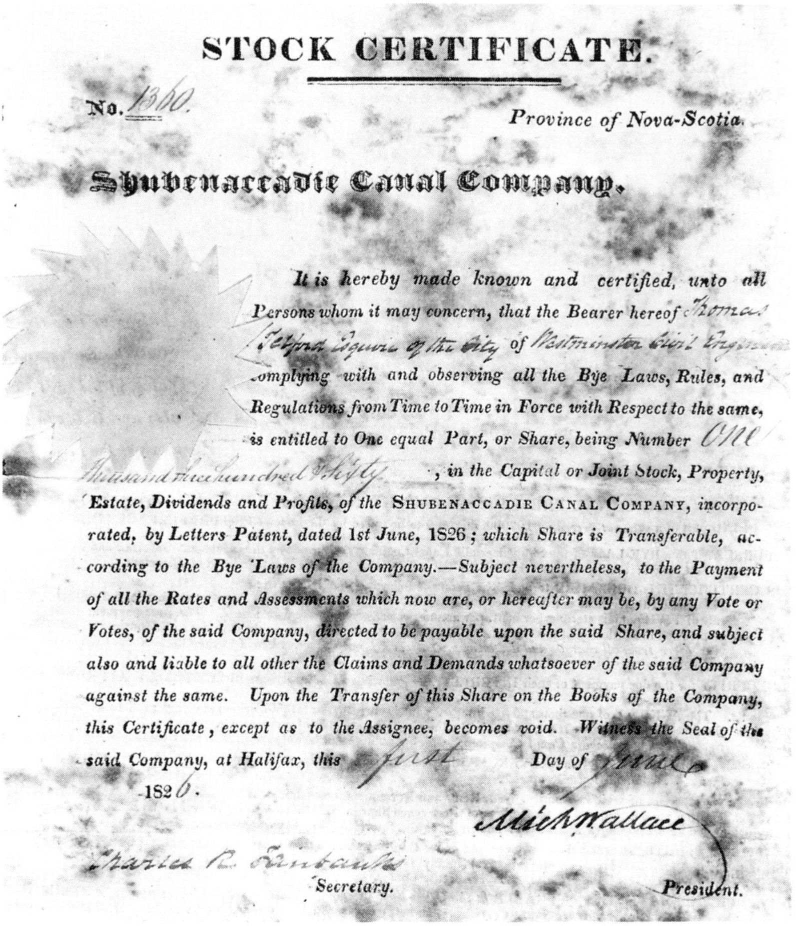

STOCK CERTIFICATE.

No. 1360.

Province of Nova-Scotia.

Shubenaccadie Canal Company.

It is hereby made known and certified, unto all Persons whom it may concern, that the Bearer hereof Thomas Telford Esquire of the City *of* Westminster Civil Engineer *complying with and observing all the Bye Laws, Rules, and Regulations from Time to Time in Force with Respect to the same, is entitled to One equal Part, or Share, being Number* One Thousand three hundred & Sixty *, in the Capital or Joint Stock, Property, Estate, Dividends and Profits, of the* SHUBENACCADIE CANAL COMPANY, *incorporated, by Letters Patent, dated 1st June, 1826; which Share is Transferable, according to the Bye Laws of the Company.—Subject nevertheless, to the Payment of all the Rates and Assessments which now are, or hereafter may be, by any Vote or Votes, of the said Company, directed to be payable upon the said Share, and subject also and liable to all other the Claims and Demands whatsoever of the said Company against the same. Upon the Transfer of this Share on the Books of the Company, this Certificate, except as to the Assignee, becomes void. Witness the Seal of the said Company, at Halifax, this* first *Day of* June *1826.*

Mich Wallace

Charles R. Fairbanks

Secretary. *President.*

Fig 96 The British engineer Thomas Telford had been consulted upon the plans for Nova Scotia's Shubenacadie Canal, and took up £450 worth of shares. This share certificate was made out to him and dated upon the actual day the company was incorporated.

Steamboat days

Steamboats began with the turn of the new century, superseding sloops on such rivers as the Hudson, and on Lake Champlain a selection of rowing-floats, sail-rigged scows for cattle, schooners and ferries propelled by horses working treadmills. Back in the 1790s John Fitch had worked a steamboat on the Delaware, but success began when Robert Fulton demonstrated his paddle-wheeler *Clermont* in 1807, which with two other steamers thereafter maintained a thrice-weekly service over the 150m between New York and Albany, except, of course, during the winter freeze. Two years later steamboats appeared upon the water sections of the New York-

Philadelphia run- from New York to New Brunswick and Trenton to Philadelphia, the 25m between being served by coaches. Another innovation was the *Juliana*, claimed as the first steam ferryboat, which began to work the New York-Hoboken route in 1811: by 1815 others were running to Jersey City and up the East River.

In 1809 the *Vermont* started a regular service on Lake Champlain, in 1810 the *Accommodation* between Montreal and Quebec, and in the winter of 1811–12 the sidewheeler *New Orleans*, designed by Fulton & Livingston, and captained by Nicholas J. Roosevelt*, steamed 1950m from Pittsburgh to New Orleans, thereafter working between New Orleans and Natchez till she sank in 1814. In 1811, the *Accommodation's* owner launched the *Swiftsure*, and by 1816 a regular steamboat service was running between Quebec and Montreal. In 1817 Capt Henry Shreve's *Washington*, built at Wheeling above Pittsburgh on the Monongahela, made a trip from Louisville to New Orleans and back in 41 days, in 1823 the *Virginia* reached Fort Snelling**, 683m up the Mississippi from her start at St Louis, and soon afterwards regular steamboat running began on the Ohio, Mississippi and such tributaries as the Monongahela, Cumberland and Wabash. The Canadian *Frontenac* appeared on Lake Ontario in 1816, the American *Walk-in-the-Water* in 1818 on Lake Erie.

By 1820,69 steamboats were working in the States: thenceforward they grew in numbers, power and size, carrying both passengers and cargo. By 1825, 125 were at work between Pittsburgh and New Orleans, encouraged by the bankside accessibility of wood or coal for fuel. Yet flatboats continued on the Mississippi system, both because the territories produced more than they consumed, so that there was traffic for both, and because the flatboatmen could now return quickly by steamboat instead of walking or by keelboat.

The War of 1812 and waterway development

After the United States had declared war upon Britain in 1812, their main attacks on Canada were made at Sault Ste Marie, where the *bateau* lock was destroyed, and on the Niagara frontier. Could they have cut the Canadian portage there, the British would have lost their line of communication to the north shore of Lake Erie and onward. But damaging attacks could have been made also upon the line of the St Lawrence itself, between Lake Ontario and Lake St Francis, the international boundary.

*His brother's great-grandson was President Theodore Roosevelt, whom we shall meet again.
**Above the confluence of the Mississippi and Minnesota rivers near present Minneapolis.

Soon after the war's end in 1814 it became clear to the British that to provide a better trading route to the west a canal should be cut across the Niagara peninsula between Lakes Ontario and Erie, and the navigation of the St Lawrence improved. Clear also that for defence reasons the St Lawrence-Niagara water line must be by-passed by a water communication between Montreal and Lake Ontario that lay further from the frontier. It became equally clear to the Americans that, the British being dominant on Lake Ontario, they needed a direct American communication between the eastern seaboard and Lake Erie. It is significant that early discussions upon what was to become the Erie Canal had assumed that it would run from the Hudson to Lake Ontario, with perhaps a canal also across the Niagara peninsula. It was now seen that if cargoes from the west got to Lake Ontario, they would militarily be at danger from the British, and commercially were as likely to continue down the St Lawrence as to pass into the Erie.

DeWitt Clinton, in his *Memorials of the Citizens of New York*, published early in 1816, saw the point:

> The most serious objection against the Ontario route is, that it will inevitably enrich the territory of a foreign power, at the expense of the United States. If a canal is cut around the falls of Niagara, and no countervailing or counteracting system is adopted in relation to Lake Erie, the commerce of the west is lost to us for ever.(6)

And so in 1817, when cutting began, it was upon a canal aimed to run parallel to Lake Ontario's southern shore to a terminus at Buffalo upon Lake Erie. The rest of this chapter will therefore consider four developments all to some extent derived from the war: in Canada the building of the Welland Canal, early improvements to the St Lawrence, and the provision of a protected military water route to Lake Ontario; in the United States the construction of the Erie Canal.

The Welland Canal and the St Lawrence

Robert Hamilton, a Canadian merchant concerned with the Niagara portage, had proposed a canal across the peninsula in 1799, but his Bill had failed. Such a canal was no easy matter. From north to south, one finds the flat lands of Ontario's southern shore, then a steep escarpment rising to some 300ft (the height of Niagara falls plus the fall of the Niagara river), a ridge above it, the Welland river running into the Niagara above the falls, and lastly the flat lands of the north Erie shore.

The canal scheme was revived in 1818 by William Hamilton Merritt, who had bought some of Hamilton's property at St Catherines on Twelve Mile Creek on the Ontario slope. The Welland Canal originated, indeed, as a means of improving the Merritt property by ensuring a steady flow of water to his mills in the summer. But it grew. The company Merritt formed was chartered in 1824 and welcomed by the Upper Canada government for defence reasons and as a counter to the infant Erie Canal. Some money was raised privately in Upper and Lower Canada*; more, curiously, in New York, where canal speculation was then strong; more still from the two Canadian and the British governments.

The Welland Canal that finally resulted was not that first planned. Originally thought of as a barge canal with a 2m summit tunnel, it was replanned as a ship canal to take mule- or horse-towed lake schooners and sloops, with timber locks 110ft × 22ft, and 7ft 6in over the sills. Then in 1826, to take the new steamboats, it was to be enlarged to have locks of 125ft × 32ft × 9ft 6in. Three locks of this size were built before economy forced a reversion to the smaller dimensions.

The first section, 16m long, was planned to run from Port Dalhousie on Lake Ontario at the mouth of Twelve Mile Creek up the escarpment by 35 locks, and then through a deep cut 1¾m long to join the Welland river, which would act both as feeder to the canal's summit-level and as its continuation for 8m to Chippewa on the Niagara river, which gave access to Lake Erie. However, late in 1828 movement at the deep cut revealed geological conditions that forced a change of plan. Because the bottom of the cut could not now be made low enough, the Grand river to the westwards was to be dammed some 5m above its mouth, and a long feeder built to carry its water over the Welland river on a wooden aqueduct to the deep cut. Two locks at Allanburg at its northern end had to be built to lift the canal's summit and so maintain its navigation depth, and another two at Port Robinson at the cut's southern end to lower it again to the Welland river. In this state it was opened on 27 November 1829 by the schooner *Ann and Jane*.

It was immediately plain that the Welland and Niagara rivers formed a most inconvenient navigation, and a direct cut onwards from Port Robinson south of the locks was decided upon early in 1830. During the next year government money was raised, and cutting began onwards to Port Colborne on Lake Erie, where a large lock,

*Ontario and Quebec respectively.

130ft × 24ft, was built. In 1833 the new main line was opened, in all 28m long and 8ft deep, with 40 timber locks. The cost had been high; to 1837 some £452,000 of which £ 288,000 had come from governments by way of grant or loan.

On the St Lawrence itself, the Royal Engineers had by 1805 replaced the two lower of the four *bateau* canals opened in 1783 by a new Cascades Canal, 1,500ft long and giving 3½ft depth, with two sizeable locks 120ft × 20ft at the lower end able to take Durham boats, and guard (flood) gates at the upper end. John By, whom we shall meet again on the Rideau, had as a Lieutenant worked at the Cascades in 1802. The Côteau du Lac cut was in turn replaced by a 400ft enlarged canal with one lock built by the Royal Staff Corps and opened in 1817. Thirdly, a private company in 1819 initiated the Lachine Canal, to replace the land portage of all goods moving upwards from Montreal to Lake St Louis past the 45ft-high rapids, and so improve access both to the upper St Lawrence and Ottawa rivers. The scheme was soon taken over by Commissioners appointed by the Lower Canada government and work began in 1821. The canal with seven locks, one a guard-lock, 100ft × 20ft giving 5ft over the sills, was finished in 1825. The St Lawrence had thus been improved for Durham and steamboats from Montreal to the upper end of Lake St Francis, but nothing had been done above. A beginning had also been made upon the Ottawa river route.

The Ottawa-Rideau defence route

As we have seen, the War of 1812 sharply brought home to Britain, responsible for the defence of the Canadian colonies, the dangers of the St Lawrence route for the carriage of military stores to the fortress and dockyard of Kingston on Lake Ontario. A new navigable line upwards from Montreal and well away from the frontier was therefore planned, by using the Lachine Canal, and then the Ottawa river as far as the Rideau river at what is now Ottawa, which could be used to reach Kingston via the Rideau lakes and the Cataraqui river.

Three obstacles prevented barge navigation on the Ottawa river route: the small rapids between Lake St Louis and the Lake of Two Mountains near Montreal, where a small wooden lock had been privately built in 1816 at Vaudreuil; the 10ft Carillon rapids; then, 4m higher and much worse, the 5ft Chute à Blondeau rapids with, a little above, the long stretches of fast-running river that fell roaring 45ft down those at the Long Sault, compelling a double portage, with at its head the village of Grenville.

Fig 97 An early postcard: the 8-rise staircase of the Rideau Canal at Ottawa, probably in 1900 when the railway was being built alongside. The second building on the left, the first stone building in Ottawa, was Colonel By's headquarters.

Surveys both of the Ottawa river and Rideau-Kingston routes were ordered in 1816, and three years later the Royal Staff Corps under Captain (later Major) Du Vernet began to build 5¾m-long Grenville Canal through uncleared land to by-pass the Long Sault on the river's north side. At its upper end was a guard-lock, then the main canal with two spaced out locks, and at the bottom end two staircase pairs close together, giving a fall of 46ft excluding the guard-lock. It was at that time intended that the Grenville should be of the same dimensions as the then proposed Lachine Canal, mainly to take Durham boats, and so the three upper locks on it were built some 100ft × 20ft.

Work went on slowly, until the arrival in 1825 of a Commission of three Royal Engineer officers headed by Sir James Smythe. It did not take them long to recommend speeding up work on the Ottawa and beginning the Rideau Canal. Their report caused the Duke of Wellington as Master General of the Ordnance to urge an immediate start. In 1826 Lt-Col John By of the Royal Engineers was sent out to begin the Rideau, and in 1827 Du Vernet was told to make a start upon the other Ottawa river obstructions.

By had now realised that with the development of steamboats, no towpath would be needed, but bigger locks, and those planned for the Rideau were enlarged to 134ft × 33ft. Roughly the same dimensions were now applied to the Ottawa, first at a single new lock and cut at Chute à Blondeau, made 128ft × 32ft 6in, then at Carillon. Here Du Vernet planned a canal just over 2m long. To avoid excessive rock-cutting, he used an odd layout. From a rising entrance lock above the rapids, it ran level (being fed with water from the nearby North river), before returning to the Ottawa through a staircase pair, the rise being 13ft and fall 23ft to overcome the 10ft fall of the rapids.

Meanwhile the Rideau Canal was being constructed 123½m long by river, lake and canal, rising from the Ottawa river at Bytown (Ottawa) for 277ft to a summit in Upper Rideau lake and then falling 162ft down the Cataraqui river to Kingston harbour on Lake Ontario. Engineered by By, with N. H. Baird (son of Hugh Baird, engineer of Scotland's Edinburgh & Glasgow Union Canal) as his clerk of works for most of the time, it had 49 masonry locks whose design was based upon those of the Lachine Canal, including the great 8-rise staircase at Ottawa, the biggest in North America, and an astonishing masonry arched dam 60ft high at Jones Falls. It was built through virgin forest and under great difficulties in five working seasons, and opened in May 1832—an astonishing feat of engineering.

Du Vernet was meanwhile being pressed to hurry up the Ottawa works. These were opened at the beginning of the 1834 navigation season, the four lower locks on the Grenville having been built, like those on the Carillon, to the larger size, but with its three upper locks left unenlarged, a bottleneck on the whole Montreal-Kingston line. The timber lock at Vaudreuil remained essential until replaced in 1843 by a masonry one at Ste Anne de Bellevue, which performed the same function but was sited on a different channel linking the Ottawa and St Lawrence. Canada's defence had been strengthened.

Trade also flowed to the new route, for, though the building of larger locks and cuts on the St Lawrence had begun in 1834, it was not finished until 1851. Until then, canoes, Durham boats, large barges and more and more steamboats used the new line; indeed, almost all the through trade between Montreal and Lake Ontario passed that way. After 1851, this activity switched back to the St Lawrence, leaving only local traffic. Nevertheless, a defence capability remained.

The Erie Canal

Only a few inland canals around the world have really caught the imagination of men: the Erie is one. There were no American precedents. When authorised by the New York state legislature in 1817, the longest canal in the country was the Middlesex, though we must remember that the Rideau waterway was then being surveyed. The state's population was not much more than a million, the country an unsettled wilderness through which it was proposed to build a

363m-long canal from Albany via West Troy (linked by a cut to the Hudson river) by way of the Mohawk river valley to Buffalo on Lake Erie. Beyond, the areas along the Great Lakes were only sparsely populated or remained wilderness, whereas most of the settled areas lay along the Ohio, Mississippi or their tributaries. In 1820, for instance, nearly 70 per cent of Ohio's population lived in the southern 36 of the state's 88 countries.

But the project had great advantages: a steady rise from the Hudson to 650ft near Buffalo, no intervening mountain range and ample water. Begun in April 1817, it was 40ft wide at top, 28ft at bottom, 4ft deep, with 83 locks each 90ft × 15ft, and a 10ft-wide towpath. New York state built it, with DeWitt Clinton as the driving force. William Weston was asked to come out of retirement to take charge; when he refused, three engineers each supervised a section. Soon, however, one of the three, Benjamin Wright, took sole charge—the man who was to become the greatest engineer of the American canal mania, as William Jessop had been of the British.

The whole line was opened in October 1825, and was at once a wild, unparalleled and extraordinary success, which opened up the north-east to industry and commerce, facilitated travel and emigration to the west, enormously reduced freight rates, and much increased property values. Steamboats supplemented sloops and schooners in the carriage of goods and passengers between Buffalo and the southern shore of Lake Erie, Michigan and beyond.

The phenomenon of the whole obscured the parts. Yet these included such engineering feats as the two aqueducts over the Mohawk, one 748ft long, the other 1188ft, the Cayuga Marsh embankment, 2m long and up to 70ft high, the aqueduct over the Genesee at Rochester, 802ft long, the twin 5-rise staircases cut through rock at Lockport near Buffalo and the deep cutting to the west beyond it. At Albany, to accommodate canal traffic, a huge mound 4300ft long and 80ft wide was built in the Hudson, open at the upper end to allow scouring but able to be closed against ice, and open below for traffic to enter and leave, the upper part being linked to the shore by drawbridges. Warehouses were then built on the mound. By European standards, of course, much of the canal work was roughly done—though the great aqueducts were of stone, not timber—and would soon need replacement. The Duke of Saxe-Weimar Eisenach, travelling the canal in its opening year, wrote: '...one who has seen the canals in France, Holland and England, will readily perceive, that the water-works of this country afford much room for improvement,' and 'it (will) be necessary to make great repairs, which...will

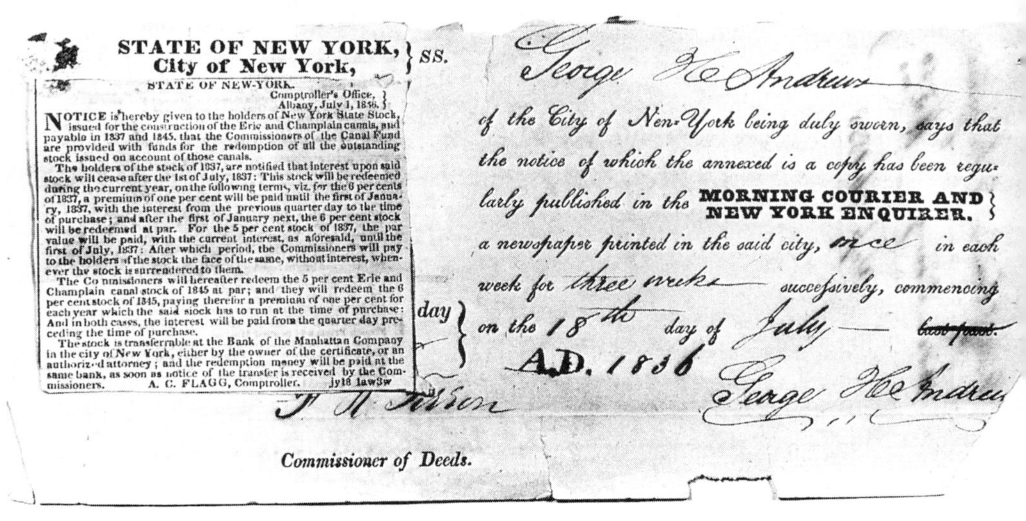

STATE OF NEW YORK, }
City of New York, } SS.

STATE OF NEW-YORK.

Comptroller's Office,
Albany, July 1, 1836.

NOTICE is hereby given to the holders of New York State Stock, issued for the construction of the Erie and Champlain canals, and payable in 1837 and 1845, that the Commissioners of the Canal Fund are provided with funds for the redemption of all the outstanding stock issued on account of those canals.

The holders of the stock of 1837, are notified that interest upon said stock will cease after the 1st of July, 1837: This stock will be redeemed during the current year, on the following terms, viz. for the 6 per cents of 1837, a premium of one per cent will be paid until the first of January, 1837, with the interest from the previous quarter day to the time of purchase; and after the first of January next, the 6 per cent stock will be redeemed at par. For the 5 per cent stock of 1837, the par value will be paid, with the current interest, as aforesaid, until the first of July, 1837: After which period, the Commissioners will pay to the holders of the stock the face of the same, without interest, whenever the stock is surrendered to them.

The Commissioners will hereafter redeem the 5 per cent Erie and Champlain canal stock of 1845 at par; and they will redeem the 6 per cent stock of 1845, paying therefor a premium of one per cent for each year which the said stock has to run at the time of purchase: And in both cases, the interest will be paid from the quarter day preceding the time of purchase.

The stock is transferrable at the Bank of the Manhattan Company in the city of New York, either by the owner of the certificate, or an authorized attorney; and the redemption money will be paid at the same bank, as soon as notice of the transfer is received by the Commissioners. A. C. FLAGG, Comptroller. jy18 1aw3w

George H. Andrews of the City of New-York being duly sworn, says that the notice of which the annexed is a copy has been regularly published in the MORNING COURIER AND NEW YORK ENQUIRER, a newspaper printed in the said city, once in each week for three weeks successively, commencing on the 18th day of July

A.D. 1836 George H. Andrews

Commissioner of Deeds.

Fig 98 Erie and Champlain Canals: a statutory notice of 1836. So successful has the Erie been that, only ten years after its opening, the State of New York is offering a premium to stockholders for early redemption of stock not yet due for repayment: '... the Commissioners of the Canal Fund are provided with funds for the redemption of all the outstanding stock issued on account of those canals'. Nevertheless, an endorsement on the back shows that the State did not pay the newspaper's $3.50 advertisement charge for over two years.

consequently require a large share of... income'.(7) But the duke saw, as we do, that speed, not permanence, was needful then. Rebuilding could, and did, come later.

Branches, some doubtfully economic, were added, to give a system length of 543m. In 1836, ten years after the Erie had opened, 3167 boats were at work on the network, and made 67,270 trips during the ice-free period of the year. They carried 1,310,807 tons and paid $1,614,335 in tolls on goods and passengers against a maintenance cost of $410,236. Goods travelling west included iron, sugar, salt, coffee, groceries, furniture and farm machinery; and before long agricultural produce began to return in larger and larger volume.

The Erie was mainly built by locals. In 1819 the canal commissioners reported that three-quarters of the workers were 'born among us' and that the majority were 'native farmers, mechanics, merchants and professional men' who lived near the canal.(8) Immigrants from Britain also worked on it*. John Richards, a Welshman, did so as a land surveyor, and wrote home (in Welsh) from 'Johnsburgh' in Warren County on 3 November 1817 that 'canals are being built in the country at present, and the people in the old country would do all

*The often repeated statement that the Irish built the Erie is incorrect, though there were Irish among the immigrants who joined locally recruited workers. Irish navvies did, however, work on the first enlargement, which began in 1836, and later.

right for this work'. He wrote again on 11 December 1818 from Utica that 'wages in the canal are one dollar per day, 13 to 14 by the month with food, with drink, with washing facilities and half a pint of whiskey every day; with own food from 22 to 23 dollars, in all weathers'. Meanwhile another Welshman, William Thomas, wrote also from Utica on 17 August 1818: 'I'm working on the cut with non-Welsh-speaking Englishmen for 13 dollars a month and a dollar is equivalent to 4s 6d of your money...I prefer to work in the old country for £8 8s od than have £20 os od here, and it's no better for the craftsmen than the labourers.' (9)

The commencement of the Erie Canal, the resolution and drive put into its building through year after year of political and practical difficulties, and above all its completion and success, set off a canal mania in the States comparable to Britain's of 1789–96. Tremendous energy and much money was put into construction, and great things were done. But because the mania occurred so close to the beginnings of America's railroad age, a great deal of what was done proved of less value than had been so ardently hoped by its promoters. Let us see what happened.

12

CONSEQUENCES OF THE ERIE

The feeling for enterprise that followed the initiation of the coal-carrying Schuylkill Navigation in 1815 and of the far-stretching Erie Canal in 1817 lay behind a Congressional resolution of 1818:

> Resolved that Congress has power, under the Constitution, to appropriate* money for the construction of post roads, military and other roads, and of canals, and for the improvement of water courses.

Thereupon federal activity was to increase. In 1824 for instance, Congress appropriated funds to survey the Mississippi, Ohio and their tributaries, also in 1824 passed the first 'Rivers and Harbors' Act, in 1828 subscribed $1 M to the Chesapeake & Ohio Canal, and in that year also made its first grants of land to help finance canal construction.

Peaking about 1825, that same feeling for waterways was to spread north, east and south as men realised that coal distribution, industrial and agricultural development all needed better water transport. In the east the money came mainly from private pockets, but to finance canals to the Ohio, and even more in the undeveloped mid-west, states had to find much of, sometimes all, the capital. We can group the results: the link to Lake Champlain and Canada; the canals of the eastern seaboard; the watershed canals that sought to reach the west and so rival the Erie; and those built in the mid-west itself, nearly all as extensions of the Erie's water line.

The Link to Lake Champlain and Canada

A trade route grew out of an Indian trail from the Hudson north to 140 m-long Lake Champlain and over the Canadian border to the Richelieu river that, past the Chambly rapids, drains into the St Lawrence. At its southern end we saw the failure of the Northern Inland Lock Navigation company to link Hudson to lake. But when the Erie was begun, so was its branch Champlain Canal of similar dimensions. It was opened in October 1823, 64 m long with 21 locks, to Whitehall at the lake's southern end. Steamboats waited for its

*In England, 'vote'.

opening to carry passengers the full length of the lake to St John's (St Jean) for onward transport to Montreal. Economic development followed up the canal line and on both sides of the lake, which itself became busy with shipping.

Back in the 1750s the French were improving the navigation of the lower Richelieu for rafts bringing lake lumber, but the Chambly rapids remained obstructive. When in 1763 Canada became British and the frontier disappeared, a heavy lumber trade developed northwards from Vermont to Quebec for export to Britain. After the War of Independence and the frontier's return the first suggestion for a canal past the rapids came from Vermont. Others followed, and Lower Canada (Quebec) was inclined to agree. But the American declaration of war on Britain in 1812 and attempted invasion of Canada changed that. Canada became less helpful to American exports, while the British defence authorities saw a Chambly Canal as making invasion easier. Though a Canadian company was authorised in 1818 to build one, a 12m line was not opened until 1843, or improved until 1860. Even then, with 9 locks 120ft × 24ft and a 6½ft channel, it was too small for Champlain Canal barges. Indeed, had not the lumber trade reversed itself in the mid-1830s, to run from Canada to the States and not *vice versa*, or Canada developed a need for Pennsylvania coal, it might not have been built at all. So it has remained.

The Canals of the Eastern Seaboard

In New England we may note three, in order of chartering. In Maine, the successful Cumberland & Oxford opened in 1827, 20m long but supplemented by lake and river navigation, carried the produce of a wide area to Portland and the sea. With the Connecticut river already made reasonably navigable, the New Haven & Northampton (or Farmington) Canal, 78m long, with 60 locks, was intended to take the river trade at Northampton and carry it to the sea at New Haven. Opened in 1828 to Farmington and in 1835 through to Northampton, short of capital, maintenance and water, it only survived until 1847. Lastly, the 45m Blackstone, part-canal, part-river, line running from tidewater at Providence inland to Worcester, was completed in 1828. Though useful until railroad days, it too suffered from poor maintenance and scarcity of water that had to be shared with millers dependent upon waterpower.

Further south lie canals as fascinating to the historian and the engineer as they were in their time successful. All had as inspiration the building of the Schuylkill Navigation, and as principal purpose the carriage of coal, much of it anthracite, from north-east Pennsylvania to New York, Philadelphia and their neighbourhoods. They were built with a power, passion, speed and engineering ingenuity that fulfilled, yet transcended, their economic motive.

Take the Delaware & Hudson Canal, intended to bring coal from the Carbondale area to the Hudson, and so to New York. In 1823 the Wurts brothers engaged Benjamin Wright to do their survey and in 1825 launched their company, the stock offered being subscribed on the first day. In 1829 it was operational. The first 17m from the collieries were by railroad, up some 950ft by five steam-operated inclined planes, then down 970ft by three more, counterbalanced because loads were downhill, to reach Honesdale. Thence the canal crossed the Lackawaxen and Delaware rivers on the level, then followed the Delaware to Port Jervis before climbing some 950ft over the intervening watershed and before falling again to Eddyville by Kingston on the Hudson; 108m long, its 108 locks and 4ft depth took 20-ton boats. It was deepened in 1842–4 to take 40-ton barges, then rebuilt by 1852 for 130-tonners. As a result, the 426,000 tons carried in 1853 jumped to a million in 1855, and dividends from 8 per cent in the 1840s to 18 per cent in 1855.

An engineering innovation interests us. The obstruction offered by river level-crossings to raftsmen caused aqueducts to be built. J. A. Roebling, who had earlier erected a wire suspension aqueduct at Pittsburgh (see p. 307), was commissioned in 1847 to build four others, the first two over the Lackawaxen and Delaware, using masonry piers and abutments with timber trunks and double towpaths.

Further south, competition from the new slackwater Schuylkill Navigation showed up the Lehigh's deficiencies. But not for long. By 1829, with Canvass White from the Erie as engineer, Josiah White and the Lehigh company had completed a part-canal, part-river navigation down to Easton with 48 locks to take 150-ton barges. The 'bear-traps' had gone for ever.

White's enterprise produced two coal-carrying canals to meet his at Easton; the Delaware Division, completed in 1832, which by-passed the Delaware river thence to Bristol, after which the river continued the navigation to Philadelphia, the other, the Morris, to which we shall return, towards New York. However, during 1835–8 White extended his own Lehigh Navigation further upwards through a difficult stretch to White Haven, using very large locks 100ft × 20ft with lifts up to 30ft: then built the 25m Lehigh & Susquehanna

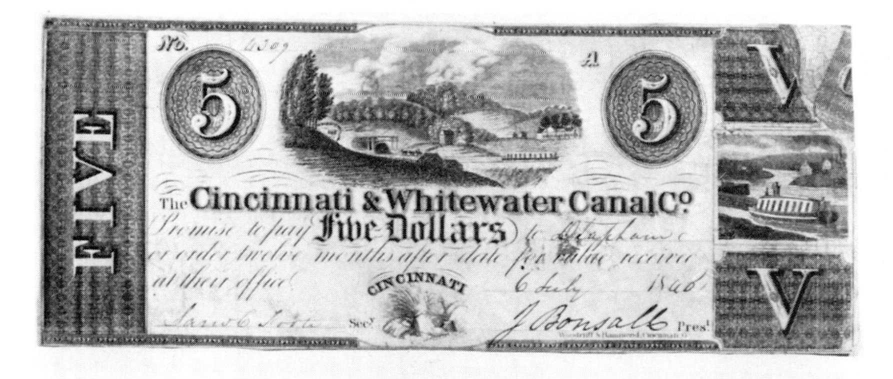

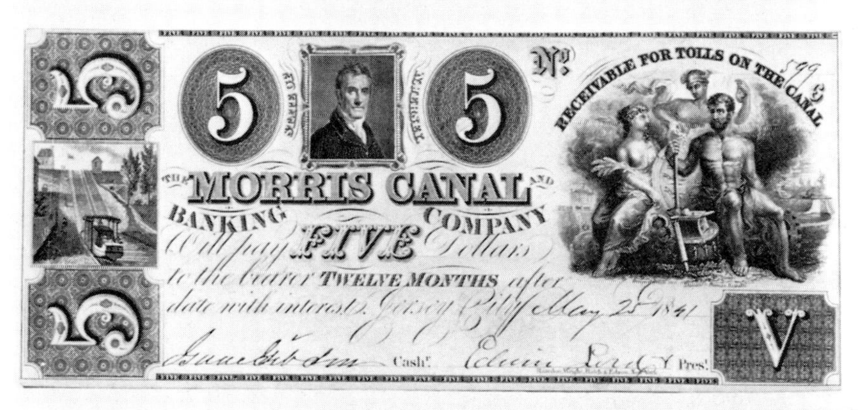

Fig 99 American canal days. Three canal undertakings issue promissory notes: (above) The Cincinnati and Whitewater Canal was chartered in April 1837 by the state of Ohio, the first boat reaching Cincinnati in November 1843. The note shows a packet boat, probably the *Express Mail*, entering the 1,782 ft long North Bend tunnel; (centre) The Morris Canal and Banking Co was authorised in 1824 with banking powers, but also to build the 102m long canal. This was opened in 1836 with 23 inclined planes, one of which is shown on this note of 1841: (bottom) packet boat and train move side by side on this note of 1836, issued one year after the company had been formed to link Richmond on the James with the Ohio by canal and/or rail.

Fig 100 The Morris Canal in 1868. The drawing shows the inclined plane at Newark (one of three with double tracks), 700 ft long at a slope of 1 in 10. Each timber framed cradle has 16 wheels and, like the boat it carries, is jointed in the middle to pass over the plane's apex. On the right overflow water from the turbine powering the plane runs down a box flume into the canal's lower pound. The toll-collector's house stands on the left at the top of the plane.

Railroad to Wilkes-Barre on the Susquehanna using three inclined planes. In the year it opened, 1841, a disastrous Lehigh flood destroyed much of the navigation. Undeterred, White rebuilt it, though the White Haven canal section was abandoned soon afterwards. Coal tonnage, 225,585 in 1840, had become 1,276,367 in 1855. This astonishing man died in 1850, but what he did lived on.

The Morris Canal, running from Easton to Jersey City in Newark Bay opposite New York, is third of the outstanding Pennsylvania coal-carriers. New Jersey authorised a feasibility study in 1822, with Ephraim Beach as engineer and Professor James Renwick of Columbia University as consultant. An Act for the Morris Canal &

Banking Co followed in December 1824, the banking side being intended partly to finance not only its own canal, but others also. Speculative excitement in the spring of 1825 was such that the $1 million of stock offered was twenty times over-subscribed. Three years later, however, one-third of that allotted had been forfeited for non-payment of assessments (calls).

Nevertheless the 90m canal from Easton to Newark was opened on 20 May 1832, and extended for 11¾m to Jersey City in 1836. It was built with 23 main-line locks and no less than 23 inclined planes to overcome a summit of 914ft. Small 18-ton boats were used, worked dry over the planes in cradles running on two 4-wheeled bogies. The biggest incline had a rise of 100ft, the smallest 35ft, and all except one (at 1 in 20) had gradients between 1 in 10 and 1 in 12. These mainly single-track inclines were designed by Renwick, and powered by turbines supplied with water from the canal's upper pounds and two reservoirs.

In 1841 the locks were enlarged to take bigger boats, consequential alterations being made to the planes. By then the Morris company was in difficulties, mainly on its banking side. Its property was sold in 1844 and a new company of the same name formed, which began an enlargement programme in 1845. The canal, widened and deepened, now took 80-ton 'hinge-boats' off the Lehigh. These resembled an ordinary boat cut in two, the open ends being closed by planks and bulkheads, and the two sections held together by latches and steadying pins. The inclines had been re-equipped with hinged timber-framed cradles running on 16 wheels. Cradles and boats could therefore bend in the middle to pass easily over their summits. Each cradle had a brake and brakesman, in case the winding rope broke or became detached. So successful was the rebuilding that traffic, 58,259 tons (of which 28,291 was coal) in the new company's first fully operational year of 1845, had in 1866 become 889,220 tons (of which 473,028 was coal and 25,833 iron ore from the Lehigh mines), figures that had been swollen by the Civil War, during which the line had been offered more coal than it could carry.

Atlantic Intracoastal beginnings

The War of 1812 brought British naval raids, and showed the States how vulnerable was her coastwise trade against a powerful fleet. After it, three components of Gallatin's ideas for a sheltered Atlantic coast passage began to be shaped: Dismal Swamp, Chesapeake & Delaware and Delaware & Raritan Canals.

The 21m Dismal Swamp Canal joined the Elizabeth river near Norfolk, Virginia, to the headwaters of the Pasquotank river opening into Albemarle Sound, North Carolina. At first used mainly for rafts taking logs from the swamp, it was rebuilt with 7 locks (one a staircase pair) and fully opened in 1823 to enable small ships to avoid the dangerous sea passage round Cape Hatteras. In 1827–9 Congress and the state of Virginia helped pay for further enlargement, with 5 bigger locks 100ft × 22ft, whereupon steamers began to supplement its schooners, sloops and barges. A charming print of 1830 shows the first regular steamer *Lady of the Lake* passing the Lake Drummond hotel, built across the state line for convenience of runaway marrying, illegal duelling and other diversions. So popular was the print that in miniature it appeared on the vignettes of many banknote issues.

Next followed the Chesapeake & Delaware Canal through the neck of the Delaware peninsula. In 1821 Matthew Carey began to seek subscriptions, more being added by Pennsylvania, Delaware and the federal government. Construction began in 1824 with Benjamin Wright in charge, and the 13½m canal was finished in 1829, able to take craft drawing 7ft—a figure soon to be increased—with two tide-locks and two others, each 100ft × 22ft (as on the Dismal Swamp). The cut had been expensive, but proved successful. Coasting vessels began to use it, together with craft from Norfolk and the James, Potomac, Susquehanna* and Delaware rivers, and passenger-boats running between Philadelphia and Baltimore. In 1834, 5438 transits were made, including the Ericsson Line's steamers using the screw propeller invented not long before by Swedish engineer John Ericsson.

A stillwater link between the Delaware and Raritan rivers, and so between Philadelphia and New York, had been in Gallatin's list. After a false start in 1825, a second company was chartered in 1830, on the same day as one for the Camden & Amboy Railroad on a similar route. In 1831 the two amalgamated and in 1834 the Delaware & Raritan Canal opened, 43m long with 14 locks 110ft × 24ft. Much of its traffic was to be coal off the Lehigh, for its dimensions were greater than the Morris, but much also came up the Delaware from Philadelphia, or from the Schuylkill. With its opening, the Atlantic Waterway became a reality, for traffic could now move through sheltered waters between North Carolina and New York.

In our consideration of the canals and developed river navigations of the eastern states, let us not forget the importance of more rough

* For the Susquehanna & Tidewater Canal and its effect on the Chesapeake & Delaware, see p. 308.

and ready navigations, some with a few locks, some with none. In Virginia alone, by 1834 the James river had 85m of navigable tributaries, the upper Appomatox another 100, the Rappahannock 55, the Shenandoah 60, and the Roanoke another 300, in this last case a sluice (flash-lock) navigation, plus the 13½m Weldon Canal round the Roanoke rapids. In all, 600m in one state.

Canals to the West

By 1825, the full achievement of those who had built the Erie Canal was measured by wild activity in eastern seaboard cities which saw their trade with the west imperilled by New York's enterprise. The new means of transport offered cheap, fast carriage west for goods and passengers, and equally good passage east for the produce of the new lands south of Lakes Ontario and Erie. Canals to the west must be built, cried the citizens of Boston and Philadelphia, Washington, Baltimore and Richmond. Roads would not do to carry cargoes now able to ascend the Mississippi in steamboats. If two ranges parted by the Connecticut river separated Boston from Albany and the Erie's magic water line, if the Allegheny mountains interposed a barrier between the other four and the Ohio, could new technology not overcome them?

All knew action was urgent if their own situations were to be saved, but another factor complicated the discussions that raged in 1825 and 1826: railroads. England's horse-tramroad system was known in America, where people had also keenly followed locomotive development. On 27 September 1824 the opening of the Stockton & Darlington Railway showed that these were practical, even when used on a line that was partially worked by inclined planes and stationary engines. For North America, railroads had one special advantage: they were not, like canals, stopped in winter, not for two or three icy weeks as normally in Britain, but for months. Yet railroads represented promise untested in the hard school America offered. What to do?

Boston's problem was to reach the Hudson-Erie line. A commission of enquiry reported that a canal to the Connecticut would be 100m long, with 1959ft of lockage, its extension to the Hudson another 78m with 1322ft more lockage and a 4m tunnel. The cost would be great, and would probably have to come from the state.

The canal party were encouraged by the opening of the nearby Blackstone Canal in 1828, while railroad supporters welcomed a charter granted in March 1826 to the 3m Granite Railway (probably America's first) from the Quincy granite quarries to the

Neponset river, a horse line. It began working in October. As more was learned about the progress of British railways, scepticism about a canal increased. Eventually the Boston & Worcester Railroad was authorised in 1831, partly to nullify the effect of the Blackstone Canal in drawing traffic south towards Providence. Finished in 1835, it was extended to Albany between 1837 and 1841 as the Western Railroad, and became highly prosperous.

In Virginia, the State Board of Works had in 1816 recommended that instead of trying further to improve the Potomac's navigation, a canal towards the Ohio should be built beside it.

The Potomac offered the shortest line to the west; it was used, and within limits it was profitable. Not unnaturally, Erie fever caused men to think that, were a canal to be built beside it from Georgetown (Washington) for 185m to the coalfields at Cumberland, and then for another 175m past the mountains to the Ohio valley and Pittsburgh, it would succeed, in spite of a summit-level of 1898ft, a 4m tunnel and 398 locks. Chartered in 1825, approved by Congress, Maryland and Virginia, and with two-thirds of its estimated cost subscribed by public bodies, the Chesapeake & Ohio Company got organised in 1828 with Benjamin Wright as engineer. Eighty feet wide and 7ft deep between Georgetown and Little Falls, 60ft and 6ft above, construction began upwards from the Potomac at Georgetown. Baltimore had proposed to join it by an extension Maryland Canal but, influenced by a merchant who had seen the Stockton & Darlington Railway at work, opinion moved towards a railroad direct to Wheeling on the Ohio, whence ran a turnpike. It was chosen, and construction of the canal and the Baltimore & Ohio Railroad, also to run beside the Potomac, began on the same day in 1828.

For the canal, there followed lawsuits with the railroad, cholera among the navvies and desertion by all public backers except Maryland, which took control in 1836. By 1831 the canal had opened for 23m to Seneca; by 1834 for 60m to Harpers Ferry, and soon afterwards for 86m to near Williamsport. Traffic now began to prosper. The line reached Hancock by 1839 and Cumberland in 1850. There it ended, with 74 locks 90ft × 15ft along its 185m, a tide-lock at Georgetown and connecting locks between canal and river at points along its length. Three years later the railroad reached Pittsburgh. The race had been won. Thenceforward the canal settled down as mainly a downwards carrier of coal mined in the Cumberland area. Among notable structures of the period were the stone-built Monocacy aqueduct, 438ft long with seven 54ft arches, and the 3118ft-long Paw Paw tunnel. Much later, the 7m Alexandria Canal was from 1843 to leave the C & O at Georgetown, cross the Potomac on a 1100ft-long wooden-trunk aqueduct, and so run to the town of Alexandria and the Potomac again.

Further south, the James River Company had made the river precariously navigable for some 200m to Buchanan, except for the 200ft Balcony falls. Fully taken over in 1823 by Virginia, a 4m part-canal, part-slackwater by-pass of the falls was then begun, and opened in 1828. Thereupon a canal was begun to by-pass the James upwards from Richmond, as the Chesapeake & Ohio was by-passing the Potomac. Simultaneously, and higher up, the road over the mountains from Covington was being improved, and the Kanawha river being made navigable from Charleston to the Ohio, whence boats could work to Pittsburgh. By 1829, four years after the Erie Canal had reached Buffalo, this route was providing a southern transport line from the eastern seaboard to the Ohio.

A pause followed, until in 1835 the James River & Kanawha company was formed, three-fifths of the capital being private and two-fifths state and city money, to connect Richmond to the Ohio by rail, or a combination of canal and rail. Immediately, the aim was to extend the canal upwards and through the Blue Ridge mountain gap to Buchanan—and then to think again. The line indeed reached Buchanan in 1851 (a year after the C & O got to Cumberland), 196½m long and 5ft deep, part canal, part slackwater, having risen 728ft by 98 locks some 100ft × 15ft. Though work was done on 15m more, the company's drive had ended.

The rival pulls of canals and railroads were exemplified by the Pennsylvania Main Line, by which not only was the Ohio reached, but also the Erie Canal and, indeed, Lake Erie itself. Until the canal mania, Philadelphia (then the nation's largest city) had relied upon the Pittsburgh Pike, opened in 1820, to link it to the Ohio. Now, in February 1825, a majority of the state's Board of Canal Commissioners recommended a canal from Philadelphia to Pittsburgh. On 4 July 1826 the first sod was turned at Harrisburg, by which time it had been realised that a mixed canal-rail line would be the likeliest outcome. The whole Main Line opened in 1834. Let us consider its components.

The first section, from Philadelphia to Columbia on the Susquehanna, whose valley had then to be followed westwards, could have utilised the Union Canal if it had been big enough. Instead, the 82m Columbia & Philadelphia Railroad was built, planned as a horse line with two easy-gradient steam-operated inclines, one near each end, but in fact worked as both a horse and locomotive line from 1834 until horses were given up in 1844.

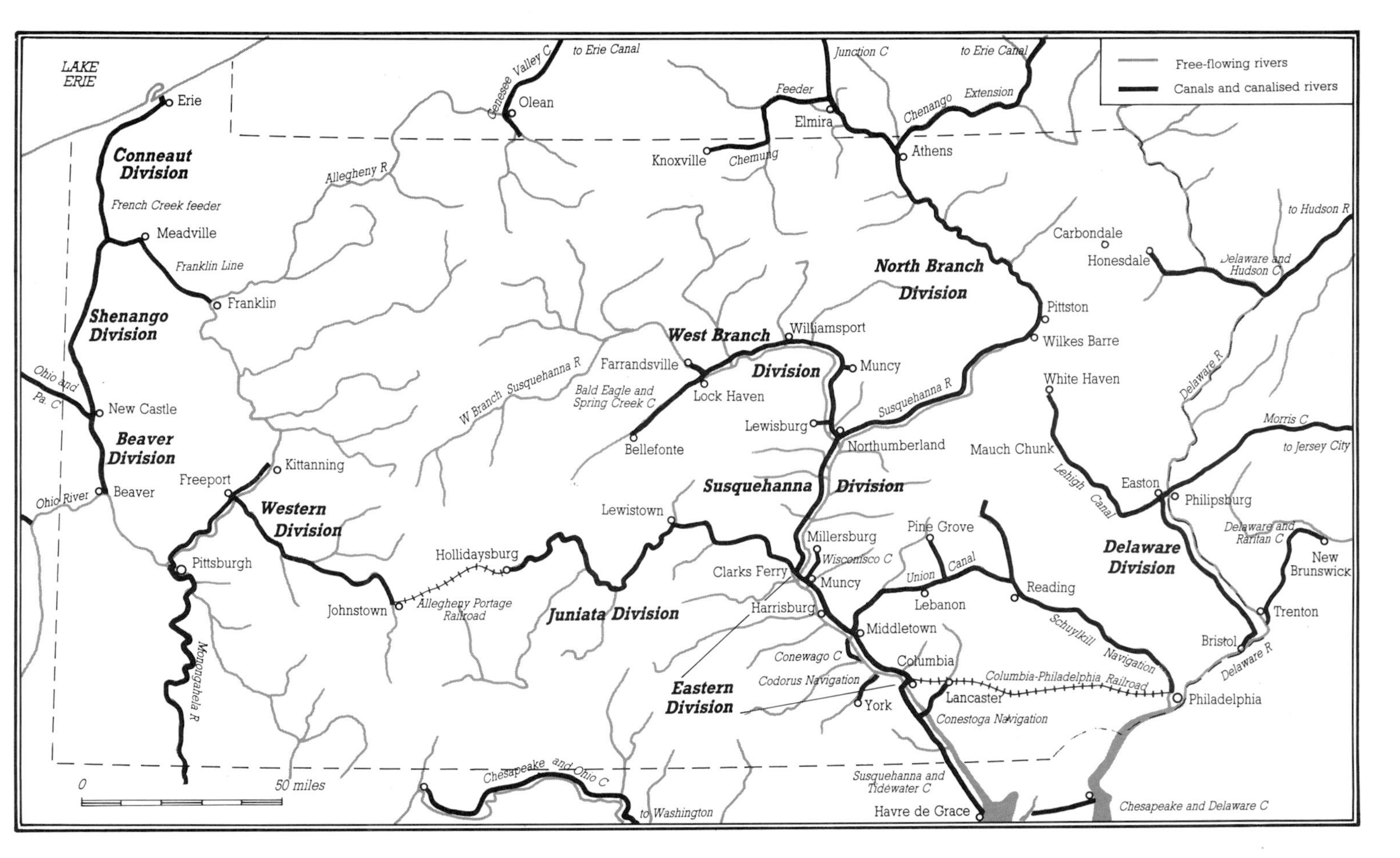

Map 13 The canals of Pennsylvania

At Columbia, convenient to the railroad and also with a river outlet, a canal began. It climbed the eastern side of the Susquehanna valley for 43m and 14 locks past Harrisburg to Duncan's Island. There barges crossed the river on the level above a dam 1998ft long, beside the Clark's Ferry wooden covered bridge with towpaths on two decks. Once across, the canal's Susquehanna Division turned north, while the Main Line followed the Juniata valley, mainly by canal but with slackwater sections, for 127m and 88 locks to Hollidaysburg at the foot of Allegheny Mountain. There the astonishing Portage Railroad began. Built by the engineer Sylvester Welch in three years from 1831, this 37m line had ten inclined planes, five on each side of the mountain, worked by steam-assisted counterbalance, and reached a summit of 2334ft.

At its other end at Johnstown, a canal with 60 locks and an 850ft tunnel ran for 105m to Pittsburgh, which it entered by a 1140ft-long wooden-trough aqueduct* over the Allegheny river. One branch ran down by four locks to the river, another with four more locks and an 810ft tunnel joined the Monongahela at the point where the Chesapeake & Ohio was planned to arrive.

The 394m of the Main Line had been opened in eight years, in many ways a more astonishing feat of enterprise and engineering ingenuity than the Erie itself. Moreover, in its opening year of 1834 a canal boat had been carried over the Portage Railroad. The innovation led to the building of sectional boats that could be divided into three or four units for carriage also over the Columbia & Philadelphia railroad. Thereafter some passengers and goods travelled without transhipment from Philadelphia to Pittsburgh.

At Clark's Ferry we noted the Susquehanna Division's canal following the river's west bank. After 41m it reached Northumberland, where the North Branch continued for 55m to loading docks at Nanticoke, the whole being finished in 1831. There huge quantities of anthracite supplied by Josiah White's mines via Wilkes-Barre were loaded into 150-ton barges, which then moved by way of Northumberland, Clark's Ferry and Columbia and the unsatisfactory navigation of the lower Susquehanna to Havre de Grace, at its estuary in Chesapeake Bay.

In the early 1800s both Pennsylvania and Maryland had built canals (for one, the Conewago, see p. 276) to by-pass sections of the

*John Roebling replaced it in 1845 with a new one of seven 162ft spans. It used the suspension principle to support the summer weight of water, though not the drained winter structure. Hence it was not a true suspension aqueduct, as were his later structures on the Delaware & Hudson.

Susquehanna. They had been useful, but now something much better was needed to carry the traffic offering: coal, but also lumber in barges and rafts, flour, and the produce of the West. So the Susquehanna & Tidewater Canal was opened in 1840 from Wrightsville opposite Columbia to Havre de Grace, and the older lines closed. At Wrightsville barges were coupled in tandem to pass together through the new canal's 28 large 170ft × 17ft locks to the estuary, whence they were towed onwards by tugs either to Baltimore or through the Chesapeake & Delaware Canal to Philadelphia and maybe on to New York. Packet- (passenger) boats also ran from Wrightsville down the S&T: it is said that at the still-standing ticket office one could buy a through ticket to London, England. Additional S&T traffic was a main cause of the enlargement in the 1850s of the Chesapeake & Delaware itself. New locks were now provided, 220ft × 24ft with 10ft over the sills, two at the Delaware City end, one with a side-pond at Chesapeake City, and an enormous steam-driven scoop-wheel that could pump 200,000cu ft of water an hour to the summit-level.

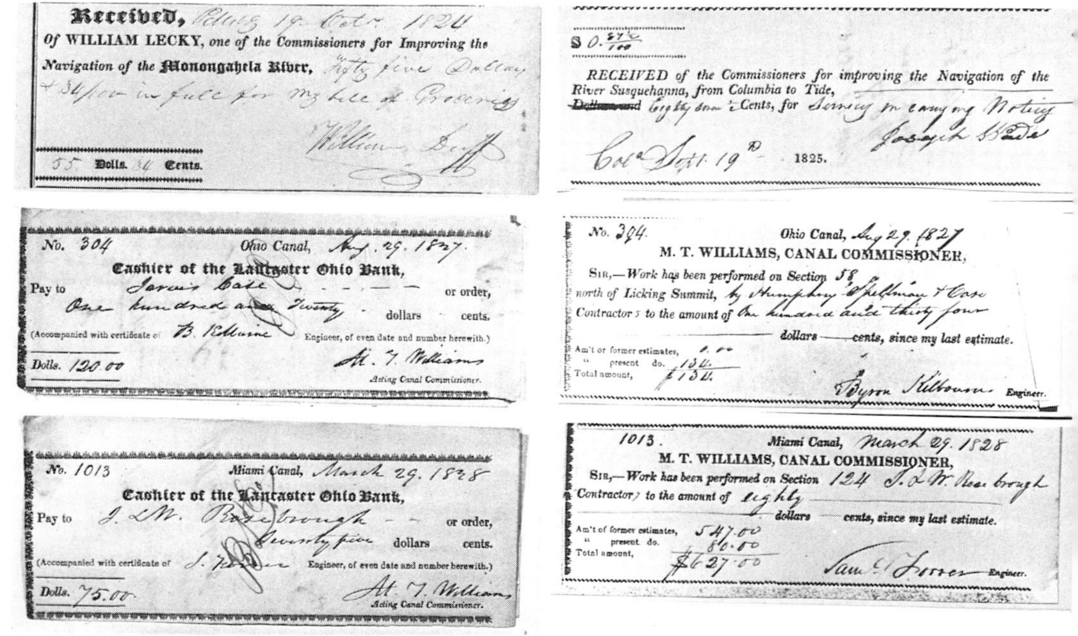

Received, 19 Octr 1824
Of WILLIAM LECKY, one of the Commissioners for Improving the Navigation of the Monongahela River, fifty five Dollars & 34/100 in full for my bill of Groceries
William Duff
55 Dolls. 34 Cents.

$ 0.87½/100
RECEIVED of the Commissioners for improving the Navigation of the River Susquehanna, from Columbia to Tide, Eighty seven ½ Cents, for Serving in carrying Notices
Joseph Wade
Col. Sept. 19th 1825.

No. 304 Ohio Canal, Aug. 29. 1827.
Cashier of the Lancaster Ohio Bank,
Pay to Jarvis Case or order,
One hundred and Twenty dollars cents.
(Accompanied with certificate of B. Kilbourne Engineer, of even date and number herewith.)
Dolls. 120.00
M. T. Williams
Acting Canal Commissioner.

No. 304. Ohio Canal, Aug 29. 1827
M. T. WILLIAMS, CANAL COMMISSIONER,
SIR,—Work has been performed on Section 58 north of Licking Summit, by Humphrey Spellman & Case Contractors to the amount of One hundred and thirty four dollars cents, since my last estimate.
Am't or former estimates, 0.00
" present do. 134
Total amount, $134
Byron Kilbourne Engineer.

No. 1013 Miami Canal, March 29. 1828
Cashier of the Lancaster Ohio Bank,
Pay to J. & W. Rosebrough or order,
Seventy five dollars cents.
(Accompanied with certificate of S. Forrer Engineer, of even date and number herewith.)
Dolls. 75.00
M. T. Williams
Acting Canal Commissioner.

1013. Miami Canal, March 29. 1828
M. T. WILLIAMS, CANAL COMMISSIONER,
SIR,—Work has been performed on Section 124 J. & W. Rosebrough Contractors to the amount of eighty dollars cents, since my last estimate.
Am't of former estimates, 547.00
" present do. 80.00
Total amount, $627.00
Saml. Forrer Engineer.

Fig 101 Canal and slackwater navigation building goes on around the time of the Erie's completion: (above left) the Monongahela Commissioners pay a bill for groceries, presumably for construction workers; (above right) the Susquehanna Navigation Commissioners serve notices, perhaps for land purchase; (below) the Ohio Canal Commissioners pay for work done; each order on the bank states for what the payment is being made, and is countersigned by an engineer, in the case of the Ohio & Erie Byron Kilbourne, in that of the Miami Canal Samuel Forrer.

Thirty-nine ft in diameter and 10ft wide with 12 buckets, it survived till 1927. By 1857 C&D tonnage had passed 600,000.

Let us now return to the Main Line. Its eventual link with the New York state canals was provided by painfully extending the North Branch from Nanticoke past Wilkes-Barre (whence White's railroad ran to the Lehigh) to Athens for two different connections to the Erie system in 1856 (see p. 314). In all, 210m of canal from Duncan's Island. The North Branch was a stupendous achievement in itself; its 169m between Northumberland and the state border included 43 locks, five river dams, 29 aqueducts and 229 road bridges. Further west, the Pennsylvania canal system made its own contact with Lake Erie by Building the Beaver & Erie Canal from the Ohio below Pittsburgh to the lakeside town of Erie. Begun in 1831, the year the first loaded canal boat arrived at Pittsburgh by the Main Line from Johnstown, it reached past New Castle to Conneaut Lake in 1836, whence a private company completed it in 1844—in all, 136m and 137 locks 80ft × 15ft, rather smaller than on the Main Line.

Perhaps in 1839 in Pennsylvania we see the American canal mania's greatest fulfilment, a year in which over 1250m of canal were built or building, 409¾m privately owned, the rest by the state.

One very different achievement remains to be mentioned. The Monongahela river runs south from the coalfields of what is now West Virginia (then Virginia) to Pittsburgh. In 1817 a private company was given 25 years to make it navigable to Fairmont. They did so with 16 locks, whereupon the river became a busy carrier of coal destined for Pittsburgh or beyond, and the company satisfactorily profitable.

Canals of the Mid-West

As Erie Canal construction moved towards Buffalo and Lake Erie, Canadians were opening the first Welland Canal to avoid it, while Americans began to consider, with the impetus of their canal mania behind them, how best to extend it. If, for instance, the southern shore of Lake Erie could be linked by water with the Ohio, the western states' isolation would be ended. Their products could move to the east by lake ship to Buffalo and then along the Eric Canal, which would bring back the goods and immigrants they badly needed. So, from the lake ports of Cleveland and Toledo, three major canal lines were constructed to the Ohio at Portsmouth, Cincinnati and Evansville. A second plan was to build an extension canal from beyond Toledo across Michigan to Lake Michigan and Chicago—and one was indeed begun. Thirdly, the traffic that would be generated upon Lakes Ontario, Erie, Huron and Michigan by the Welland and Erie Canals, as well as the rapid growth of lake steamships, suggested a link south from Chicago at the foot of Lake Michigan towards the Mississippi—and so a canal was created.

The cutting of the Mid-western canals has elements of classical tragedy, for we with hindsight know that, however hard engineers drove ahead, however sketchy the navvies' work, however desperate state expedients to raise money to keep going, railroads would inevitably overtake them, sometimes even before they were finished. Immensely long canal lines (308 m for the Ohio & Erie, 468 m for the Wabash & Erie) were launched into the hardly settled interiors of newly created states with only a fraction of New York's financial resources. Yet out of enterprise, courage, foolhardiness, desperate engineering, inevitably poor administration, every kind of financial trouble—and always hissing steam coming closer and closer—two successful canals did in fact emerge, the Ohio & Erie from Cleveland to Portsmouth, and the Illinois & Michigan south from Chicago. Indeed, as the Illinois Waterway in much enlarged form, the last flourishes today to offset some of the tragedy of the old canal men.

In the state of Ohio, irreconcilable economic interests made it necessary to build more than one canal from Lake Erie to the Ohio river. A state Canal Commission reported in 1825. It recommended the Ohio & Erie line from Portsmouth on the Ohio by way of Columbus, Newark and Akron to Cleveland, and a second, the Miami Canal, from Cincinnati lower down the Ohio north to Dayton, with a view to later extension. Both were authorised the following month. The first boat, *State of Ohio,* reached Cleveland from Akron on 4 July 1827, and the whole Ohio & Erie, now with a branch to the state capital, Columbus, was completed in October 1832. With its two summits 489 ft and 419 ft above the Ohio and 150 solidly built stone locks, the canal was a great engineering feat comparable to the Erie itself. Indeed, many Erie men had helped to build it. Now people could, and did, travel entirely by water the 2700 m from New York to Portsmouth by way of Albany, Buffalo and Cleveland, the packet-boat journey through the O & E taking 80 hours.

By 1840 large tonnages of wheat, flour, corn (maize), pork and coal were arriving at Cleveland, with back cargoes mainly of salt and merchandise from the east. In that year also, the Ohio & Erie was connected to the Pennsylvania system at New Castle by the 91 m Pennsylvania & Ohio (Mahoning). This quickly became busy, partly because it brought coal from collieries in the Mahoning valley to Cleveland, that enabled industry to develop and lake steamboats to be

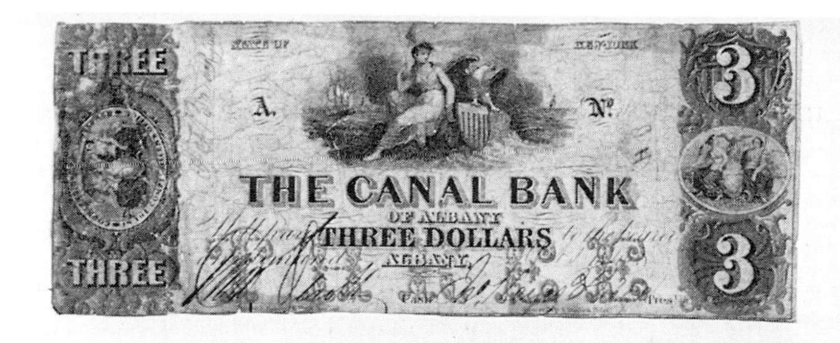

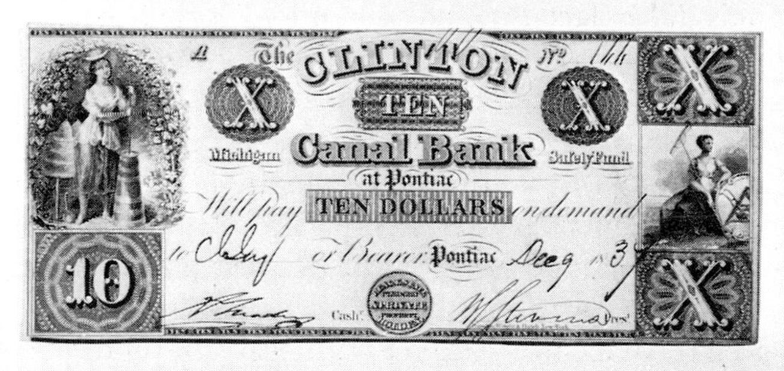

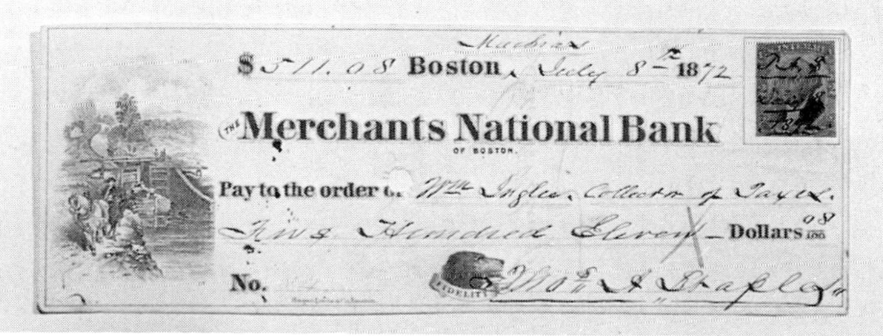

Fig 102 Canal banks sometimes provided finance for construction; in other cases they merely adopted a fashionable name: (above) the Canal Bank of Albany, which issued this note in 1844, was incorporated in 1829, and probably held deposits from the state's Canal Fund. It failed in 1848; (upper centre) the Chemung Canal Bank was founded in 1833, the year the canal, an extension of the Erie's branch to Seneca Lake, was opened. The canal closed at the end of 1878, but the bank's descendant, the Chemung Canal Trust Co, is still active in Elmira; (lower centre) Michigan's Clinton Canal Bank was founded in December 1837, the year the Clinton-Kalamazoo Canal project was authorised. Both had short lives, the bank's only until October 1838, the canal's to 1842; (below) though Boston's canal, the Middlesex, had been abandoned in 1860, a canal scene still illustrates a check in 1872. The scene was popular; it appears again, for instance, on a Delaware & Hudson Canal Co's share certificate of 1896.

fuelled, partly because it offered passengers a water route between Pittsburgh, Cleveland and Lake Erie ports.

Indiana's canal boom started too ambitiously and too late, though Congress had in 1827 granted half a million acres of land to help finance by sales what became the Wabash & Erie Canal from Toledo, on Lake Erie and in Ohio state, to Indiana's Evansville on the Ohio, at 468m the world's longest after the Grand Canal of China. As neither citizens nor state had much cash, canal commissioners were not appointed until January 1832. Land was then bought by Indian treaty, cutting began, and a commissioner travelled for ten days by steamboat and Main Line to New York to raise a loan.

Meanwhile, Ohio's Miami Canal from Cincinnati had reached Dayton in 1832, and been extended towards Toledo. Building also began at the Toledo end backwards to Junction near Defiance to meet the Wabash & Erie, and then along the latter's line, but at Ohio's expense, to the state border. The whole Cincinnati-Toledo route, completed in 1845, became known as the Miami & Erie. Meanwhile the Indiana length of the W & E was being slowly constructed south-west and then south past Fort Wayne, Peru, Lafayette and Terre Haute to Evansville, reached in 1853.

It was an achievement as pertinacious as it was seemingly hopeless. The ravages of nature and vandalism, the frequency of local stoppages and the cost of maintenance, were all as persistent as money and traffic were scarce. Within a few years the Evansville-Terre Haute section was unnavigable; that to the north survived into the 1870s.

The Ohio river had another Indiana canal connection, the Whitewater, built from Lawrenceburg on that river upwards to Cambridge City by 1845, whence a private extension took it to Hagerstown: 76m in all. A branch, the Cincinnati & Whitewater, begun in 1836, ran to the Ohio state border and onwards for 25m to Cincinnati past the 1782ft Harrison or North Bend tunnel. In 1849 an engaging diarist recorded a packet-boat journey from Hagerstown to Cincinnati:

> (From Laurel) Start at Sundown, run aground 2ce. Pilot got mad, swore...snorted, about to leave, concluded to stay, all glad of it...Boat, made poor progress, last night run aground 7 times...(1)

The Whitewater suffered so badly in 1852 from floods that it became unusable. The Cincinnati branch was sold to a railroad in 1862, the Whitewater itself in 1866. A small section of the Whitewater, including a lock and a covered aqueduct at Metamora, Indiana, has been restored.

Yet Indiana's canals were economically successful: ton-mile freight rates began lower than road transport, and were reduced as traffic increased. Canal-building also encouraged Indiana's development. The creation of canal ports, the impetus given to corn shipments, population growth and land values, has caused a present-day economic historian to write that the Wabash & Erie contributed 'the single most important development in the political and economic life of Indiana during the mid-nineteenth century'.(2)

Michigan's canal effort began late, with an Act of 1837 that proposed to connect Lakes Erie and Michigan by a Clinton-Kalamazoo Canal. Cutting began in 1838 from Mount Clemens on the Clinton river and in 1842 had reached westwards to near Rochester, 12m away. Then, Michigan's finances being in even worse case than Indiana's, money gave out and work ended.

Illinois did better. A canal from Lake Michigan to the Illinois river, tributary of the Mississippi, had been suggested in 1810 as an addendum to Gallatin's proposals. Needed to develop the country, it became possible after Illinois' statehood in 1818, and likely after the Erie's opening. In 1827 Congress donated alternate sections of land for 5m on either side of the proposed route, land sales beginning in 1830 in Chicago, then with a hundred inhabitants. After long discussions on dimensions, and on whether a railroad would not be better, construction began in 1836 of the lake-fed Illinois & Michigan Canal, 60ft wide at surface and 6ft deep, big enough to take river and lake steamers.

By 1841, however, Illinois, Indiana and Michigan were all in similar situations. They had tried to give their settlers a trading infrastructure by building basic public works financed by loans. But their cost was outrunning the states' ability to pay interest. Therefore the more that was spent, the higher the interest rate asked for new loans, or the greater the discount sought upon the issue prices of state scrip, because of the increasing risk of default. Yet could the works be completed, the states' economies would receive an impetus that would make interest payments secure. Faced with this situation, Illinois decided to finish the canal somehow. It was restarted in 1845 under the management of three state commissioners, on a smaller scale to take 150–75-ton craft loaded to 4ft 8in. To save money, it now began in the lowest reach of the Chicago river near its mouth in the lake, and rose 12ft to its summit-level, being supplied with water not by the lake as had been intended, but by four river feeders. It was opened in April 1848, 96m long from Chicago to La Salle on the Illinois, a notable engineering feature being the long rock-cut north of

Lockport. The value of unsold land grants now appreciated, and much debt was paid off.

New York trade now began to move by the Erie Canal and the Lakes to Chicago and then by water to the Illinois river and St Louis on the Mississippi, instead of by New Orleans. In return, grain, flour and farm products, local and from Mississippi ports, passed back for Chicago and Lakes destinations. Packet-boats ran between Chicago and La Salle to connect with river boats to St Louis, while steam-tugs towed long lines of canal barges. Chicago's population grew, from the original hundred of 1830 to 20,000 when the canal opened, and 74,500 in 1854, when the Chicago & Rock Island RR began to compete for Mississippi trade. Nevertheless the canal continued to prosper. For 1860–2, tolls averaged $207,080 pa, with $142,745 of that figure surplus to expenditure, while 529,441 tons were carried. The state of Illinois, and Congress who had made the land grants, could be proud of their achievement.

Back to the Erie

Having traced some of its consequences, let us return to their origin. New York state, proud of their great canal, in 1825 ordered surveys for feeder routes. One, the Oswego Canal, joined the Erie to Lake Ontario in 1828. By 1834 a substantial amount of traffic was between the Ohio canals and American ports on the Lakes to New York by way of the Welland Canal and the Oswego branch. Another, the Black river canal from Rome to that river, running also into Lake Ontario, finished in 1851, was for a time successful. But a number of optimistic branches towards the south were more costly than profitable.

A 5m canal ran to Lake Oneida (1835). Another, the 97m Chenango, from Utica to Binghamton on the Susquehanna, begun in 1833 and completed in 1837, was in 1856 linked to the Pennsylvania's North Branch. Another, by way of Seneca lake and the Chemung Canal (1833), with 53 locks, continued to Elmira, while its navigable feeder at Corning had in the 1830s been linked by railroad to the Pennsylvania mines at Blossburg. Not, however, till 1856, by the privately owned Junction Canal, did this line also cross the state border to join the North Branch. For a time boats were busy moving north towards the Erie with Pennsylvania coal and with lumber—but not for long. Later, the Genesee branch was built south from Rochester to Olean in 1856 and the Allegheny river in 1862. With side-cuts, well over 100m long, having 114 locks, expensive to build, unproductive of revenue, enthusiasm had outrun discretion in its building.

Meanwhile, enlargement of the Erie's main line was authorised in 1835, only ten years after opening. The decision perhaps marked the end of the country's first national improvement policy as state and private enterprise on both sides of the Alleghenies took over. The canal was now to be 70ft wide (40ft) and 7ft (4ft) deep, with duplicated locks each 110ft × 18ft. Trade leaped up as sections of what amounted to a new canal were opened, the last in 1863, though congestion at the most critical points had been relieved long before that. Toll receipts, $1,010,290 for 1830–2, were $2,947,653 in 1848–50. By now, through traffic on the canal was predominating over local, as Lakes trade increased. In early years, too, the main flow of traffic had been westwards, but from the late 1830s eastern shipments steadily increased.

The reconstructed line included new and duplicated lock staircases at Lockport, a new aqueduct at Rochester (1841) with seven 52ft spans that was 45ft wide, and could take boats passing while on the structure, the great 31-span Richmond aqueduct (1856) at Montezuma near the Seneca Canal junction, and that of 14 arches at

Fig 103 An early postcard shows the 7-span Rochester aqueduct, wide enough to pass two boats, of the reconstructed Erie Canal.

Fig 104 An early postcard. The twin combines or 5-rise staircases of the Erie Canal at Lockport near Buffalo. That on the left was later removed to accommodate the staircase pair of locks of the New York State Barge Canal; that on the right, though weired, can still be seen.

Schoharie Creek which replaced a previous level-crossing across which boats were hauled above a dam by rope and windlass.

Canadian Developments

A year before the military waterway route to Lake Ontario by way of the Ottawa river and Rideau Canal had been completed, the Welland had been opened in 1833 along its improved route. The Canadian provinces of Upper (Ontario) and Lower (Quebec) Canada immediately employed the American engineers Benjamin Wright and John B. Mills to recommend improvements to the St Lawrence. They proposed cuts round rapids, with a 9ft channel and locks 200ft × 55ft. It is ironical that the Erie Canal's chief engineer was here proposing the beginnings of a Seaway which would one day replace it. Meanwhile, the start of work in 1834 on the first cut, the Cornwall Canal round the Long Sault rapids, 11m long with six locks 200ft × 45ft × 9ft, must have influenced New York state's 1835 decision to enlarge the Erie.

Meanwhile the Welland Canal company was in bad financial trouble, which led to Upper Canada appointing a majority of directors from 1837, and taking it over in 1841, the year the two Canadas were united. In 1838, however, N. H. Baird (of the Rideau) and H. H. Killaly* had been asked to recommend improvements. They reported that, had not their instructions limited them to schooner locks 110 ft × 24 ft, they would have proposed those 180 ft × 45 ft. Their foresight was supported by Lt-Col George Phillpotts, engineer of the Cornwall Canal, who clearly envisaged the Welland becoming part of a through line that would better the Erie Canal when he wrote of the 'communications from Lake Erie to the sea by the Welland Canal, Lake Ontario and the River St Lawrence'. He recommended a small ship canal to take:

> large freight steamers, capable of conveying a cargo of at least 300 tons, without any transhipment before they arrive at Montreal or Quebec.(3)

In spite of these proposals, the provincial Board of Works chose locks 150 ft × 26½ ft with 8 ft 6 in over the sills, able to take schooners, but too narrow for the side-paddle steamers which the Cornwall Canal would be able to accommodate. The two entrance locks and Lock 2 were, however, to be bigger, so that steamships could enter the canal and reach St Catherines. Reconstruction began in 1842, along much the same winding route, and the second Welland Canal was opened in 1851. There were 28 locks on the main line (25 between Port Dalhousie and Thorold, two deep locks at Allanburg and Port Robinson, and one at Port Colborne); in addition, one at Welland to reach the Welland River and one at the junction of the Port Maitland feeder and the main channel. Later, the feeder's enlargement to a fully navigable branch added two more, one at Port Maitland a guard-lock (later a pound-lock) at Dunnville. Not long afterwards work began upon deepening to provide a 9 ft channel as on the St Lawrence, and was finished in 1855.

On the St Lawrence, the Cornwall Canal had been completed in 1843, and was followed in 1845 by the Beauharnois Canal, 11¼ m long, 9 ft deep, with nine locks 200 ft × 45 ft, the reconstructed Lachine in 1848, and the Williamsburg canals in 1851, all with similar dimensions. The new St Lawrence route was finished, as was the new Welland Canal in the same year, and Canada had a 9 ft waterway (the standard depth of the Mississippi system today) from Montreal to Lake Erie and on to Lake Huron. It was bigger than the reconstructed but not yet finished Erie Canal in depth and lock dimensions, though not yet big enough to compete successfully against railroads, given its annual closure of five months because of ice. In 1854 a Reciprocity Treaty gave American vessels the right to navigate the St Lawrence;

*Son of John Killaly, who engineered many miles of Irish canal.

Map 14 Three Canadian river systems

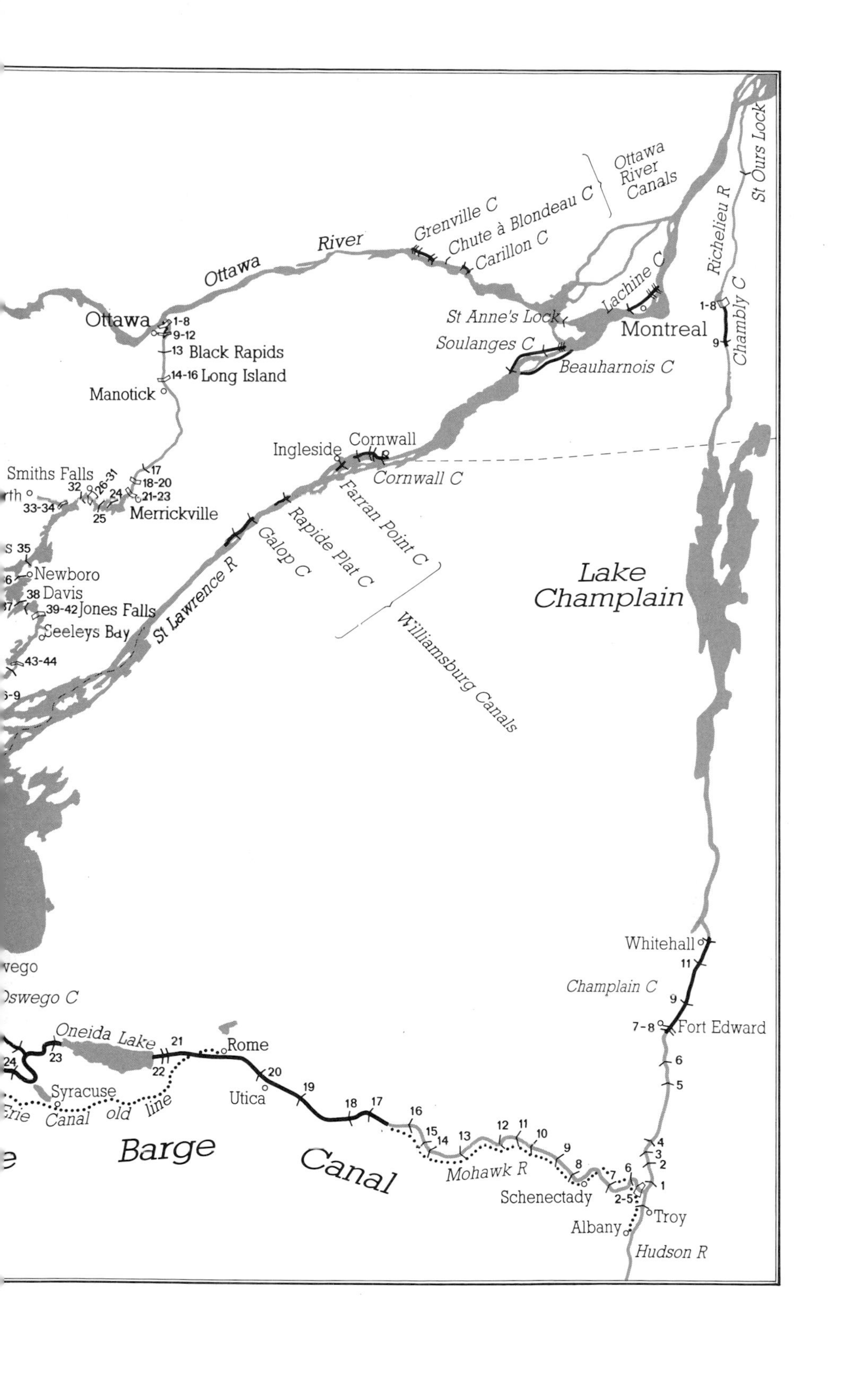

Ottawa River Canals
Grenville C
Chute à Blondeau C
Carillon C
Ottawa River
Ottawa
1-8
9-12
13 Black Rapids
14-16 Long Island
Manotick
St Anne's Lock
Lachine C
Montreal
Soulanges C
Beauharnois C
Richelieu R
St Ours Lock
Chambly C
1-8
9
Ingleside
Cornwall
Cornwall C
Farran Point C
Rapide Plat C
Galop C
Williamsburg Canals
St Lawrence R
Smiths Falls
17
18-20
21-23
24
25
26-31
32
33-34
Merrickville
35
Newboro
38 Davis
39-42 Jones Falls
Seeleys Bay
43-44
Lake Champlain
Whitehall
11
Champlain C
9
7-8
Fort Edward
6
5
4
3
2
1
Oneida Lake
21
22
23
24
Rome
20
Utica
19
18
17
16
15
14
13
12
11
10
9
8
7
6
Syracuse
Erie Canal old line
Barge Canal
Mohawk R
Schenectady
2-5
Troy
Albany
Hudson R

previously they had been barred by the British navigation acts. It was for many a better route than the Erie. Traffic on the St Lawrence canals, 288,103 tons in 1850, was 733,596 in 1860.

The Great Lakes

By this time the Great Lakes had a network of passenger- and cargo-steamer and sailing-ship services. Peter Stevenson, for instance, says of Lake Erie that between

> forty and fifty splendid steamboats, and many sailing vessels, were in 1837 employed in its trade...and several harbours with stone-piers have been erected on its shores for their accommodation. (4)

Indeed, by 1839 eight boats were working a regular service between Buffalo and Chicago, though ice stopped navigation from early December to early May. Steam-towing of barges, however, common on the Hudson river and on the St Lawrence below Montreal, was only partially successful to and from Buffalo and the Erie Canal, because of handling difficulties in storms.

Steamers flourished on Lake Ontario, Canadian vessels running mainly between Niagara, York (Toronto) and Kingston, American between Ogdensburg, Oswego, Rochester and Lewiston. From 1837, also, they began running through the Thousand Islands to Prescott at the head of the St Lawrence. Thence a small steamer, *Dolphin,* left daily for the top of the Long Sault rapids, whence by stage, steamer, stage, steamer and coach passengers could by-pass the rapids and reach Montreal the same day. When, however, the St Lawrence canals (except those at Williamsburg) had been opened, followed by the rebuilt Lachine in 1848, steamers from Lake Ontario began to run the rapids downstream, returning through the locks.

From the early 1840s to the 1860s the railroads were thrusting west. Just ahead of them ran a boom in lake transport also heading westwards, away from connections with the Ohio and Indiana canals, soon to decline, and towards establishing direct communication from the St Lawrence, the Erie Canal at Buffalo, and eastern Lake ports to Detroit, and places round Lakes Huron and Michigan. Occasionally the drive to the west developed fantastically, as when in 1849 the barquentine *Eureka* left Cleveland on Lake Erie for San Francisco and the Californian goldfields, via the St Lawrence and Cape Horn. The first ship had arrived at Chicago from the lower lakes in 1834, yet by 1860 the city was a base for eleven major railroads. In between, freight-carrying rose, and fine passenger-steamers were built, culmi-

Fig 105 The excursion steamer *Lady Elgin* at Chicago on 7 September 1860, the day before she was lost in a gale on Lake Michigan after being rammed by the schooner *Augusta*; 287 lives were lost. Note the hogframe curving from forward to aft. Such ships had hulls lightly built for the great weight of boilers, engines, paddle-wheels, etc, they had to bear. The hogframe therefore provided external stiffening to a hull which without it would have been pulled apart.

nating in the 2000-ton *Western World* of 1854 with her 300 staterooms. By 1857, 107 sidewheel and 135 propeller steamers were recorded on the lakes alongwith 1006 sailing craft.

In 1852, however, the railroad reached Toledo from Cleveland, and in 1856 the Grand Trunk was opened from Montreal to Toronto. A panic among shipowners followed in 1857, big passenger-steamers being laid up and smaller, cheaper ships being built instead. The start of the Civil War in 1861 hit passenger-carrying again, though freight transport increased, with the peak of the sailing-ship era in 1868, the growth of tug-towing and then of bulk freighters, the first of these on the Lakes in iron, the *Onoko*, being launched in 1882. Railroads and the war had ended one era of passenger-carrying. Another was soon to begin.

Ships now needed to penetrate beyond Lakes Erie, Huron and Michigan into Lake Superior, blocked by the 18–24ft falls at Sault Ste Marie (the Soo) on the St. Mary's river that joined Superior to Huron. Especially was this so after the St Lawrence river canals had been completed, so enabling ships to reach the sea.

In 1844 one steamer, the *Independence*, and in 1846 another, the *Julia*

Palmer, had been hauled overland to Lake Superior. Then in 1852, Congress granted 750,000 acres of Michigan land as a carrot to attract an entrepreneur. It did so, and the Sault Canal, 1800yd long, was built by a private company under contract to the state; though it cost twice the estimate, the company was left with much rich copper-bearing land. Opened in 1855 by the steamer *Illinois,* this short canal had a staircase pair of locks (the State lock) each 350ft × 70ft × 12ft. Thus Lake Superior was given water transport of iron ore and grain, and a line lay open for 2342m from Duluth to the sea.

The Canal Age in North America

An Englishman, comparing the North American canal age with his own of a few decades before, notices the differences before the similarities, and one above all, that in North America some three-quarters of construction finance derived from the public purse, whereas in Great Britain (though not in Ireland) almost all of it was subscribed privately.

In England, a major canal project involved negotiation, valuation, argument and sometimes arbitration with hundreds of separate owners, each occupying land that had long been settled, together with such problems as building substantial accommodation bridges to rejoin divided farms, and carrying well-used roads over or under and streams beneath the proposed line. North American canal commissioners and engineers, except near the long settled eastern seaboard, had a less complicated job. Most land was either unsettled or recently settled; some was virgin and still owned by the federal government. Nevertheless, the problems were there, each landowner naturally arguing that nowhere in the New World was there more valuable land than his.

Surveying in uncleared land, often rocky or thickly wooded, was necessarily slow, difficult and liable to some error; therefore even after construction had begun, changes in the line were likely. Du Vernet on the Grenville Canal wrote in 1820:

> a good deal of the ground through which the Canal has to pass is still un-cleared and the wood very thick... I may on a better knowledge... alter the direction a little.(5)

There was far more tree-felling, stump-pulling and dangerous rock-blasting, far more building across marshland.

The winter made a great difference to construction. Du Vernet's men had to be taken about 40m by *bateau* to Montreal each winter.

Afterwards came spring floodings. In 1822 he could not restart work until 12 June, when the river level had fallen far enough. In the year following, his working season was 102 days. The further south, however, the less interruption. Another practical difference was the lack of local transport services and supplies. In England, local roads were everywhere, and villages and towns never far away could provide them: this was not the case where distances between populated areas were so much greater.

North American engineers and British seem not very different—indeed, some were British; military engineers in Canada, and also men there with British experience such as Francis Hall, N. H. Baird and H. H. Killaly; and such as William Weston in the States. A very few professionals, the rest men who learned their job by practising it. In the States, as in Britain, the earlier canals were built mainly by local labour plus a minority of navvies, some of them immigrants. Sometimes skilled men were imported, as the masons brought from Scotland to work on the early Shubenacadie Canal locks. The more the canal was cut through virgin land, however, the more professional navvies there were. Again, by its very nature it was more difficult to supervise and get good work in uncleared country; so the need for repairs, work to be done again, the endless effort to keep a canal watertight and operational, became greater in the mid-west.

Compared with British, American canal construction was rough and ready—sensibly so, in a new country of great distances, virgin territory and plentiful timber. Peter Stevenson notes

> undressed slopes of cuttings and embankments, roughly built rubble arches, stone parapet-walls coped with timber, and canal-locks wholly constructed of that material...

but he recognises these expedients are

> to meet the wants of a rising community, by speedily and perhaps superficially completing a work of importance, which would otherwise be delayed...and although the works are wanting in finish and even in solidity, they do not fail for many years to serve the purposes for which they were constructed...(6)

Timber was used for most things that in England, where timber was becoming scarcer, would have been of brick or stone. Wooden lock-houses or warehouses were common; so were wooden aqueducts.

Rivers were often crossed on the level, barges moving above a dam, and with a wooden towpath bridge (sometimes a double-decker) for the towing mules, or occasionally a floating wooden towpath bridge, as on the Middlesex Canal at Concord. But the obstruction caused to raftsmen by such dams and bridges made aqueducts preferable.

Masonry aqueducts were built on some of the great lines, like the Erie and the Chesapeake & Ohio. Iron troughs, however, were not needed when wood was so cheap. America's first important canal aqueduct, at Black Brook on the Middlesex, 110ft long, had a wooden trough on wooden piers upon stone foundations. Much later, a trestle aqueduct at Bolivar carried the Sandy & Beaver Canal over the Tuscarawas river to join the Ohio & Erie. But most had timber troughs on masonry piers, like the 'vast low wooden chamber full of water'(7) over which Charles Dickens entered Pittsburgh; it was 1140ft long, 16½ft wide at top and 8½ft deep, supported on six piers. This, like many others, was roofed. Another slightly smaller such aqueduct was that over the Potomac from Georgetown to Alexandria, its trunk 1100ft long, 17ft wide and 7ft deep, borne on eight stone piers some 30ft above tidal water. We have already noted J. A. Roebling's use of the suspension principle along with wooden trunks at Pittsburgh and on the Delaware & Hudson.

Locks, especially on the main canal lines built with public money, were usually of masonry upon timber foundations, and with timber flooring. Poor masonry work at the sides, however, might then be covered by wooden planking, as on the later locks of the Sandy & Beaver Canal, or, as on the Chemung, the 'locks were constructed of wood, supported at the sides with braces, with a stone wall of masonry at the head, and a dry wall at the sides, resting on the foundation timbers...'(8), the wall's task being to separate woodwork from earth. A somewhat similar construction had been used in the 1790s on some of the Middlesex Canal locks: an inner wooden lock chamber was separated from walls of rough masonry by wooden braces. Sometimes, however, lock chambers were entirely of timber, as the locks of the first Welland Canal, opened in 1829, or entirely of masonry, as some of the 1826 locks of the Shubenacadie Canal, based on those of Scotland's Forth & Clyde.

The so-called 'bear-trap' lock used mainly on the early Lehigh Navigation was a simple form of inclined plane, built where the traffic was all one-way, and disposable craft could be used. The device consisted of a chute or flume cut through a dam and then down a rapid below it, having at the top end 'V' shaped piers within which the 'arks and rafts' would be brought to the head of the chute. There an ingenious one-way lock utilised water pressure first to back up the river water, and then release it to provide a flash. One built in the early 1830s on the Chemung river had a timber-framed chute 47ft wide and 156ft long to drop lumber-rafts and boats 7½ft, an apron being carried for some distance downstream from the bottom of the chute to prevent craft suddenly hitting deep water and breaking up. Later this chute was found too steep for boats to use during the spring freshets. A better one, at Towanda on the Susquehanna, having been quoted as an example, a new one was built 450ft long with a slope of 1 in 100.

Most locks were of the conventional early Continental or English pattern, with mitre-gates and balance beams, and various types of paddle-gear. Because most canals had plenty of water, being river-fed, such economising devices as side-ponds were seldom built. One often used pattern, however, is thought to have been first introduced by Josiah White on the Lehigh. Lower gates were of normal mitre pattern, but those above were replaced by a drop-gate which, when water-levels on both sides equalised, fell outwards to lie on the canal bed, it being raised by winches and chains placed above. On the Delaware & Hudson Canal, however, the lock-tender could operate both sets of gates from a bridge built over those at the lower end. Rods and chains ran along the bottom of the lock to the upper gate, while the balance beams of the lower mitres were replaced by gate arms worked by winch and ratchet gear from the bridge.

Staircase, combine or flight locks were not uncommon—fine examples were the parallel 5-rises at Lockport on the Erie Canal, while 2- and 3-rises were often to be found, eg on the Schuylkill canals and the Ohio & Erie. In Canada staircases were common: the Rideau, for instance, had the great 8-lock staircase at Ottawa, and a selection of 3-rises and pairs. Today, the great parallel 3-rises of the Welland Canal and of the Panama at Gatun form perhaps the world's most impressive lock flights.

No vertical lifts were built during the canal age, and, with the glorious exceptions of the Morris Canal and the Portage Railroad (with which I include the two planes on the Philadelphia to Columbia railroad that was part of the Pennsylvania Main Line) four inclined planes, an early one at South Hadley falls, two on the Shubenacadie Canal, and one at Georgetown on the Chesapeake and Ohio, which replaced two overworked locks connecting the canal with the Potomac river. With a single track, this last had a fall of 39ft. The caisson, 112ft× 16ft 9in× 7ft 10in, rode on six 6-wheeled trucks running on four rails on a 1 in 10 slope. It was intended to carry a 115-ton barge afloat, but the weight proved too great for the track, and craft had to be carried dry. Turbine-powered at first, later by steam engine, the incline, opened 1876, in its day the world's biggest, was in 1886 destroyed in a flood.

North America's few and short canal tunnels, none as long as a mile, cannot compare with Britain's 42 m of tunnelling or France's similar achievement. Instead, its engineers built some fine reservoir dams. The South Fork dam and reservoir was begun in 1838 to supply the Johnstown-Pittsburgh section of the Main Line, its wall 931 ft long and 72 ft high. As a canal reservoir it was only needed for eighteen years. Much later, when in use for other purposes and after it had been raised, it was to collapse in the Johnstown disaster of 1889. Most remarkable for its period was the Jones Falls masonry dam on Canada's Rideau Canal, built between 1826 and 1832 by Col John By of the Royal Engineers. It is 350 ft long on the curved crest and 62 ft high, the first arched dam in North America, when built the highest, and still a wonder to the visitor.

Most canals closed during the winter freezing, when channels and feeders were drained to prevent ice damage to structures. English canals might have to close for days or weeks, very occasionally for a month or more, but this was incidental to all-the-year operation. The whole elaborate business of the annual closing and reopening, as described for instance in W. D. Edmunds' *Rome Haul*, is strange. Usually however, repairs could be done while the canals were drained, so obviating England's two-week summer stoppages.

Throughout the canal period in the States, animals were used for towing, mules or occasionally oxen for freight-carriers, horses for packet-boats. One finds the occasional oddity: boats on Maine's Cumberland & Oxford Canal would probably be towed (on canal sections), sailed (cross lakes) and poled (on a river). Steam-tugs were used on tidal waters, on rivers such as the Mississippi, on ship canals like the Chesapeake & Delaware, or in slackwater (canalised) navigations, but American saw little steam-towing on ordinary canals and few steam self-propelled canal boats. Towing animals usually had a stable on board, for distances were too great for a system of bankside stables, usually at pubs, to grow up. Therefore a spare team was sometimes carried.

Britain saw the creation of a number of canal villages and towns. But in America the effect was multiplied in space and time. On a prosperous canal, hamlets would spring up round stores to sell liquor and groceries, with a warehouse by the lock-house. Except at flights, every lock had its lock-tender. Towns would grow as basins were dug, turnpikes built to bring goods to the boats, packet services begun, warehouses built and boatyards opened. With the boats came cheaper goods, more population, wider culture. With them left all that the town and its neighbourhood could grow or make.

One other feature of American canals must be mentioned, the

packet-boat services. Britain had such services over short distances, some run at high speeds, but only on the Forth & Clyde were there overnight sleeper boats. Many American canals had long-distance passenger-boats running regularly, with a captain and crew of maybe six. Some also carried freight. The Erie Canal's packet-boat services are well known. Here are two descriptions from other canals one east, the other west of the Alleghenies.

One described at the opening of the Eastern Division of the Pennsylvania Main Line in 1832 carried 25 passengers* and 30 tons of freight. It was 79ft long, 12–15ft broad, drew 12in when loaded, and was pulled by three horses on 10m stages using a 150ft towline, at a speed of 4–4½mph.

> The apartments are these: A ladies' cabin in the bow of the boat, calculated for eight persons. This cabin is handsomely decorated, and has tables, chairs and beds for that number of persons, and is as neat and comfortable as such rooms usually are in steam boats. The next room is what is called the 'midships', containing the freight. Next is the gentlemen's room, large enough for all passengers; this room besides a bar with the choicest liquors, is calculated for a table, at which all the passengers breakfast, dine and sup, and contains beds and bunks for all the male passengers. The last room is the kitchen, at the steerage, where cooking is done in superior style.(9)

Peter Stevenson describes travelling on a packet-boat as experienced west of the Alleghenies—quite different, he says, from the east:

> About eight o'clock in the evening, every one is turned out of the cabin by the captain and his crew, who are occupied for some time in suspending from the ceiling two rows of cots or hammocks, arranged in three tiers, one above another. At nine, the whole company is ordered below, when the captain calls the names of the passengers from the way-bill, and at the same time assigns to each his bed, which must immediately be taken possession of by its rightful owner on pain of his being obliged to occupy a place on the floor, would the number of passengers exceed the number of beds... I have spent several successive nights in this way, in a cabin only 40ft long by 11ft broad, with no fewer than forty passengers; while the deafening chorus produced by the croaking of the numberless bullfrogs ...was so great, as to render it often difficult to make one's-self heard in conversation, and, of course, nearly impossible to sleep. The distribution of the beds appears to be generally regulated by the size of the passengers; those that are heaviest being placed in the berths next the floor... At five o'clock in the morning, all hands are turned out in the same abrupt and discourteous style, and forced to remain on deck while the hammocks are removed and breakfast is in preparation. This interval is occupied in the duties of the toilette... A tin vessel is placed at the stern of the boat, which

*Other packet-boats carried as many as 60–70.

> everyone washes and fills for his own use from the water of the canal, with a gigantic spoon formed of the same metal; a towel, a brush, and a comb, intended for the general service, hang at the cabin door... The breakfast is served between six and seven o'clock, dinner at eleven, and tea at five.(10)

And yet Charles Dickens found compensations

> ...there was much in this mode of travelling which I heartily enjoyed at the time... Even the running up, bare necked, at five o'clock in the morning, from the tainted cabin to the dirty deck; scooping up the icy water, plunging one's head into it, and drawing it out, all fresh and glowing with the cold; was a good thing...the fast, brisk walk upon the towing-path... the exquisite beauty of the opening day... the lazy motion of the boat... the gliding on at night... the shinning out of the bright stars... all these were pure delights.(11)

The canal age's justification lay in its low-cost transportation. In the 1840s, a ton hauled 100 miles by waggon on the best roads generally ranged from $10.00 to $15.00. The corresponding charge on the Erie Canal was 91 cents, and on the Ohio & Erie Canal $1.00. With the latter's opening in 1832, wheat which formerly was worth 20 to 30 cents a bushel at a northern Ohio farm began to yield the farmer 50 to 75 cents. There let us leave it.

13
RIVERS, RAILROADS AND WAR

The Western Rivers to the Civil War

The first western rivers steamboat, the *New Orleans,* had reached its namesake from Pittsburgh in 1812. A rapid growth of steamboating followed, from the early 1820s linked with oceangoing ships and packets running to and from New York and the eastern seaboard.

Design quickly moved from the small ship form of Fulton and Livingston's *New Orleans* to Henry Miller Shreve's *Washington* of 1816. Steamboats were soon to have a long, narrow, flat-bottomed, shallow-draught hull, upon and not within which engines, cargo and passengers were to be carried. Propulsion was to be by side paddlewheels in huge casings, each wheel driven by a small, light, high-pressure engine. As design developed, draught was further reduced by building longer and wider hulls, weight by lighter construction and materials.

Charles Dickens travelled on such a one from Pittsburgh to Cincinnati in 1842:

> There is no visible deck...: nothing but a long, black, ugly roof, covered with burnt-out feathery sparks; above which tower two iron chimneys, and a hoarse escape valve, and a glass steerage-house. Then, in order as the eye descends towards the water, are the sides, and doors, and windows of the state-rooms, jumbled as oddly together as though they formed a small street, built by the varying tastes of a dozen men: the whole is supported on beams and pillars resting on a dirty barge, but a few inches above the water's edge: and in the narrow space between this upper structure and this barge's deck, are the furnace fires and machinery, open at the sides to every wind that blows, and every storm of rain it drives along its path.(1)

Peter Stevenson, visiting Pittsburgh in the mid-1830s, wrote of a visitor:

> ...in the very heart of the continent...the appearance of a large shipping port, containing a fleet of thirty or forty steamers moored in the river, cannot fail to surprise him; and his astonishment is not a little increased if he chances to witness the arrival of one of those steamers, whose approach is announced long before it makes its appearance by the roaring of its

Fig 106 Sidewheeler *Magnolia* loads cotton bales from an ingenious covered shoot on the Alabama river. 'At a given signal from below a thousand-pound package of the staple was started at the top of the slide, two hundred and fifty feet perpendicular above the level of the water. Slowly it moved at first, but, gaining momentum as it proceeded, the pace quickened – quicker, quicker, quicker – till at last it fell like a thunderbolt on the deck, knocking the bales of the barricade in every direction. In one moment a dozen black fellows were upon the new arrival, dragging it out of the way with instruments resembling boathooks, or busying themselves with reconstructing the barricade'.

> steam, and the volumes of smoke and fire which are vomited from the funnels... this same vessel has come direct from New Orleans... and ...fifteen days and nights have been occupied in making this inland voyage, of no less than two thousand miles.(2)

Steamboats were then basically freight-carriers, but as their passenger business became more competitive, so did the amenities that they offered increase for those who could pay well, while deck passengers travelled uncomfortably but cheaply. By 1837 there were nearly 400 steamboats on the Ohio-Mississippi system, most of them built at Pittsburgh or Cincinnati, and in 1855 there were 727, with a tonnage of 170,000. In early days, steamboating was highly profitable, but thereafter competition kept freight rates, passenger fares and profits generally low, though all fluctuated with traffic offering, boats available, and river levels. Freight between New Orleans and Louisville, some $5.00 per 100lb in 1815, had fallen to 25 cents by 1860.

About 1850 came a major design change, the substitution of a stern paddlewheel for two sidewheels. Once boats could be built strong enough to take a sternwheel's weight, it offered safer shallower-draught navigation, a wider useful hull without increasing total width, and the possibility of lashing barges alongside. Multiple rudders were now fitted forward and aft of the sternwheel, to make the craft manoeuvrable in either direction.

As steamboats multiplied, keelboats and sailing barges disappeared from the trunk routes, though they survived for a time on shallow tributary rivers. Early steamboat trade, along with passenger-carrying, was mainly in high-value, low-bulk upstream cargoes. Flatboats, carrying bulk agricultural cargoes downstream only, survived into the late 1840s, then fell rapidly away. They had been helped to survive by the ease with which their crews could return upstream by steamboat, and also by river improvements which reduced losses from snags and enabled them to travel by night. Only in the last two decades before the Civil War of 1861–5 did steamboats take over most downstream bulk agricultural cargoes.

Until the Erie Canal opened in 1825, the Mississippi was the main outward carrier from the farmlands of the West. Thenceforward the Erie route caught up, until by 1860 each line was transporting some 2M tons a year, with carryings on the third, Pennsylvania, route much less at some 300,000 tons. For inbound traffic, however, except in passengers, the Mississippi's importance decreased after the 1830s because of the Erie, and after the 1850s because of railroads also. Thenceforward the river trade became more regional in character, and so little affected by railroads except in the Cincinnati-Pittsburgh sector.

At first, steamboats kept largely to the main trunk routes of the Ohio and lower Mississippi. Later, they began slowly to penetrate the upper Mississippi and tributary rivers in spite of their navigational difficulties. Most were basically cargo-carriers: nevertheless, crowds of passengers—many of them seeking one of the land trails to the West—found them preferable to any possible alternatives. What happened on the Missouri may stand for the tributaries. Though its first steamboat appeared in 1819, few ran until about 1840. Numbers then rapidly increased to 1858, when more than 700 were at work, and packet lines ran from the Missouri's mouth to Miami, Kansas City, St Joseph, Omaha and even Sioux City, bringing prosperity to the valley. Then railroads came, and traffic fell away—in spite of the 1862 gold rush to Montana which temporarily took steamboats some 2500m up the river to Fort Benton—to nothing at all.

Snags (sunken trees) were and long remained the rivers' worst navigation hazard. Dickens experienced them on the Mississippi between the Ohio confluence and St Louis:

> For two days we toiled... striking constantly against the floating timber, or stopping to avoid those more dangerous obstacles, the snags, or sawyers, which are the hidden trunks of trees that have their roots below the tide. When the nights are very dark, the look-out stationed in the head of the boat, knows by the ripple of the water if any great impediment be near at hand, and rings a bell beside him, which is the signal for the engine to be stopped; but always in the night this bell has work to do, and after every ring, there comes a blow which renders it no easy matter to remain in bed.(3)

Some boats, indeed, had false bows fitted to protect them from snags. Boat losses were reduced, however, between 1827 and 1832 by the introduction of snagboats, hand-operated until 1829, increasingly steam-powered thereafter, which worked to remove sunken trees.

By the General Survey Act of 1824, Congress gave to the US Army Corps (then Board) of Engineers the responsibility for inland waterway improvements for navigation, a responsibility greatly enlarged by later enactments. Thereafter, surveys and channel improvements continued. The Louisville & Portland Canal was one.

Along the Ohio's 981 m from Pittsburgh to the Mississippi, with a fall of 460 ft, there was one serious obstacle to navigation, the 25 ft-high falls at Louisville, round which, except for a few weeks during floods, all cargoes and passengers had to be transhipped. The 2-mile Louisville & Portland Canal, much of it through rock, by-passed them. Built with private and Kentucky state money, it was opened in December 1830 with three locks 185 ft × 50 ft, at that time America's biggest. In 1833 it carried 1585 boats of 170,000 tons capacity, but by 1852, so much had steamboats been enlarged, only 57 per cent of the craft in use could pass it. In 1860, the canal being then under *de facto* federal control, lock enlargement and partial realignment began, only to be temporarily stopped by the Civil War.

In this period also, efforts were made to improve channels by cutting through shoals and sand-bars, building wing-dams to force the current towards the centre, and dredging. But all three were local and sporadic. Levees (flood-banks) also began to be extended upwards from New Orleans to St Francisville, and then at intervals to Natchez.

Improved track, better steamboat engines, faster transits, much quicker turnround times and a longer navigation season combined to make Western river steamboats an outstanding example of transport productivity, and the rivers themselves the dominant factor in the development of the West before the Civil War.

The Western rivers saw the rise of the showboat. The first of the famous showboats was that of an emigrant English family, the Chap-

Fig 107 Steamboat days: (above) a letter marked for, and carried on, the *General Lafayette* from Louisville on the Ohio to New Orleans in 1852; (centre) the *Edna* ran on the Black River (the letter is dated 1847) and the Ouachita; (below), Robert M'Kinney sends 37 sacks of wheat from Steubenville on the Ohio to Pittsburgh in 1857; (bottom) a sidewheeler from the note of an Omaha City, Nebraska, insurance company; a printer's proof vignette of an early rail-steamboat interchange point; and a mailcarrying sidewheeler from a check of the Boatmen's Saving Bank of St Louis. The wheel casing of this last is inscribed 'St Louis & New Orleans Packet. U.S.M. City of St Louis'.

mans, who launched it at Pittsburgh in 1831; nine Chapmans, with two others, began in that July to drift down the Ohio and Mississippi. They poled the boat, attracting audiences as best they could to their little floating theatre. When they reached New Orleans, the boat was broken up, the Chapmans returning to Pittsburgh by steamboat to build a new one. By 1836, however, they had done well enough to build a steamboat theatre.

Others followed, such as Sol Smith who bought the Chapmans' *Floating Palace* in 1847, or Spaulding & Roger's huge *Floating Circus Palace*, some 200ft × 35ft, built in 1851 and bigger than any river steamer. Boats also appeared on the canals: *Dixie* on the Wabash & Erie, *Huron* on the Miami & Erie, and Henry Butler's on the Erie. And then for a time, the Civil War ended the showboats, for riverside folk had other things to think of than floating entertainment. The tradition was continued in 1985 with the delivery of the Nashville-based snowboat *General Jackson*, which includes a 1000-seat theatre.

Though steamboats carried freight, and sometimes lashed a barge or two alongside, or towed one behind, the practice was disliked by crews as making the boats harder to handle, passengers who suffered delay, and insurance interests. Demand for more cargo capacity than flatboats and keelboats offered came from the Pittsburgh coal industry, which as early as 1814 began sending coal to riverside towns down to New Orleans. At first it was shipped in 'coal boxes', flat-bottomed and square-ended wooden barges some 80ft to 175ft long that were lashed together, floated downriver on the floods, and broken up for lumber at journey's end. However, losses of boats were about 10 per cent annually, while towns' fuel supply remained uncertain.

So experiments began in the 1840s with purpose-built barges, at first small, then larger, and also with the new technique, first recorded in 1845 and accepted by 1851 of pushing barges in front of a towboat with vertical timbers (towing knees) to press against the nearest barges, these latter being firmly lashed both to the towboat and to each other, ahead and sideways. Sternwheelers, which steered well in reverse when flanking (see p. 341) were ideal for push-towing. By 1855 10-barge tows were being used: in 1857 a million tons of coal from the Pittsburgh area moved by river, and by 1866 some 80 towboats were handling its output.

Railroads and Canals in the Nineteenth Century

It is difficult to grasp how quickly railroads were built in the United States. Much of their mileage being through virgin land that presented no ownership problems, they were constructed cheaply, often more so because of land grants, but with immense energy. Whereas in 1830 the United States possessed 1277m of canal and 73 of railroad, in 1840 there were 3326 of canal and 3328 of railroad, and in 1850 3698 of canal but 8879 of railroad. Ten years later the railroad figure was 30,636.

Because states had helped to finance most important canals, there was at first some resistance to railroad building on the reasonable ground that it would depreciate assets which had only begun to yield revenue. A railroad paralleling the Erie Canal, for instance, was initially authorised only for passengers; then goods carrying was allowed only when the canal was closed by ice; and not until 1851 generally, on condition that charges were at least equivalent to the canal tolls.

From about 1834 the impetus of canal-building and enlargement, especially in New York, Pennsylvania, Ohio, Indiana and Illinois, kept it moving forward. Maybe North America's canal age can be considered ended when in 1848 both the Illinois & Michigan Canal and the St Lawrence's locks were completed. Thereafter to the Civil War was a time of finishing, improving, adding to what had gone before. During the Civil War, rivers and canals played a notable part, especially because of the North's control of the tributaries of the Mississippi and possession of the vital Erie Canal, just rebuilt in time. Especially was this so in the Confederate States. A canal such as North Carolina's Weldon could have expected a short life, for after some initial active years, most traffic had left it for railroads. But because of Northern destruction of Southern railroads, a brief time of furious use returned before peace brought it to bankruptcy.

Passenger traffic left the canals for the faster and more comfortable railroad cars; high-yielding freight followed, while bulk freights often remained because early railroads were not sturdy enough and had insufficient facilities to carry them. The figures (4) for states with appreciable canal mileages tell their eloquent story:

	1830		*1840*		*1850*		*1860*
	Canals	*Rail-roads*	*Canals*	*Rail-roads*	*Canals*	*Rail-roads*	*Rail-roads*
New York	546	—	640	453	803	1409	2682
Pennsylvania	230	70	954	576	954	900	2598
Ohio	245	—	744	39	792	590	2946
Massachusetts	74	3	89	270	89	1042	1264
Virginia	—	—	216	341	216	341	1731
Indiana	—	—	150	20	214	226	2163
Maryland	10	—	136	273	136	315	386
New Jersey	20	—	142	192	142	332	560
Illinois	—	—	—	26	100	118	2799
	1125	73	3071	2190	3446	5273	17129*

*The United States canal total can be taken as 4254m in 1860, after some 350m had been closed.

The earliest were to go were those barely economic eastern tidewater canals and river navigations which were in no case to resist railroad competition: such were the New Haven & Northampton (1847), Blackstone (1848), Connecticut river locks and cuts (from 1849), Santee (1850) and Middlesex (1852). Others, though on a smaller scale than in Britain, were transferred to railroads, especially in Pennsylvania.

In 1846 the Pennsylvania Railroad had been authorised to build a direct rail link from Harrisburg to Pittsburgh: by 1852 it and its partners could offer continuous rails from Philadelphia to Pittsburgh via the Portage Railroad and in 1854 by its own line. Passengers and freight were at once attracted, and Main Line canal/rail traffic fell. The state reacted by in 1855 rebuilding the Portage Railroad without inclined planes, and putting a tonnage tax on goods carried by the PRR except when the canal/portage line was closed. Carriers by the old line intensified competition, one advertising:

> Since the completion of the New Portage Railroad, avoiding the Inclined Planes, Boats meet with no detention at the Allegheny Mountains, Trucks being in readiness at all times to carry Boats over immediately upon their arrival at Hollidaysburg or Johnstown. Ours is the only through line on the Pennsylvania Canal, and being composed entirely of portable boats, only one transhipment is required.(5)

After heavy rate-cutting, the whole Main Line was sold to the Pennsylvania Railroad in 1857, the Columbia-Philadelphia railroad being taken over, the Portage closed.

Also in Pennsylvania the Susquehanna, North and West Branch, and Delaware Divisions were all sold in 1858 to the Sunbury & Erie RR, later itself to form part of the Pennsylvania RR, and in 1859 the Pennsylvania Canal Board was abolished. Some of the purchases were closed, as the Main Line from Johnstown to Pittsburgh in 1863–5. Others were for a time expanded, for in 1867 the Pennsylvania RR formed the Pennsylvania Canal Company as a subsidiary, and this widened and deepened the Juniata and Susquehanna Divisions together with the North and West Branches to take 260-ton boats. For a few years they did well, but after 1875 business fell off, disastrously so after the 1889 Susquehanna flood. The North Branch north of Wilkes-Barre to the New York state line closed in 1872, most of the West Branch by 1891. After 1890 only 144m of the central canals remained: by 1903 not one. Another flood of 1894 caused the Reading RR, who operated the Susquehanna & Tidewater Canal, to close that also.

Further south, the Chesapeake & Ohio Canal had reached Cumberland in 1850, and stopped. It maintained a declining existence until 1924 when, badly damaged by flood, it was closed. In 1938 the federal government bought it, and restored a 22m section above Georgetown, which became the nucleus of the present national park (see p. 401). In 1851 the James River & Kanawha Canal reached Buchanan. In 1859 a French company proposed to complete it to the Ohio, a scheme halted by the Civil War. After it Edward Lorraine, seeing the canal's value in keeping down railroad rates, re-surveyed the unbuilt line, proposing a 9m tunnel greatly to reduce the necessary lockage and improve water supply, after which the Corps of Engineers themselves put forward a modified version of Lorraine's scheme. However, serious flood damage in 1870 and 1877 caused the canal's sale in 1880 to the Richmond & Allegheny RR, which used the towpath for some of its track. Thus after 1880 only one east-west waterway line, the Erie, remained within the United States, and another, the St Lawrence and Welland, in Canada.

What of those ambitious canal lines further west that had been built to join the Ohio to Lake Erie, or to connect one to another? First to go was the longest and economically shakiest, the Wabash & Erie, which in sections roughly from south to north became disused between 1865 and the 1870s. With it went the Miami & Erie. The Beaver & Erie closed in 1871, the Pennsylvania & Ohio the year after. Only the great Ohio & Erie was still busy in 1886. Indeed, new locks were built in 1913, and bank repairs done. Serious flood damage followed, and this time the state decided to abandon.

Westwards again, the Illinois & Michigan remained prosperous, so much so that by 1871 the whole capital debt had been repaid. The originally planned method of finance, by land sales and loans secured on the tolls, had worked. Though the opening of a competing railroad in 1854 removed the canal's passenger business and encroached on freight-carrying, tonnages increased until in 1882 they just topped the million. Locals benefited from low freight rates offered by both. After 1882 carryings fell away; but by then new ideas were about.

The Erie Canal resisted railroad competition better than most. Its limitations began to show in 1863 with a decline in the flour traffic, and became clear to all in the depression years after 1873, when total tonnages began to fall. Carryings on the Erie system in 1876, at some 4.2M tons, were less than two-thirds those of 1872. They then rose to some 6.46M in 1880, a figure better than it looks because of branch closings, among them the Genesee Valley, Chenango and Chemung Canals in 1878.

In 1882 canal tolls were abolished with almost no effect on traffic. Tonnages began to fall again in the 1890s, to 3,541,000 in 1906. Meanwhile New York state railroads, which had carried under 15 M tons in 1876, moved over 100 M in 1906. The Champlain Canal had best withstood competition: traffic, 1,120,000 tons in 1868, was still 800,000 in 1906. Indeed steamboats ran on Lake Champlain into the new century, the last one, *Ticonderoga*, being built in 1906.

Professor J. A. Fairlie in 1907 wrote:(6)

> ... it seems clear that the decline of canal traffic has been largely aided by the failure to improve the canals to keep pace with the railroads.

By then, however, the New York State Barge Canal had been begun (see p. 357).

The Coal-carriers of the East

In the eastern states, things were different for two groups of waterways: those that carried coal and those that formed part of the Atlantic Intracoastal Waterway.

One of the leading coal-carriers, the Delaware & Hudson, flourished until the early 1860s. Then, beginning to suffer severely from competition, the company retaliated by itself going into the rialroad business, by way of lease and new construction. By the 1870s, canal traffic had become unimportant, and in 1898 the canal was disused.

In 1862, when demand was at its height on the Lehigh, due to the Civil War, a flood destroyed the works above Mauch Chunk and damaged those below. The upper section was never rebuilt. Nevertheless, though the Lehigh Valley RR's parallel line competed with it, the company emerged full of energy from the Civil War. It bought new coal lands, absorbed and built railroads, and acquired the Delaware Division Canal in 1866. Thenceforward, however, its canal interests lessened. In 1908 electric-locomotive towing was introduced on two experimental 2 m sections, but probably ended when the navigation was badly flooded in 1912. Nevertheless, the lower part of the navigation and the Switchback Railroad that succeeded the Gravity at Mauch Chunk (now Jim Thorpe) worked until 1932, the navigation finally closing after the 1942 flood.

At Easton, the Lehigh's line met those of the Delaware Division and Morris Canals. The former, having been bought by the Sunbury & Erie RR in 1858, finished up in Lehigh Navigation ownership. After 1867 the latter began to suffer badly from railroad competition in coal and iron, until in 1871 it was leased to the Lehigh Navigation's competitor, the Lehigh Valley RR, along with its important Jersey City terminal facilities. Traffic fell and became more local, until in 1922 the Morris Canal passed to the state of New Jersey, and was closed in 1924.

Finally, serious flood damage in 1850, added to the expense of the 1845–7 improvements, forced reorganisation of the Schuylkill in 1852. Thereafter the navigation remained active until it was leased to its competitor, the Philadelphia & Reading RR, in 1870. It remained a modest carrier, mainly of anthracite, for many years before closure in 1931.

The Atlantic Intracoastal Waterway

Between 1840 and 1856 the Dismal Swamp Canal had been improved by new approach channels. But while these were being built, three developments gave reality to the idea of an Intracoastal Waterway, and new life to its components: the completion of the Susquehanna & Tidewater Canal in 1840, of the Chesapeake & Ohio to Cumberland in 1850, and the increased efficiency of steam-towing. Down the S & T came anthracite from the Lehigh, down the C & D bituminous coal: both were needed up and down the Waterway. Small sailing ships, instead of making their solitary way along the coast, could now be hauled, as coal barges were, by tugs along a protected passage.

So a chain effect occurred. It began on the Delaware & Raritan, whose locks were lengthened to 220ft in the winter of 1852. The Chesapeake & Delaware followed, building in 1854 220ft locks, two at the Delaware City end, one with a side-pond at Chesapeake City. Then came a new waterway, the 70m Albemarle & Chesapeake, begun in 1855 and opened in 1859, to provide a more efficient alternative to the Dismal swamp. Part river, part canal, part sea sound, it ran from Hampton Roads and the Elizabeth river to the North river and Albemarle Sound. Early steam pile-drives prepared the foundations of its one lock, also 220ft long but 40ft wide. Built with gates facing in either direction, its purpose was to stop tidal surge from the Elizabeth river. The 1850s therefore saw the coastal route become truly an Atlantic Waterway.

The consequences for the Delaware & Raritan were shortlived: the canal and its parallel railroad were leased by the Pennsylvania RR in 1871. Five years later the Reading RR built its own coal-carrying line and canal business quickly fell away. The process had begun which was to take the D & R out of the modern Waterway.

On the Chesapeake & Delaware, tonnage had passed the 600,000 mark by 1857. To the older grain and lumber trades were now added increasing quantities of coal. Then came the Civil War, when the canal played a unique part, being for a time the Union government's only line of communication between Washington and the rest of the North. It now achieved its maximum toll revenue, and shared prosperity with canals and railroads elsewhere. After the war, business increased for a time, tonnage reaching a million in 1869, but receipts did not, as much of the Cumberland coal trade was lost to rail. Passages and tonnages, 16,394 and 1,312,816 in 1871, had by 1890 fallen to 6769 and 686,067.

Further south, the Civil War had affected both the Albemarle & Chesapeake and the Dismal Swamp. The latter failed to recover from destruction, depression and competition. Deserted by the Federal government and Virginia, both large shareholders, it struggled on with few improvements other than lock lengthening, but with a fair amount of freight and passenger-traffic, until the 1890s. Then new owners, the Lake Drummond Canal & Water Co, rebuilt it between 1896 and 1899, lowered the summit-level and provided two new locks 250ft × 40ft with 9ft on the sills, and widened and deepened the channel. It remains substantially in this state today. Rebuilt, it regained much traffic from the Albemarle & Chesapeake, itself soon to be in trouble, steam-tugs now hauling long trains of sailing ships and barges, while steamers worked independently.

Steam on the Western Rivers, 1865-1900

Railroad competition came later to the western rivers than to the eastern canals. In 1854, for instance, the first of them reached the Mississippi, the Chicago, Rock Island & Pacific at Davenport, Iowa. Another year, and three more had gained the banks across from Iowa's Dubuque, Clinton and Burlington. Three years more, and Quincy (Illinois), Prairie du Chien and La Crosse (Wisconsin) had rail links. St Paul (Minnesota) followed in 1867. Its severity was mitigated while railroad lines ran east and west, the Mississippi north and south, for interchange of goods and passengers took place. But not for long, for north-south lines followed in the 1870s. By December 1873, a through train had run from Chicago to New Orleans.

Immigrants and cargoes brought by the new railroad and older Mississippi steamboats to St Paul were taken north by waggon to where the newly founded Red River Transportation Co's boats could take them to Winnipeg. Supplemented by construction materials for the Canadian Pacific Railway, this route remained busy until rails from the south reached Winnipeg in 1878. Similar connections were

No. 45

SHARES

Demoine Navigation & Rail Road Company

Endowed with the Lands and Franchises accruing from grants made by the Governments of the United States and of the State of Iowa

For the Improvement of the Valley of the Demoine from the junction of Illinois, Missouri and Iowa, through Iowa, into Minnesota

MISSOURI, IOWA AND MINNESOTA

CAPITAL STOCK $ 3,000,000

INCORPORATED 1854.

THIS IS TO CERTIFY that

is entitled to Shares of One hundred Dollars each in the Capital Stock of the DEMOINE NAVIGATION & RAIL ROAD COMPANY. The said Stock is transferable only on the books of said Company at the Office of their Treasurer in the City of New York or at any transfer Agency at the pleasure of the holder by the said in person or by Attorney on the surrender of this Certificate.

IN WITNESS WHEREOF the said Company have caused this Certificate to be signed by the President and Treasurer. Dated at NEW YORK, this day of 185

Treasurer Orville Clark President

Fig 108 An early effort at river-railroad cooperation to develop the valley of the Des Moines river.

improvised elsewhere, notably by way of the Arkansas and Missouri rivers.

The war over, steamboating had returned and maintained itself for some twenty years before serious decline began. Maybe its high point was the 1870 race between the 300ft sidewheel packets *Natchez* and *Robert E. Lee* from New Orleans to St Louis, won by the *Lee* in 3 days, 18 hours, 14 minutes—a record still unbroken by a commercial craft.

The future, though a difficult one, lay with the towboats, pushing lumber rafts, or barges carrying coal or grain. Only in the coal trade were very large tows, now consisting of standard width and stronger barges than the old 'coal boxes', used. Known tows of 1868 took 16,000 tons in 24 barges, and of 1878, 22,200 tons in 32 barges, though normal tows were smaller. A towboat usually had a sternwheel, but also up to five balance rudders, which extended both fore and aft of the rudder post to increase leverage. When about to enter a bend, 'flanking' was used. That is, as the front of the tow entered the bend, the towboat's engines were reversed. This checked the tow, so allowing more time for the turn, and also increased the turning movement by making the sternwheel drive water against the rudders. Once round the bend, engines were again put ahead. Later in the nineteenth century a British engineer, Charles Ward, was foremost in introducing water-tube boilers, multiple-expansion engines and screw propulsion on towboats, though sternwheelers were to survive well into the twentieth century.

In 1857 coal shipments on the Ohio had passed the million figure, then fallen during the war, then risen to 3,361,934 tons in 1880. In early days, barges had loaded little except coal, but later began to carry raw materials, manufactured goods and agricultural products in direct competition with steamboats. Expensive towboats and barge fleets, however, required capital resources which after the war came increasingly from the mine operators—in 1886 nearly all the coal-trade towboats were so owned. The tendency culminated in the formation of the Monongahela River Consolidated Coal & Coke Co in 1899, absorbing some ninety firms to own 80 towboats and 4000 barges and other craft, 63 per cent of all tonnage capacity on the Mississippi and its tributaries. Theirs was supplemented by other barge lines carrying traffic other than coal. On the upper Mississippi, towboats and craft tended to be owned by the lumber and sawmill companies, while lower down three barge lines formed the Mississippi Valley Transportation Co, to concentrate the grain trade.

The lumber trade was (though with foresighted reafforestation need not have been) temporary, but in its time important. In the second half of the nineteenth century enormous numbers of logs were floated down the upper Mississippi from Wisconsin and Minnesota. The logs were then collected into rafts, to be taken by towboats to the milling towns. When push-towing lumber-rafts, lines from the stern corners of the raft to a separate engine on the towboat enabled the stern of the raft to be moved from side to side. The sawmills' products were then loaded on barges and pushed by towboats to ports down to St Louis: sawn logs, planks, shingles and laths. By 1892, 100 towboats were working in the timber business, and in 1900, 32 M ft of logs, 79 M of boards, 25 M shingles and 18½ M laths had passed through the Mississippi's Des Moines cut. Most of it ended in 1914, when the trees ran out.

Third of the great trades was grain. Railroads, shifting it from west to east, had mechanised their handling by using grain storage elevators, special cars for carrying it in bulk, and gravity-fed loading of cars from the elevators. After the war, however, the waterways counter-attacked. The Transportation Co (The Barge Line), based in St Louis, organised itself to carry grain to New Orleans, using towboats, grain barges, and elevators for loading and unloading. By 1881 the company had 13 towboats and 98 barges; by 1885 some of the latter held 1600 tons, and 10,000-ton tows were being handled. In addition, the Barge Line carried merchandise, both downstream and up. Another grain-towing trade was based at St Paul on the upper Mississippi, a third upon the Illinois river, steamers towing canal boats to the Illinois & Michigan Canal and Chicago.

The increasing size of tows was opposed by the steamboats, who found them a danger; towboat owners in turn attacked railroads for building obstructive river bridges, and some of them saw proposals to build locks on the Ohio as likely to delay the tows.

Towboats and barges, penalised by spells of low water and frost, battling against the railroads, hardly held their own in a time of great industrial growth: in 1870 there were 325 towboats on the Western rivers, in 1900, 304. As for steamboats, they increased in numbers but halved in tonnage between the same dates before collapsing in the next decade. Indeed, barge transport played a comparatively small part in the economic life of the post-Civil War years of the river valleys; only in the bulk transport of coal and some raw materials, mainly lumber, was it important. As Louis Hunter in *Steamboats on the Western Rivers* points out:

> Its development was hampered by the narrow limits of the commercial as well as the physical framework within which it had to operate. Already the great movement of western farm products had been largely diverted to the new West-East channels, and the rapidly developing industrial economy was oriented with reference to the railway network rather than the river system.

Water transport was in 1890 far less important to its area than it had been forty years earlier.

Changes in Canada

The Montreal-Ottawa-Kingston route by the Ottawa river and Rideau Canal was transferred from military to Government control in 1857. Nevertheless, defence significance remained for much of the century; certainly during the *Trent* affair in the American Civil War, and in 1865 when a first study was made of the possible extension of the Ottawa river navigation by way of Lake Nipissing to Lake Huron, later to be called the Georgian Bay Ship Canal project. Local replaced through traffic from 1851, when the major St Lawrence improvements had been done. Comparatively small freight tonnages were, however, supplemented by a considerable steamboat traffic between Montreal and Ottawa, though from 1854 passengers avoided lock delays by using a portage railway from below Carillon to above the Grenville Canal. It survived until 1910.

In 1870 a Canal Commission was appointed to review the whole Canadian canal system, and in 1871 recommended the enlargement of the Ottawa River canals to the later St Lawrence size: 200ft × 45ft

Fig 109 The *James Swift*, built in 1893, was renamed the *Rideau King* in 1900 after refitting.

locks and 9 ft depth. Action followed. The Grenville Canal locks were rebuilt by 1875 as recommended with five locks instead of the original seven, thus eliminating the Grenville bottleneck. At Carillon a new canal was built with two locks and a dam (it incorporated a slide to carry cribs of timber) which backed the river up sufficiently to do away with the need for a lock at Chute à Blondeau. These works, finished by 1882, were followed in 1886 by a new lock at Ste Anne. Railway competition had, however, now arrived, though the river line held the downstream timber trade and good passenger-steamboat business.

Mainly because of political pressure rather than as part of a serious plan, some work was done above Ottawa. At Chats Falls a 3 m ship canal with six locks was begun through hard rock in 1854. Cutting stopped in 1856 after considerable expenditure. Higher up a staircase pair of 200 ft × 45 ft locks were built in the Culbute channel not far from Pembroke, and finished in 1876. In 1879, however, the Union Forwarding & Railway Co, main users of the upper river, gave up passenger services in the face of railway competition, so that in 1885 only 32 lock transits were made. They were abandoned in 1889.

St Lawrence to the Soo

Three years after the dominion of Canada was born in 1867, a royal commission was appointed to study waterways. It worked fast to report in 1871, recommending enlargement of the Welland and St

Lawrence canals to a uniform 12ft depth with locks lengthened to 270ft, and a Canadian canal at the Soo.

Work began on the St Lawrence, though with depth increased to 14ft, to include a new Lachine Canal, a 14m Soulanges Canal with five locks to replace the Beauharnois, and new Cornwall and Williamsburg canals. The programme was not, however, finished until 1904, because successive Canadian governments thought it likely that railways would render river expenditure useless, a view that slowly changed as the value of waterway competition in keeping down rail rates was realised, change being encouraged by the freeing of the Erie Canal from tolls in 1882. Thus the work mostly fell into the 1891–1903 period. The result, even before completion of the whole route, was a rise in traffic by 'canallers', small ships of some 1000 tons that could navigate the 14ft canals before transferring their inward cargoes to the bigger lakers. With it came rising prosperity for the communities along the waterway. Some passenger-steamboats, however, deliberately shot the rapids instead of using the locks: so did the great downward log-rafts, controlled by oars or a tug.

Enlargement of the Welland Canal began in 1873, and was completed in 1881 to 12ft depth and in 1887 to 14ft, part of it, between Port Dalhousie and Allanburg, 13m long, having been built on a new route: There were now 25 locks, 270ft × 40ft 4in, their working being electrically powered and night use electrically lit. Though a towpath was provided, most traffic was now towed by tugs taking a sailing vessel or several barges—or was self-propelled.

In 1903 tolls were abolished on all Canadian canals. They were only to be reintroduced with the coming of the Seaway.

The American canal at the Soo having been opened in 1855 (see p. 322), the great iron-ore deposits of Lake Superior could be reached. Five years later, 128,982 tons passed the locks, to supply about a tenth of US pig-iron production. In 1870 the state of Michigan began to build an additional lock, the Weitzel, 515ft × 60ft × 17ft, its gates hydraulically powered, and carry out corresponding channel deepening. It was opened in 1881*, the year the canal was transferred to the Corps of Engineers and made toll-free. Tonnage rose phenomenally, coal and lumber upwards, iron ore, copper and grain downwards. In 1870, 690,826 tons, it was 3.7M in 1886 (in 6203 craft, 3880 of them steamers) and 13.2M in 1894. Meanwhile, between Lakes Erie and Huron, Dredged channels were

*The Weitzel must be one of the earliest examples in which water was admitted to the chamber through openings in the lock floor, so avoiding turbulence. Three years later it and the canal were lit by electricity.

needed where the St Clair river joined Lake St Clair, and at Lime-Kiln Crossing on the Detroit river. Here the ruling depth, 12 ft in 1866, had become 16 ft by 1882 and 18 ft by 1891.

The completion in 1862 of the enlarged 7 ft Erie Canal had been followed in 1871 by the beginning of work on the St Lawrence canals to 14 ft depth, and in 1873 also on a new Welland Canal. Transhipment of Lakes cargoes was therefore necessary at Buffalo for Erie traffic in vessels requiring over 7 ft, and at Port Colborne for those needing over 12 ft (1881) or 14 ft (1887). The opening of the Weitzel lock in 1881 at 17 ft, therefore, and corresponding deepenings of the Soo and Detroit channels, turned Lakes Superior, Huron, Michigan and Erie into a self-contained shipping area for craft requiring over 14 ft. Within it, harbour and channel depths and ship size and design tended to develop roughly in parallel.

The Americans showed they had taken the point when in 1886 the Corps of Engineers began their Poe lock at the Soo, 704 ft × 100 ft (later 95 ft) × 21 ft, on the site of the original staircase pair, destroyed in 1888, and completed it in 1896. Its start was followed by the Rivers & Harbors Act of 1890 to authorise the cutting of a ship channel uniformly 21 ft deep and 300 ft wide through the connecting waters of the four lakes. Completed in 1897, it was able to take 5000-ton craft moving between Duluth, Chicago and Buffalo. The Canadians also took the point. In 1887 they began another Soo lock 900 ft × 60 ft × 18¼ ft on their own territory (it had a military as well as a commercial motive), and completed it in 1895. There were now three locks at the Soo, one Canadian and two American*.

By 1907 the Great Lakes from Erie to Superior were carrying an enormous traffic during their open season of roughly nine months, the larger part of it in coal westbound and iron ore, grain and flour eastbound. In all, 70 M tons of freight and 25,000 ships passed Lime-Kiln Crossing, of which 51,751,080 tons in 22,155 craft transitted the Soo locks—80 per cent of it eastbound. There were at this time 2572 Lakes craft, of which 1844 were steamers (mainly bulk carriers up to 13,000 tons), 519 sailing vessels (many no longer sailed, but towed), 209 barges, the latter averaging some 350 tons and being usually steamer-towed, and also 480 canal boats of some 100 tons, presumably mainly off the Erie. Iron-ore-ship ownership was largely in the hands of American steel companies, headed by the United States Steel Corporation; the carriage of other commodities mostly rail-controlled, except for that through the Erie Canal.

*A fourth (American) still larger lock, the Davis, 1350ft × 80ft × 23ft, was completed in 1914 and a fifth (also American), the Sabin, the same size as the Davis, in 1919.

Fig 110 The four American locks at Sault Ste Marie (the Soo). From right to left, MacArthur, Poe, Davis, Sabin. The fifth, (Canadian) lock lies out of the picture to the left.

Between 1895 (before the new locks at the Soo were fully operational) and 1910, great changes had taken place in the business done by Lakes ports above the Welland Canal. In 1895 Chicago-Calumet was handling the greatest amount, at some 10M tons, followed by Buffalo with 8½M, Duluth-Superior with 6M and Cleveland with 5M. But by 1910 Duluth-Superior had far outstripped the others, with nearly 37M, Buffalo running second at nearly 15M, Cleveland having come up to 13M, while Chicago-Calumet* had remained nearly stationary at 11½M. Two other notable ports in 1910 were Milwaukee (8M) and Toledo (6½M). With these increases in tonnages handled went technological improvements at ports, especially for handling bulk cargoes; Jerome K. Laurent gives us an example:

> Grain was gravity-fed from an elevator, utilising one or more long spouts direct into the vessel's hold; unloading utilised a leg with an endless line of metal buckets lowered into the hold. The moving buckets carried the grain upward into the elevator, and steam-operated shovels were lowered

*I assume that these Chicago-Calumet harbour and port figures exclude those of Indiana Harbor and Gary, which are integral to metropolitan Chicago, and became of increasing importance for the reception of Lake Superior iron ore after the US Steel Corporation had been formed in 1901.

into the hold to move grain to the leg as needed. This efficient system enabled one elevator in Buffalo to unload more than 300,000 bushels of grain from vessels during one day in 1891.(7)

Watershed

In Chapter 11 we saw that in Canada after 1841 financial control had moved from company to province to nation, but that in the United States east of Pittsburgh much had been done by private subscription, more by majority or total state participation, but with Congress help only on the Chesapeake & Ohio. Further west, however, federal help by way of land grants had been vital to such canals as the Illinois & Michigan, as it was by way of the work of the Corps of Engineers to the Mississippi river system from 1824 onwards.

By 1850 the importance of federal action in support of national transportation development had been reaffirmed, but with the emphasis on railroads, for whose construction, especially in the west and south, 130M acres were given. Beneath this federal encouragement, supported also by local authorities and states, the drive of private investors was towards railroads, with state governments, except perhaps for New York, moving to divest themselves of what had become canal liabilities. Then slowly, as it was realised that where waterway competition still existed, railroad rates were lower, a contrary movement began.

A precursor of change was the 2000-delegate waterway convention that met at Chicago in 1863 at a time when there was interest in rebuilding the Illinois & Michigan Canal, enlarging the Atlantic coast route, and improving the Ohio. Out of the movement then initiated was to come federal purchase of the main waterways and their subsequent enlargement. It occurred at a time when the formerly great canals were approaching their peak tonnages ahead of decline: 6,673,000 for the Erie and 1,318,800 for the old Chesapeake & Delaware, both in 1872, and the second highest for the Delaware & Raritan, at 2,838,000. Ahead lay fierce railroad competition or in some cases control, obsolescent equipment and falling revenues. But a time, too, when the future for really big waterways was brightening year by year, in the United States and Canada, in Latin America, and in Europe.

The change in American public sentiment was marked by the 1874 report of the Senate's Select Committee on Transportation Routes to the Seaboard (the Windom Committee), which recommended river and canal improvements as a way to check railroad rates. In 1878 Congress authorised a 4½ft channel on the upper Mississippi, which would involve by-passing of rapids by cuts and locks, and in 1879 set up the Mississippi River Commission, to make surveys, plan improvements, build levees, prevent floods, improve navigation and assist commerce (it still exists, but since 1928 as adviser, not executant). By the 1880s, it was being widely realised that not only did waterway competition restrain railroad rates, but that modern, large-sized waterways were in themselves cheaper carriers. Moreover, they were part of modern progress, for the decade saw work begin on four great ship canals around the world, the (French) Panama, Kiel, Corinth and Manchester.

Two American laws of the 1880s were especially important. The Rivers & Harbors Act of 1882 substituted a general prohibition on tolls on federal waterways for previous piecemeal enactments:

> That no tolls or operating charges whatsoever shall be levied or collected upon any vessels, boats, dredges, craft or other water craft passing through any canal or other work for the improvement of navigation, belonging to the United States.

The 1887 Interstate Commerce Act tackled the problem at the other end, for the first time establishing the principle of federal regulation of railroads, and including among its objectives the prevention of rate-cutting aimed at eliminating competition.

Meanwhile the federally financed canalisation of the Kanawha had begun about 1875 (it was completed with 10 locks and dams by 1901), and in 1885 a first experimental Ohio lock was completed at Davis Island*, 5m below Pittsburgh, under the direction of Col William E Merrill of the Corps of Engineers. It was remarkable indeed, as being big enough at 600ft × 110ft to take coal-boat tows, having sideways-rolling gates running on ground rails, and being accompanied by a Chanoine movable dam (see p. 86), larger than those being built on the Kanawha, which would enable tows to pass without using the lock at times of high water. The lock width, at the time probably the greatest on any inland navigation, set the dimension of all future river canalisations. Thus a 6ft-deep stretch was provided where coal boats could wait for 'a coal-boat rise' to enable them to move downstream. These coal boats came mostly from the Monongahela river, whose navigation company was in 1880 paying 12 per cent and carrying over 3M tons of coal. Lock- and dam-building on that river to give 6ft depth was completed in 1890, and in 1897 its company was taken over by the federal government through the Corps of Engineers.

The 1880s and 1890s saw *The Waterways Journal* founded in 1888,

*It was replaced in 1922 by Emsworth lock.

and work begun in 1892 on the Chicago Drainage Channel and the federally financed Hennepin Canal. The Ohio Valley Improvement Association was founded in 1895, the year also in which the United States and Canada jointly agreed to set up a Deep Waterways Commission to study improved routes between the Lakes and the Atlantic. In 1896 Congress authorised a 9ft-deep, 250ft-wide navigation channel on the Mississippi downwards from Cairo, in 1899 Governor Theodore Roosevelt of New York set up a committee to produce a state policy for canals, to be the precursor of the New York State Barge Canal, and in 1902 the Corps set up the Board of Engineers for Rivers & Harbors, to review all projects independently of local 'pork-barrel' politics.

William R. Willoughby writes of the 1900s:

> With the possible exception of the twenty-five years following the close of the War of 1812, at no time in the history of the American people has interest in rivers and canals been quite so intense as during the first ten or fifteen years of the present century. They wrote about waterways, talked about them, created dozens of associations to promote their development, and persuaded their congressmen to vote millions of dollars for their improvement.(8)

These associations quickly became basic to waterway development, for local support was such that they could bring considerable political pressure to bear. Among them were the Atlantic Deeper Waterways Association, the Lakes-to-the-Gulf Deep Waterway Association, and the Gulf Intracoastal Canal Association.

Generally promoting waterways was the National Rivers and Harbors Congress, formed in 1901 at Baltimore to bring all concerned individuals, corporations and pressure groups together in one organisation. This body pressed for annual appropriations for waterway development (they had been triennial from 1896 to 1905, though Congress enacted another in 1907) and demanded a national commission to study waterways at home and abroad and advise Congress upon them, and a federal department of transportation to develop waterways, highways and railroads. At that time there was no Department of Transportation: waterway improvements fell under the War Department, the regulation of common carriers to the Interstate Commerce Commission.

This energy of the 1900s was backgrounded by three events: the United States' purchase of French rights in the Panama Canal and a new beginning of construction; the start of work on the New York State Barge Canal; and the opening of the Chicago Ship Channel.

14
WATERWAYS DEVELOP: PANAMA TO THE LAKES

The Isthmus

The quest to find, build and then control transit routes across the Central American isthmus is as old as European discovery. In the early 1800s individuals, Englishmen among them, and governments got involved in schemes from Darien to Mexico, which culminated in the formation of an American ship canal company proposing a Nicaraguan route. Because of its likely international importance, the greatest Old and New World powers of the time, Britain and the United States, by the 1850 Clayton-Bulwer treaty agreed to combine to maintain free and uninterrupted passage across the isthmus. Both countries were foreshadowing what later came about at Suez, a canal open to the ships of all nations on equal terms, and not controlled by one alone.

Until 1869 access from the eastern United States to California was by one of several waggon trails, or around Cape Horn, or across the Central American isthmus. The Accessory Transit Co was formed in 1851 to operate what was for a time a successful river, lake and road route across Nicaragua, until the Panama Railroad was opened in 1855. The idea then entered men's minds that railways carrying ships might be used to cross isthmuses. It was an idea especially associated with four engineers, the English James Brunlees, the Americans W. F. Channing and J. B. Eads and the Canadian H. G. C. Ketchum. Brunlees, indeed, in 1859 suggested one at Suez to Napoleon III.

Though Honduras offered a longer route and a higher summit than Panama, the sailing distance to American ports at either side was much less. In 1864 the British Honduras Co* was formed in association with Channing to build a ship-railroad, but could not raise money. Then in 1872 Brunlees proposed one to take craft of at least 1200dwt, by using six rails altogether carrying 60 4-wheeled bogies.

*Its seal showed a ship-railway carrying a sailing ship.

Propulsion would be by steam rack-locomotives, and ships would be got out of the water and returned to it by hydraulic lifts. His estimated cost for 235 m of ship-railway rising to almost 3000 ft was some £30 M. No action followed.

J. B. Eads, engineer of the St Louis bridge and Mississippi channel improvements, after having tried to interest de Lesseps in a ship-railway at Panama, was in 1881 granted a concession to build a line across the isthmus of Tehuantepec in Mexico. He planned a railway for 3000-ton ships, to be carried in water-filled caissons hauled by steam-locomotives. However, difficulties that had been experienced in the 1870s in providing a stable roadbed for the 400-ton caisson of the Chesapeake & Ohio Canal's Georgetown incline (see p. 325) made investors nervous, and no action resulted.

Brunlees had proposed to use hydraulic lifts in Honduras. It was, indeed, this invention by the British Edwin Clark in the 1860s that made practicable the only scheme actually begun. Clark had used such a lift at a graving dock on the Thames and later for the Anderton canal lift in England, opened in 1875. H. G. C. Ketchum, who had worked with Brunlees, thereupon proposed a ship-railway across the 17 m of the Chignecto peninsula that separated Canada's Bay of Fundy from the Baie Verte leading to the Gulf of St Lawrence and New England ports, so saving some 500 m of sea passage. Such a railway presented special problems. Because of the Bay of Fundy's great tidal range, to haul ships out of the water on an incline, as in the Suez and Tehuantepec schemes, would have been impracticable. Clark's invention provided the answer, as it did for Brunlees' later Honduras scheme; they could be lifted vertically by hydraulic power to ship-railway level. Eventually the Chignecto scheme was accepted by the Canadian government, and a 20-year subsidy promised once the line was completed.

In 1883 the Chignecto Marine Transport Railway Co was formed in England, Brunlees and Edwin Clark being associated with it, and construction started in 1888. The track was built straight and almost level with four heavy toughened steel rails* upon which ran an articulated cradle with 192 wheels able to take a ship and cargo of 2000 dwt, up to 235 ft long, 56 ft wide and of 14 ft draught. At the Bay of Fundy end, the ship would enter a tidal and then an enclosed lifting dock, and be positioned over a gridiron upon which rested the carrying cradle. The gridiron would then be raised by 20 hydraulic rams and presses to trackle-level, after which two locomotives would haul the cradle off the

*Each pair of 4 ft 8½ in gauge, set wide apart.

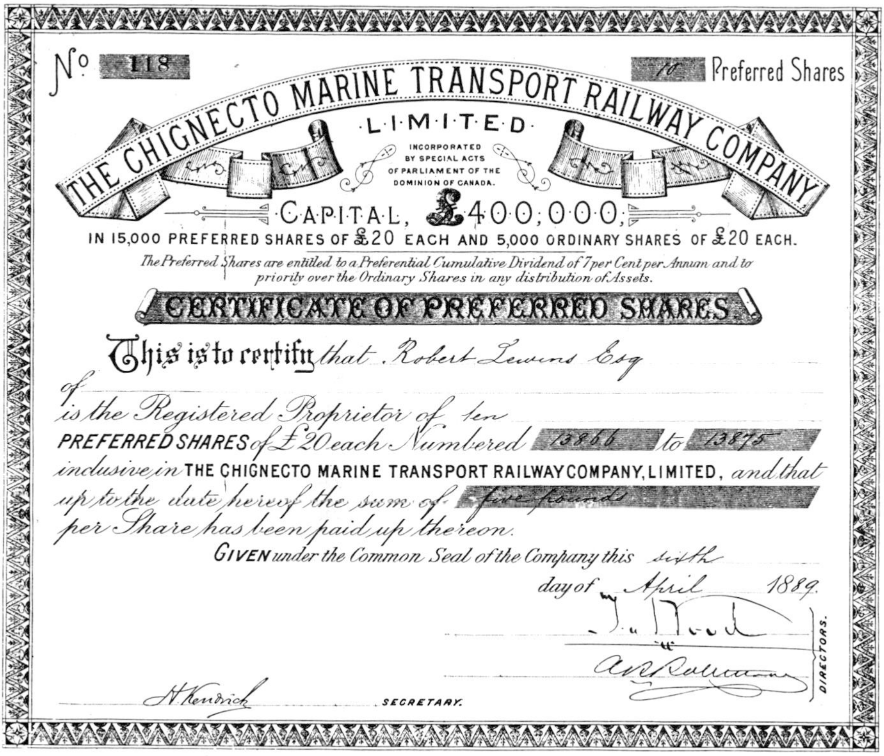

No 118 — 10 Preferred Shares

THE CHIGNECTO MARINE TRANSPORT RAILWAY COMPANY

LIMITED

INCORPORATED BY SPECIAL ACTS OF PARLIAMENT OF THE DOMINION OF CANADA.

CAPITAL, £400,000;

IN 15,000 PREFERRED SHARES OF £20 EACH AND 5,000 ORDINARY SHARES OF £20 EACH.

The Preferred Shares are entitled to a Preferential Cumulative Dividend of 7 per Cent per Annum and to priority over the Ordinary Shares in any distribution of Assets.

CERTIFICATE OF PREFERRED SHARES

This is to certify that Robert Lewins Esq
of
is the Registered Proprietor of ten
PREFERRED SHARES of £20 each Numbered 13866 to 13875
inclusive in THE CHIGNECTO MARINE TRANSPORT RAILWAY COMPANY, LIMITED, and that up to the date hereof the sum of five pounds per Share has been paid up thereon.

GIVEN under the Common Seal of the Company this sixth day of April 1889.

DIRECTORS.

H Kendrick SECRETARY.

Fig 111 A preferred share certificate of the Chignecto Marine Transport Railway Co, issued in April 1889. The seal bears the company's name and the words 'Incorporated by Special Act of Parliament of the Dominion of Canada'. An endorsement on the back, also signed by Kendrick as secretary, notes the hopeful transfer of the fully-paid up shares on 4 October 1895.

gridiron and along the track at 5–10 mph to the other end, where another gridiron would lower the ship to sea-level. By July 1891 over half the work had been done. Then money ran out at a bad time, more could not be raised, the Government of Canada—perhaps persuaded by those interested in threatened ports—refused an extension of time, and construction stopped after some £700,000 had been spent, with half as much again still needed. The works decayed, and little can now be seen of the world's only large-scale ship-railway.

In 1875 the 70-year-old Ferdinand de Lesseps, whose Suez Canal had been opened in 1869, first declared interest in a Panama Canal. In 1879 the international congress he had summoned approved the idea, in 1880 he visited Panama, and on 3 March 1881 the Compagnie Universelle du Canal Inter-océanique de Panama was incorporated to build a sea-level cut through the mountainous, rainy, fever-ridden isthmus, operating under a concession from Colombia. By December 1888 it was bankrupt, having raised mainly from Frenchmen three and a half times the cost of Suez. Between 1880 and 1888, the French

THE PANAMA CANAL.

THE UNIVERSAL INTER-OCEANIC CANAL COMPANY
(COMPAGNIE UNIVERSELLE DU CANAL INTER-OCÉANIQUE).

PARTICULARS AND STATISTICS OF THE UNDERTAKING.

A.—THE CANAL CONCESSION.

On the 20th March, 1878, an agreement or concession for the formation and working of an inter-oceanic canal through the territory of the United States of Colombia in Central America, was entered into between Senor Salgar, on behalf of the Colombian Government, of the one part, and Mr. Wyse, on behalf of the International Company then intended to be organized of the other part. This concession was ratified by an Act of the Colombian Congress, on the 17th May, 1878, containing (*inter alia*) provisions, of which the following is an abstract :—

> Art. 1, Sec. 1. The privilege to continue for 99 years from the date of opening. Sec. 2. The Colombian Government cannot grant or undertake the construction of any other canal. Sec. 5. The canal shall be completed in twelve years, or, in case of *vis major*, within eighteen years from the formation of the company. Sec. 8. A strip of land 200 metres (about 220 yards) wide on each side, is granted for the use of the canal. Art. 4. 500,000 hectares (about 1,250,000 acres) of land, with the mines they may contain, shall be granted to the company, in such places as the company may choose. Art. 14, Sec. 3. The navigation duty is not to exceed 10 francs for each cubic metre.* Art. 15. The Government shall collect 5, 6, 7 and 8 per cent. respectively out of the gross receipts during each period of twenty-five years. Art. 18. No tax shall be imposed by the State upon the agreement, or upon the shares, bonds, or other documents of the Company.

M. Ferdinand de Lesseps assigns to the present Company the above mentioned Concession, together with the benefit of all agreements relating thereto (including the purchase of the Panama Railway, subject to an annuity of 3,800,000 francs net), in consideration of 10,000,000 francs (5,000,000 of which are payable in 10,000 Canal shares).

B.—THE CONSTRUCTION OF THE CANAL.

The length of the Canal will be 73 kilometres (about 44 miles), being thus less than half the length of the Suez Canal (166 kilometres or about 100 miles.)

The Canal will take about six years to complete, or, according to the estimate of M. Couvreux, the contractor, five years and a few months. The estimated number of labourers will be 9,000, but this number can with facility be increased to 12,000 if required, in order to expedite the opening.

Messrs. Couvreux and Hersent, who constructed the Suez Canal, have contracted to carry out the works for 512 million francs, with a proviso that the net resultant profit shall be divided between them and the Company. This profit, it is anticipated, will be very considerable : the recent report of M. Sosa, the Company's engineer at Panama, shows a saving of 22,738,000 francs on a portion of the construction, as compared with the estimate on which the contract was based. It is confidently asserted by competent authorities that 450,000,000 francs will be the extreme limit of expenditure on the works of construction.

C.—THE CAPITAL OF THE COMPANY.

The total cost of the Canal is estimated at between 550 and 600 million francs, viz. :—

Construction	500	million francs
Purchase of Concession	10	,, ,,
Incidental Expenses	15	,, ,,
Interest on Capital during Construction	75	,, ,,
Total	600	,, ,,

From this has to be deducted the share of profit on the Couvreux contract (without taking into account the before mentioned saving of 22,738,000 francs), estimated at from 25 to 50 millions, thus leaving a net estimated cost of from 550 to 575 million francs.

* This would make 40 francs per ton but M. de Lesseps has reduced the tariff to 15 francs per ton.

Fig 112 The first page of the English version of the original 1880 prospectus for the French Panama Canal.

had done an immense job, in excavation, the building of docks, offices, living quarters and hospitals, the assembly of technical data, the purchase and running of the Panama Railroad, and land acquisition.

Their effort has been superbly described by David McCullough in *The Path Between the Seas, 1870-1914*. It is a story of 'ifs'. If de Lesseps had not been convinced that nineteenth-century inventiveness and technology, faced with any problem, could solve it; if therefore he had not been sure that a sea-level canal, as at Suez, was practicable; if from the start he had faced the problem of how to control the floods of the Chagres river, and so had paid more attention at the 1879 congress to the experienced engineer Baron Godin de Lepinay*; if his had not been a personality so inspiring that it compelled men to do what their better judgement told them was imprudent—then de Lesseps might indeed have built and opened a canal in his lifetime, in spite of the appalling geological, geographical and medical problems.

But by the time he had admitted that locks were necessary, and commissioned Gustave Eiffel to build them, it was too late. Financial confidence collapsed under the company's weight of debt and endless need for money, and the French nation turned against Ferdinand and his son Charles with as much energy as previously it had supported them. It lost its faith along with its money. De Lesseps declined to senility, and died in 1894, aged 89.

In 1898 came the Spanish-American war, during which the American battleship *Oregon* had to voyage from San Francisco to Cuba round Cape Horn. An American sea-to-sea canal was needed, such as Germany had completed in 1895 between the Baltic and North Seas. But prominent Americans like Theodore Roosevelt, then governor of New York, would have nothing to do with the idea of an undefended, international waterway as envisaged in the Clayton-Bulwer treaty. Such a canal, they thought, would enable foreign warships to attack the States as easily as it would allow American warships to pass from coast to coast. The United States must control it. Britain, unable to think that she and the States would ever again be at war, agreed, and a new treaty was signed. By then Roosevelt was President.

Nicaragua was the favoured route. The choice of Panama was due mainly to the splendidly devious and devoted Philippe Bunau-Varilla. A boyhood hero-worshipper of de Lesseps, he had joined the French company from Corps des Ponts et Chaussées in 1884, to become, at 26, its chief engineer and, later, contractor for the Culebra Cut. Out of the ruins of the bankrupt concern a new Panama Canal Co was created in 1894, whose object became to sell its rights in the unfinished canal. An Isthmian Canal Commission, appointed by President McKinley, recommended Nicaragua, and valued the French company's rights at a maximum of $40 M. The canal company then set out to influence public opinion against Nicaragua; so did Bunau-Varilla, helped by the fortunate eruption of a Nicaraguan volcano. In 1902, therefore, Congress authorised negotiations with Colombia, of which Panama was then a part, for a concession.

*He proposed flights of locks leading to a summit-level formed of two artificial lakes that would absorb the Chagres water, roughly the way the American canal was later to be built.

Fig 113 The Culebra cut as seen probably in early 1888, and published in June. The *Illustrated London News* artist wrote: 'Here there are over thirty excavating machines, sixty locomotives, two thousand railroads cars and four thousand workmen employed. Fifteen thousand tons of earth and other matter are actually excavated, loaded in cars, and transported far away in every twenty-four hours'.

Colombia played her hand badly, to the extent that Bunau-Varilla helped to organise an almost bloodless revolution in 1903 that created an independent Panamanian state. Whereupon he, now its ambassador to Washington, signed a treaty granting a canal zone which the United States could defend. The French company got its $40 M and the States began to build a new lock canal, with Col Goethals in sole charge from January 1908.

The prospect of ships soon being able to pass through the canal to and from the west coasts of North and South America gave some of those associated with the Mississippi hope of a far greater transhipment trade through New Orleans. The president of the Missouri River Valley Improvement Association summed it up in 1908:

> There is a close relation between the improvement of our rivers and the building of the Panama Canal. If the United States is to realize what it should from this great undertaking, it is absolutely necessary to improve the waterways of this country. If we do not do so, we are practically building that great canal for the use of foreign nations... (1)

Presciently, Roosevelt had early realised the danger of a Big Navy Germany. It was appropriate, therefore, that the American Panama Canal opened on 15 August 1914, a few days after WWI had begun. Philippe Bunau-Varilla, in himself a link with de Lesseps, was there. The canal was to be a success, and especially an American success. In 1936–7, of 28,108,375 tons carried through it, 22.5 per cent was between two US ports, and another 38.5 per cent to or from a US port.

The New York State Barge Canal

In 1884, when the French were working at Panama the American Society of Civil Engineers discussed whether the now-declining Erie Canal should be replaced by a small ship canal, maybe 18ft deep, with locks 450ft × 60ft, from Lake Erie to the Hudson. Nothing resulted, but an idea had been born. In 1896 a Rivers and Harbors Act ordered a study by the Army Corps of Engineers which, seeking partly to head off a further improvement of the Welland-St Lawrence route, in 1897 recommended a 1500-ton barge canal.

Then in 1899 Governor Theodore Roosevelt of New York appointed a commission to formulate a state waterway policy. Its recommendation of 1900 was to replace the Erie Canal with a new one, toll-free, to take 1000-ton barges, at the western end to follow much the same route, but at the eastern a different line by way of Lake Oneida and the Mohawk river. These ideas were subsequently expanded to provide locks 328ft × 45ft × 13ft, to accommodate two 1500-ton or one 2500-ton barge. The project was approved by referendum in 1903; by then Roosevelt had become President. Contemporary opinion saw its purpose as mainly to help safeguard the trade of New York, assist the iron, steel and general industrial development of western New York state, and hold down railroad rates.

The second Erie Canal was finally closed in 1917, and the Barge Canal, with three reconstructed branches, opened in the following year. The main line was now 348m long, and the Champlain, Oswego and Cayuga & Seneca branches 68m, 24m and 92m respectively, the last having two termini, Ithaca on Lake Cayuga and Watkins on Seneca lake. There were 35 electrically operated locks on the main line, 57 on the whole system. In 1924, not long after opening, Albany on the Hudson, hoping for transhipment as well as new trade, announced its intention of dredging the Hudson to 27ft to allow access by oceangoing ships.

Roosevelt in 1907

The important part Theodore Roosevelt played in New York state and in launching the Panama Canal, give him a key place in the movement to improve American waterways. In 1907 he appointed an Inland Waterway Commission, which the following year recommended that waterways should be system-artically improved and a National Waterways Commission be established. Set up in 1909, the Commission produced a report in 1912 with proposals for encouraging waterborne trade.

On 4 October 1907 the flamboyant president, having travelled by steamboat from Keokuk, Iowa, addressed a Deep Waterway Convention at Memphis, saying that 12,000m of river were 'now more or less fully navigable' in the Mississippi valley. Yet, he added, much of the system was 'practically unused for commerce'. 'We cannot afford', he said, 'any longer to neglect the great highways which nature has provided for us... Year by year transportation problems become more acute, and the time has come when the rivers really fit to serve as arteries of trade should be provided with channels deep enough and wide enough to make the investment of the necessary money profitable to the public. The National Government should undertake this work.'(2) He blamed the situation on the granting of haphazard appropriations* instead of on a define and continuous plan, stated

*In England, vote provisions for expenditure.

that the rivers should begin to supplement the railroads', and called for the linking of navigation, irrigation, flood-prevention and hydroelectric power development. In October 1909, President Taft followed up Roosevelt's work by travelling in a steamboat convoy from St Louis to New Orleans with many who were attending the Lakes-to-the-Gulf Deep Waterway Convention there.

Behind this activity was a general realisation that unregulated railroad competition had greatly damaged water-borne trade, that viable waterways held down railroad rates, and that the railroads were congesting themselves with long-haul, low-toll bulk freights to the detriment of more profitable and suitable traffics. Therefore waterways, by taking those freights, would be complementary rather than competitive. A realisation also that large waterway transportation, whether on the Lakes or on rivers, was cheaper, a desire by mid-western riverside towns to return to their earlier prosperity, and a feeling that there was something to be learned from others. Here is Senator Francis G. Newlands in 1907

> Germany has perhaps the most perfect system of transportation in the world. Her rivers have been artificialized from source to mouth and they are supplemented by a system of canal, rail, and ocean transportation which, combined, give that country a transportation machinery unequaled anywhere in the development of domestic and foreign commerce. (3)

It is against the background of events taking place at Panama between 1902 and 1914 that we should consider what followed as well as what went before.

The Chicago Ship Channel

Even before the Illinois & Michigan Canal had opened, it was realised that the Illinois river needed to be deepened and preferably locked to give an efficient communication between the canal's terminus at La Salle and the Mississippi. In 1866 Congress authorised a survey, which recommended that the canal be widened to 160 ft and deepened to 7 ft from Chicago to Lockport, whence a similar-sized channel should be built along the Des Plaines and Illinois rivers. Locks were to be 350 ft × 75 ft. Though the federal government did not act, the state of Illinois did, in 1867 authorising two Illinois river locks of the recommended size, and a 7 ft channel, works finished in 1877. Two further locks were later built by the federal government to complete the river improvement in 1893.

In 1890 the federal government also began to build the 77 m

Hennepin (Illinois & Mississippi) Canal, a steamboat canal-slackwater route from the Illinois River below La Salle to Rock Island on the upper Mississippi that could take 600-ton craft. Completed in 1907, and notable as the first in America in which concrete instead of cut stone was used for locks and dams, and canal proved too small for the twentieth century. Closed in 1951, it was handed over for recreational uses.

Given the dimensions of the Illinois river, enlargement of the Illinois & Michigan Canal was urgent to do away with transhipment at La Salle. Back in 1871 Chicago had eliminated two locks and lowered the canal's summit to lake level, so that city sewage would be carried, not into Lake Michigan, but through the canal to the Des Plaines river. The rapidly growing city quickly over-polluted the old canal. In 1892 therefore, work began on a new Chicago Ship Channel, 28m long. This commenced in the Chicago river, the flow of which was reversed, and then left it for a new cut, 24ft deep, the first part through heavy rock-cutting, which was completed to Joliet in 1901, the old Illinois & Michigan Canal then becoming disused for traffic. A 600ft × 110ft lock with a 41ft lift was built where the canal joined the Des Plaines river at Lockport, to control the volume of water taken from Lake Michigan.

The Channel and the Chicago river were the beginnings of an enlarged waterway to the Mississippi; argument then began upon the dimensions of its extension. In 1906 a Lakes-to-the-Gulf Deep Waterways Association was founded to promote a 24ft (20ft navigable depth) ship channel from Chicago to New Orleans, but finally in 1915 the Illinois Waterway Commission was established to build a waterway onwards from Lockport to Utica, 91m from Chicago, in the hope that the federal authorities would continue it. In fact, it was largely built by them, with a 9ft channel, to Utica in 1930 and the mouth of the Illinois river in 1933.

The Great Rivers to 1914

After the Civil War steamboats and towboats on the great rivers flourished for a time, and then declined in losing rivalry with the expanding railroads. They struggled and almost died, except for a small and still-surviving excursion trade.

Freight-carrying tonnages on the Mississippi in the eighty years from the end of the Civil War to that of WWII in 1945 reached a peak in 1880, kept nearly level to 1890 (with some 6M tons hauled), then fell away till 1918. Railroads caused the decline, indirectly by developing east-west traffic flows where waterways could not compete, directly by building lines into grain- and cotton-growing areas away from water carriage. Then came a change. From 1918 tonnages began to grow and have steadily increased ever since, the immediate reason being a transport shortage due to WWI, an underlying cause probably the amount of infrastructure that had by then been provided. There had naturally been a long time-lag from the beginnings of waterway development to the time when, channels dredged and buoyed, craft and towboats built and terminals provided, results began to show in traffic figures. Let us return to those beginnings.

On the Ohio the tolls of the Louisville & Portland Canal were a handicap in facing transport competition. Lock enlargement and realignment began in 1860, but was stopped by the Civil War. Work began again in 1870 by the Corps of Engineers, already improving the Ohio itself, and the canal was reopened in 1872, with two locks now 390ft × 90ft. Tonnage tripled within five years, but traffics had changed. The agricultural trade to the east had gone mainly to railroads, and instead it passed bulk cargoes of coal, iron ore and salt. In 1874 the federal government took over the canal, and in 1880 made it toll-free. Then came Davis Island lock in 1885 (see p. 349).

In 1890, Congress authorised study of an extension downwards of the Ohio canalisation; this favoured more locks and dams and a 6ft

Fig 114 Steamboat days: a dramatic picture that tells its own story.

channel. An appropriation for a second lock was made in 1890, but it was not until the Ohio Valley Improvement Association (OVIA) was formed in 1895 that appropriations became more generous. In 1902 the Association began to press for a 9ft channel—it was agreed in 1905—and by 1908 six locks and dams were built or building, with four others being planned. In 1905–6 the Ohio carried $4\frac{1}{2}$M passengers and 15M tons of freight and indeed supplied most of the Mississippi's traffic between Cairo and New Orleans. Hopefully the OVIA considered that its efficient canalisation would make it 'the greatest of all feeders to the Panama Canal'. (4)

The effort after river revival brought work upon the design of high-capacity towboats able to cope with a still largely unimproved river. Outstanding was the *Sprague*. Designed by Captain Peter Sprague, this biggest steam sternwheeled towboat ever built become operational late in 1902, and in 1907 pushed 64 wooden coal boats* and barges carrying 67,307 tons, a tow covering six acres of river and the biggest handled till then. Sadly, she lost several when out of control near Memphis. The lesson was learned. Future tows did not exceed 50 pieces, and the *Sprague*, after performing more modestly until 1948, eventually became a tourist attraction. Burnt in 1974, I saw her later, sunken in the Yazoo river near Vicksburg.

In 1910 Congress authorised 54 locks on the Ohio to provide a 9ft channel from Pittsburgh for 981m to the Mississippi, such depth being thought essential if deep-laden barge-tows were to be operated. It was to be the precursor of today's widespread 9ft system. By 1929 modification had resulted in the work's completion with 50 dams and 600ft × 110ft locks. The event was celebrated when President Hoover took a river trip in autumn 1929, a year in which the Ohio carried 23M tons; by 1953 it was to be 62M.

Traffic had increased on the Ohio, but on the Missouri steamboating ceased completely. Railroad rates rose, and in 1890 local businessmen started a river transportation company. Though it failed against railroad rate cutting, in 1906 the Missouri River Valley Improvement Association was formed. In the next year the Kansas City Transportation & Steamship Co put on a boat service to St Louis, and in the same year federal funds financed snag boats and the placing of navigation lights. This time the boats stayed, their backers hoping for the building of a Lakes-to-Mississippi deep waterway.

By 1907 it was clear that the Ohio canalisation was going ahead; therefore tonnages would be moving on the Mississippi from Cairo

*In 1907 new coal boats carried 1000 tons or so.

Fig 115 Two steamboats are moored to one bank of the Monongahela river at Pittsburgh, a medley of coal boats to the other.

downwards much greater than the 2,388,000 carried thence to Memphis in 1905. No question of lock-building arose. The Mississippi River Commission was trying to maintain a channel at least 9ft deep and 250ft wide down to New Orleans and improvement was a matter of reducing the silt coming in from tributaries, building levees, dredging and maintaining a wandering channel. The same applied to the section of river upwards from Cairo to St Louis and the mouth of the Missouri, where also the Commission was maintaining a 9ft channel, at a time when tonnage had fallen from 1,260,000 tons in 1897 to 417,000 tons in 1905.

No wonder that Senator Newlands, descending from St Paul to Memphis in 1907,

> travelled for miles without seeing a single boat...there were a few tow-boats, but the river towns were neglected, the wharves rotting, and the river fronts largely occupied by the tracks of the railroads...(5)

Such was the effect of the current interest in big waterways, however, that the Rivers & Harbors Bill of 1907 gave enabling powers for investigation of a 14ft channel from St Louis to the Gulf.

One must remember that middle-west grain from St Louis had alternative routes to the Atlantic: by rail, lake and canal (or rail) to

New York, or down the Mississippi to New Orleans and onwards. Waterway improvement, therefore, looked not only at the downward run to New Orleans, becoming more important as the Panama Canal neared completion, but also upwards to Chicago and the Lakes by improvement through to the Chicago Ship Channel. Above St Louis, therefore, traffic was much greater, 4,089,000 tons in 1905, the government having spent a good deal on this section in order to restrain railroad rates.

The Great Rivers, 1914–45

When war came in 1914, little waterway revival had taken place. Resultant pressure on railroads, however, caused the federal government to set up a committee to assess possibilities for the upper and lower Mississippi, Ohio, Kentucky and Tennessee rivers. They decided that, given good terminals, much rail-carried freight could move by water, and recommended that private enterprise be encouraged to begin services, and that the federal government should itself take action. Consequently, in 1917 $3.9M was allocated to provide barges and towboats for use on the St Louis-St Paul stretch of the upper Mississippi.

After many vicissitudes, an Inland Waterways Corporation under the War Department was set up in 1924, a year in which no private common carrier barge company was operating on the Mississippi, as a wholly-owned government body, with two partly-contradictory objects: to run a business-like organisation that would show inland waterway transport to be economically feasible; and yet develop new routes, new methods and new equipment. When created, it took over operations from New Orleans to St Louis, and from New Orleans via the Inner Harbor Canal and Lake Pontchartrain to the Gulf Intracoastal (see p. 377) and Mobile and then up the Tombigbee, Warrior and Black Warrior rivers to Birmingham, Alabama, using 25 powered craft and 69 barges, some of them out-dated.

The Corporation, operating as Federal Barge Lines, did well enough to be given more capital in 1928, and also a policy. It was to operate until four conditions had been met: navigable channels existed adequate for reasonably dependable and regular transportation services; there were terminal facilities reasonably adequate for joint water and rail services; joint tariffs had been negotiated with rail carriers which would make joint water-rail transportation generally available; and private enterprise was ready and willing to operate common carrier services.

Among the Corporation's actions were the purchase in 1926 of a 23m railroad linking the Warrior river to Birmingham's railway system, so making through transport available between Birmingham and the Gulf ports. In that year also operations began on the Mississippi up to Minneapolis and St Paul, and in 1931 up the Illinois river and through to Chicago, in spite of the route's comparatively poor and in some places congested state.

The Corporation, up to 1939, was indeed remarkably successful. It made profits in 11 out of its 15 years, and finished up in the black, having reconciled satisfactorily its two conflicting objectives. It did this in spite of the cost of providing regular merchandise traffic services mainly to small shippers, a task few private operators would tackle, relying instead upon bulk freights*, and of avoiding direct competition with independent barge lines. More, it did much development work on towboats for use on different types of river around the 1920s when diesels with screw propulsion were beginning to take over from steam sternwheelers, and upon barge conformation at varying channel depths. By the late 1930s, new construction was mainly of compact, powerful diesel towboats. Michael C. Robinson(6) indeed says that 'During the 1920s and early 1930s, the IWC was virtually the sole source of technological innovations in this field'. The IWC also did much to encourage states, municipalities and private business to build waterway terminals, itself providing some of the necessary capital, and then leasing them. Twelve had been built with IWC participation by 1935.

During this time also it succeeded in negotiating many through routes and rates with the railroads, though as waterway traffic began to build up, so did rail opposition and obstructiveness. Private barge lines, too, while admitting the pioneering role of the IWC, now began to object to its continuance—notably when American Waterways Operators, Inc were founded in 1944, and took an anti-IWC stand. Nevertheless, the IWC had done a good job. In 1953 it was sold to a private company.

In terms of channel engineering on the lower Mississippi, the great flood of 1927 was a turning point. Until then navigation and flood-control interests had agreed that the lower river should be slowed down by preventing bends being cut off, and its banks then made secure by levees raised as high as necessary. Thereafter, thanks largely to the persistence of W. E. Elam, an experienced Mississippi engineer, the opposite view began to be adopted: that natural cut-offs

* Bulk freight accounted for 93 per cent of Mississippi carryings in 1939.

should be encouraged and artificial ones built in order to speed up the flow of the river, and that an additional outlet to the sea, the Atchafalaya river, which left the main stream to the west of and above Baton Rouge, should be utilised.

The result was a partial success: flood-levels were lowered, and navigation benefited by a deeper and shorter channel: indeed the Atchafalaya and Port Allen routes to the western Gulf Intracoastal (see p. 378) owe their origin to these measures. From the 1950s, however, policy swung back to stabilisation, using wing-dams to deepen existing channels and form banks, and revetments to maintain them. Important to the prevention of major floods are the numerous flood-control reservoirs scattered throughout the Ohio and Mississippi basins on small tributaries above the heads of navigation. These retain water during the rainy season and release it during the dry.

One year after he had opened the 9ft Ohio canalisation in 1929, President Hoover approved Congressional authorisation of a 9ft channel from St Louis to the twin cities, to enable the deep-draught towboats of the Ohio and lower Mississippi to work upwards to the Illinois river junction above that of the Missouri, along which a 9ft channel was at that time being made on its way to Chicago, and to Minneapolis. Mostly finished by 1940, and supplemented later, it consisted of the two St Anthony locks at Minneapolis and 27* below. No 27 at Chain of Rocks just below the Missouri mouth, on a 10m lateral canal, is the only one not in the river itself, or with an accompanying dam. The St Anthony locks were 400ft × 56ft, the others were (and are) 600ft × 110ft, except for Nos 19 and 27 (1200ft × 110ft), and No 26 at Alton now being rebuilt to that size. Some have smaller auxiliaries. In justification of it all, in 1940 the upper Mississippi carried 3,495,028 tons; in 1976, 68,420,307 tons.

This last figure has taken us beyond our period. On the Mississippi generally, the largest component of the whole river system, from 1925 to 1934 the tonnage figure kept level. Then a rise began, until by 1945 it had tripled. Helpfully, during the 1935–9 period, upbound and downbound traffic balanced well, the latter being slightly greater. Operationally, two innovations of the 1930s were the tunnel stern, to enclose the screws and so direct the force of the water to the entire radius of each, and radiotelephonic communication with other towboats and the shore.

Back in 1827 Corps of Engineers officers were surveying a by-pass canal to the Muscle Shoals rapids in Alabama, as part of a canal-river

*When relating locks to their official numbers, it should be noted that there is a Lock 5A, but no Lock 23.

plan to link the Tennessee to the Atlantic. The later Windom Committee again recommended such a scheme, and for fifteen years from 1875 the Engineers incorporated the old work done into a new canal at Muscle Shoals: one of them was Goethals, later of the Panama Canal. Then in 1933 the Tennessee Valley Authority was set up at the beginning of Franklin Roosevelt's New Deal, its objects including flood control, electricity generation and navigation. By 1945 a 9ft channel with nine locks had been established from the Ohio for 648m to Knoxville. The work done then has in turn led to the Tennessee-Tombigbee Waterway of our own day.

The Atlantic Intracoastal Waterway achieved

Against a background of national interest in waterways, local feeling grew in favour of a Chesapeake & Delaware ship canal, to be a rebuilding of the existing line by the federal government for commercial and defence reasons. The latter were given point by Germany's not long opened Kiel Canal, an object lesson in how to increase naval effectiveness, the former by the existing water carriage of the two bays, then estimated at some 90M tons. The result was President Theodore Roosevelt's commission of 1906 to investigate the case for a ship canal able to take the biggest vessels afloat. Recommended in 1907, bills for government purchase of the old canal failed.

At the time of the Atlantic Deeper Waterways Conference of that year, the Waterway was thought of as extending from New Jersey southwards to Beaufort Inlet, North Carolina, and consisting of four canals, the Delaware & Raritan, Chesapeake & Delaware, Dismal Swamp and Albemarle & Chesapeake, together with the Core Creek cut then being made between Beaufort harbour and Pamlico Sound. All were basically towed-barge canals having less than 10ft depth at low water. Here are dimensions:

	Length (miles)	*Width (ft)*	*Depth (ft)*	*No of locks*	*Dimensions (ft)*
Delaware & Raritan	44	80	7	14	220 × 24
Chesapeake & Delaware	14	66	9	3	220 × 24
Dismal Swamp	28	60	6	7	100 × 16½
Albemarle & Chesapeake	14	80	7½	1	220 × 40

Railroad competition had brought declining traffic. The Chesapeake & Delaware, carrying 1,318,772 tons in 1872, had fallen to 639,543 in 1900, while on the Delaware & Raritan the 2M tons of 1870 had become 623,751.

There seemed to be two choices: to develop the Waterway for barge traffic, or enlarge it to ship-canal standard. Engineers considered either to be geologically feasible, and locks anywhere on the route unnecessary. The Association chose to press for a large barge channel, to be extended northwards by way of New York's East River to Long Island Sound and by a Cape Cod canal to Boston, and southwards to Florida, with the possibility also of a Cross-Florida waterway to link with Mississippi traffic. They were successful. Of twelve canal links between Boston and Florida, eleven had been built by 1941: only an enlarged Delaware & Raritan across New Jersey was missing. Let us glance at some of its components.

A canal through the Cape Cod peninsula had long been considered, and twice begun. Finally a company, chartered in 1899 and led by the nostalgically-named De Witt Clinton Flanagan, got necessary banking support, and with W. B. Parsons as engineer began work in 1909 on a cut to be 7.7 m long, 13 m including approach channels, and 25 ft deep, at a time when some 30,000 craft a year were thought to be making the time-consuming passage round Cape Cod. The discovery of scattered glacial boulders weighing up to 100 tons, each of which had to be dynamited, slowed work, but the canal opened, though only to 15 ft depth and with one-way traffic, on 29 July 1914. A feature was its fast-running current, due to differing tidal heights and times at each end. Two months earlier, the rebuilt Kiel Canal had been completed; less than a month later, the Panama Canal opened.

The Cape Cod Canal was soon deepened to 25 ft, and traffic increased steadily, though the company lost money. However, the war showed its value, and in 1918, after a U-boat attack on a coal-barge tow off Cape Cod in July, the government took over its operation and improved it. Returned to the company in 1920, it was bought by the US government in 1928 and put in charge of the Army Corps of Engineers.

Federal interest in the whole Intracoastal line became active in 1910 when the Albemarle & Chesapeake company went bankrupt in face of competition from the rebuilt Dismal Swamp. Because in any developed Atlantic Waterway the A & C had the better line, the Rivers & Harbors Act of that year authorised purchase. It was transferred to the Engineers in 1913 and made toll-free, whereupon the Dismal Swamp lost customers until it, too, was taken over in 1929 as a secondary 9 ft route for the Waterway. On the Albemarle & Chesapeake, a big new guard-lock, 600 ft × 75 ft, was built in 1932 to replace the older one removed in 1917. On the Dismal Swamp, new locks 300 ft × 52 ft with 12 ft on the sills were opened in 1940 and 1941.

The purchase of the Albemarle & Chesapeake, followed by the outbreak of war, led the federal authorities towards taking responsibility for creating and maintaining a Waterway very much as the 1907 conference had foreseen it. They did so, and by WWII had succeeded.

From 1900 the Chesapeake & Delaware's tonnage had increased to reach a million in 1915. But war had come, and the canal was too small for modern 1500–2000-ton barges. So the government stepped in to buy it. Sold in 1919, work began in 1922 upon a sea-level barge canal 12 ft deep. It was opened in 1927. So quickly, however, had the size of ships requiring passage increased that in 1935–9 the line was again rebuilt as a ship canal (of 26.1 ft controlling depth, later increased to 29.6 ft) just in time for the heavy use it received in WWII. Here are tonnages:

	tons
1921	490,000
1926	610,000
1931	990,000
1936	1,300,000
1941	4,062,000

The Cape Cod Canal also proved its value in WWII, when convoys of cargo-ships bound for Iceland, Greenland and the UK passed through it. In 1933 the Engineers had begun to widen the channel, straighten the approaches, and rebuild its three bridges, and after the Rivers & Harbors Act of 1935 to cut a much bigger canal 32 ft deep and able to take two-way traffic. The work was done in June 1940, just in time. Cargo tonnage, 2,627,000 tons in 1935 and 9,901,000 in 1940, reached a maximum of 18,851,000 in 1944—though the war killed the old-established passenger-ship transits.

The federal government had worked the Delaware & Raritan during WWI for military and naval traffic. Thenceforward, however, it resumed its decline. The government did not consider it essential, and it was abandoned at the end of the 1932 season. Locks have been weired, but the canal has been kept in water for supply purposes. Since then craft passing north on the Waterway from the Chesapeake & Delaware have had to turn down Delaware Bay, pass through the small Cape May Canal if dimensions permit, and then follow the unprotected New Jersey coast to New York.

Further south the Waterway was extended from the Albemarle & Chesapeake by rivers, canals and lagoons to south Florida, by way of Albemarle Sound, the Alligator river and the Alligator-Pungo Canal to Pamlico Sound and so to Beaufort. Further south, a difficult piece of cutting had to be done from Little river by the North Carolina border through high ground to the Waccamaw river, before the

Waterway continued to Charleston and beyond. Some of it was finished just in time, for during WWII German submarines operated so effectively off the Atlantic coast that in 1943 a pipeline was built from the Gulf to Jacksonville, Florida, whence oil could be carried northwards by barge. Over 1½M short tons of gasoline had been so hauled by 1945.

The shallow-draught southern section of the Atlantic Waterway links with the deep-draught Gulf Intracoastal by a smaller waterway from Stuart north of West Palm Beach to Lake Okeechobee, then by a lock and the River Caloosahatchee to the west coast. (For the Cross-Florida Canal, see p. 380.)

Georgian Bay plans and Welland action

In 1904 a 14ft waterway, entirely within Canada, was completed from Lake Erie and beyond by way of the Welland and St Lawrence canals. On the Lakes the distinctive 'laker' developed, similar to an oceangoing bulk carrier of grain, coal and ores, the last increasingly important with the steel industry's growth, but operable only upon the Lakes and not the sea. At Port Colborne, where the Welland began, cargoes had to be transhipped to 'canallers' to navigate onwards towards Montreal. Given increasing trade (2399 vessels carrying over a million tons transitted the Welland in 1900), thought began to be given to an alternative route to both the St Lawrence and Welland Canals that would take lakers to Montreal, bring oceangoing ships inland, and reduce the length of the existing route to Lake Huron and beyond.

A steamboat canal line from Montreal by way of the Ottawa and Massawa rivers, Lake Nipissing and the French river to Georgian Bay and Lake Huron had been suggested by Col By in 1829, and we have glanced at later developments on the Ottawa. The then worldwide interest in big waterways, however, led to a Georgian Bay ship canal company being chartered in 1894, a favourable Canadian Senate committee report in 1896, and a survey published in 1908 by the noted Canadian engineer Arthur St Laurent, who proposed a 22ft waterway, to be built in ten years, to take the largest lakers of 600ft × 60ft with 20ft draught.

Meanwhile surveying had been going on in 1906–8 for a much bigger Welland Canal, also to take the largest lakers. The Georgian Bay scheme went down before these Welland proposals, the defeat of Laurier's favourable government in 1911, and the onset of war in 1914, though the company's charter did not lapse until 1927.

Fig 116 A ship occupies each chamber of Lock 4 on the Welland Canal's 3-rise lock staircases.

A Welland enlargement would be a lesser commitment that yet would open Lake Ontario to big lakers and move the transhipment point for Montreal traffic from Port Colborne to Prescott on the upper St Lawrence. Work began in 1913, but was delayed by the war, the new canal being opened in August 1932. This fourth Welland Canal, 27½m long, with only seven locks 820ft (859ft between gates) × 80ft × 30ft, a 1380ft guard-lock at Port Colborne and a 27ft channel, was built on a straighter route from a new terminus at Port Weller, replacing Port Dalhousie. Three locks led the canal to the foot of the escarpment, whence a 3-rise staircase of locks in parallel took it uphill and to the seventh near the summit to complete a rise of 327ft, whence the line ran level to Port Colborne. There was no towpath, and the provision of a tug service for barges was no longer thought necessary. Average transit time now became twelve hours. The change from Port Dalhousie to Port Weller left the former together with the old canal through St Catherines, as interesting industrial archaeological sites. In 1935, soon after its opening, 5092 vessels and over 8M tons passed through. The canal was then toll-free, as all Canadian canals had been since 1903.

What was to become the St Lawrence Seaway can be dated back to 1895 and the Deep Waterways Commission set up jointly by Canada and the United States to study all possible routes between the Lakes and the Atlantic. In 1897 they reported in favour of improving the St Lawrence. Wide opposition showed itself, mainly from railroads and from eastern and Gulf ports, which crystallised in a National St Lawrence Project Conference that campaigned against the proposal for half a century. The St Lawrence improvement had, however, a fanatical supporter in Narcisse Cantin, who sought to publicise it by asking for charters between 1898 and 1904, and who in 1914 formed the Great Lakes & Atlantic Canal & Power Company.

A new argument in favour had appeared, power generation. In 1914 the International Joint Commission of the USA and Canada* was asked to study the possibility of both improved navigation and electricity generation on the St Lawrence. The war interfered with the work, but in 1932, the year the new 27ft Welland Canal was opened, agreement was reached upon a 27ft waterway and upon power projects. In the same year the two countries signed a treaty to give effect to the IJC's proposals, only to have the opposition block them in the American Senate. So matters remained until after WWII.

At the Soo, however, the Corps of Engineers had in 1943 replaced the nineteenth-century Weitzel lock with the new MacArthur, 800ft × 80ft × 31ft, dimensions much the same as the Welland's locks.

Steamer Days on the Lakes

Passenger-steamers had once been the most efficient means of long-distance transport. When railways came to compete, they did not eliminate the role of the steamer but, because the two often worked together, changed it. A new steamer age therefore began, first with sidewheelers, later with propeller-driven craft, which lasted from the early 1880s to the 1930s. It was based on scheduled summer sailings, comfort, enjoyment, cheapness, and the substitution of coal for wood as fuel, so enabling longer non-stop journeys.

The Canadian Pacific Railway was still being built along Lake Superior's northern shore when the CPR's *Algoma*, *Alberta*, and *Athabasca* began to run from a railhead at Owen Sound, Georgian Bay (from 1912, Port McNicoll, some 55m away) to Port Arthur (now incorporated in Thunder Bay). Able to carry some 240 first-class passengers and a thousand altogether, as well as considerable freight:

> No such vessels have ever been seen on the Great Lakes but their excellence lies not in the gorgeousness of their furniture or the gingerbread work of

*A body that since 1909 has been vital to discussions of traffic on the St Lawrence.

decoration but in their superiority over all other lake craft in...construction and equipment...(7)

Later came the *Assiniboia* and *Keewatin* to provide trans-Continental railway passengers with two nights in a good cabin and food from a good saloon to break their long journey, or holiday-makers with an attractive round-trip. Not till 1965 did the CPR passenger service on the Lakes end.

CPR passenger-steamers were a few among very many. The Anchor Line ships *India, China* and *Japan*, earlier built than those of the CPR, were also directly linked with a railway, the Pennsylvania RR. Carrying 120 passengers each as well as freight, they worked over the 1065 m run between Buffalo and Duluth. After the century's turn they were replaced by the 4330-ton *Tionesta, Juniata* and *Octorara*, fine white ships carrying 595 passengers, lit by electricity, and able in their dining saloons to take 250 at a sitting. Also on the Buffalo-Duluth as well as the Buffalo-Chicago run in the 1890s were the *North West* and *North Land* of the Northern Steamship Co, 5000-tonners carrying 500 first-class passengers.

The steamers of the Detroit & Cleveland Navigation Co ran overnight between those cities and Buffalo, while such ships as the

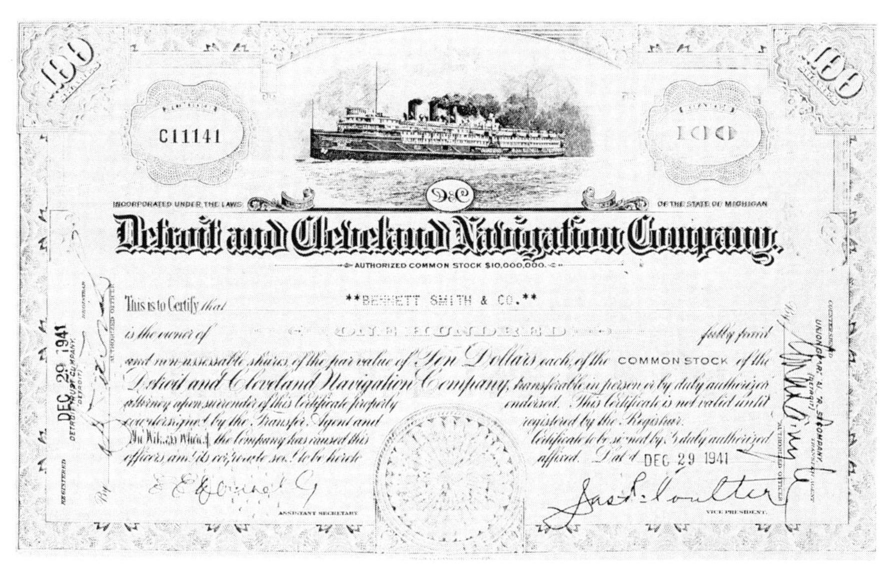

C11141

100

INCORPORATED UNDER THE LAWS OF THE STATE OF MICHIGAN

Detroit and Cleveland Navigation Company.

AUTHORIZED COMMON STOCK $10,000,000.

This is to Certify that **BENNETT SMITH & CO.**

is the owner of ONE HUNDRED fully paid and non-assessable shares of the par value of Ten Dollars each, of the COMMON STOCK of the Detroit and Cleveland Navigation Company, transferable in person or by duly authorized attorney upon surrender of this Certificate properly endorsed. This Certificate is not valid until countersigned by the Transfer Agent and registered by the Registrar.

In Witness Whereof the Company has caused this Certificate to be signed by its duly authorized officers and its corporate seal to be hereto affixed. Dated DEC 29 1941

ASSISTANT SECRETARY

VICE PRESIDENT.

REGISTERED DEC 29 1941 DETROIT TRUST COMPANY, DETROIT

Fig 117 In the last days of the great ones. A share certificate of the Detroit and Cleveland Navigation Co, issued in 1941, shows the *Greater Detroit*. She and her sister Ship, *Greater Buffalo*, built in 1924, made the overnight Lake Erie run between Detroit and Buffalo until the *Greater Buffalo* was taken over by the U.S. Navy in 1942. Each had sleeping accommodation for 1700 passengers and carried a crew of 275.

Put-in-Bay of the Ashley & Dustin Steamer Line, well known for her dining room, ballroom and observation parlour, catered mainly for excursionists, carrying 2800 day passengers from Detroit to Sandusky. Long-distance cruises could be taken on craft of the Chicago, Duluth & Georgian Bay Line, or, by changing ships more than once, of Canada Steamship Lines running from Lake Erie to the Saguenay river below Quebec, shooting the St Lawrence rapids on their way. On Lake Ontario the paddlewheelers *Chicora*, *Chippewa* and *Corona* ran day trips between Toronto and Niagara, and the propellers *Macassa* and *Modjeska* others from Toronto to Hamilton.

By the 1930s the spread of the automobile and the road truck was affecting passenger- and freight-carrying. Then, at sea off the New Jersey coast, came the burning of the *Morro Castle* with the loss of 124 lives. As a consequence, safety regulations for ships were so tightened that many owners could not afford compliance, and laid up their ships.

A new demand, however, appeared. There had been railway-car ferries on the Lakes for a long time: back in 1858 one had begun work between Buffalo and Fort Erie, Ont, and soon afterwards another between Detroit and Windsor, Ont. Bigger ones arrived on Lake Michigan in 1892, carrying some passengers also, and saved freight transhipment costs. Now came automobile ferries, as the number of motorists increased. Among the busiest were those across Lake Michigan or from mainland Michigan to Mackinac Island. On the latter run, indeed, almost a quarter-million vehicles were carried in 1938.

The Trent-Severn Waterway

This river-lake-canal route from Trenton, Lake Ontario, to Port Severn, Georgian Bay, Lake Huron, had been initiated back in 1833, and thereafter built slowly and piecemeal as a carrier of lumber, and of settlers and their freight. In the 1860s, however, a Toronto & Georgian Bay ship canal was proposed. This, turned down by the Canal Commission in 1871, resulted in a movement to develop and complete the Trent route as a grain carrier by barge from Lake Huron to the St Lawrence. A side result was the opening in 1889 of the Murray Canal to continue a sheltered barge route from east of Trenton to Lake Ontario.

This activity having produced a favourable Commission report in 1890, the government decided in 1896 to complete the route at 203 m, with 13 more locks 134 ft × 33 ft and three vertical lifts. C. E. Rogers, the canal superintendent, after visiting European lifts, and working

Fig 118 Peterborough hydraulic lift on the Trent-Severn Waterway.

with W. J. Francis, then designed one with a 65ft rise at Peterborough. It followed the general pattern of its hydraulic ram prototypes, using two water-filled caissons each 140ft × 33ft × 7ft. Opened in 1904, it was followed in 1907 by Kirkfield lift at 48½ft rise.

The third lift of 53ft was to have been at Healey Falls on the Trent, but by Kirkfield's completion, it had become clear that grain would move by lakers or rail, not Trent barge, and that a completed waterway would be mainly useful for pleasure craft. Two inclined planes (marine railways) were added, to take cruisers up to 50ft × 13ft 6in × 4ft, and on 3 July 1920 the launch *Irene* left Trenton to arrive on 12 July at Port Severn. She initiated a 240m-long cruising route (see p. 398) that has become increasingly popular.

The Steamboats' Work is Done

In their great days, steamboats worked everywhere in North America where a need was shown and a boat would float, from the Rio Grande to the Far North, and their stories have been told in words and pictures no steam enthusiast should miss. In their time they were not only essential to the transport of goods and people, but indeed to settlement itself—as in Ontario's Muskoka lakes region a little north of the Trent-Severn Waterway. But when railways came they drove most steamboats off their accustomed waters. Some survived, however, because railways never reached their distant shores.

In 1886, for instance, the Canadian Pacific Railway's transcon-

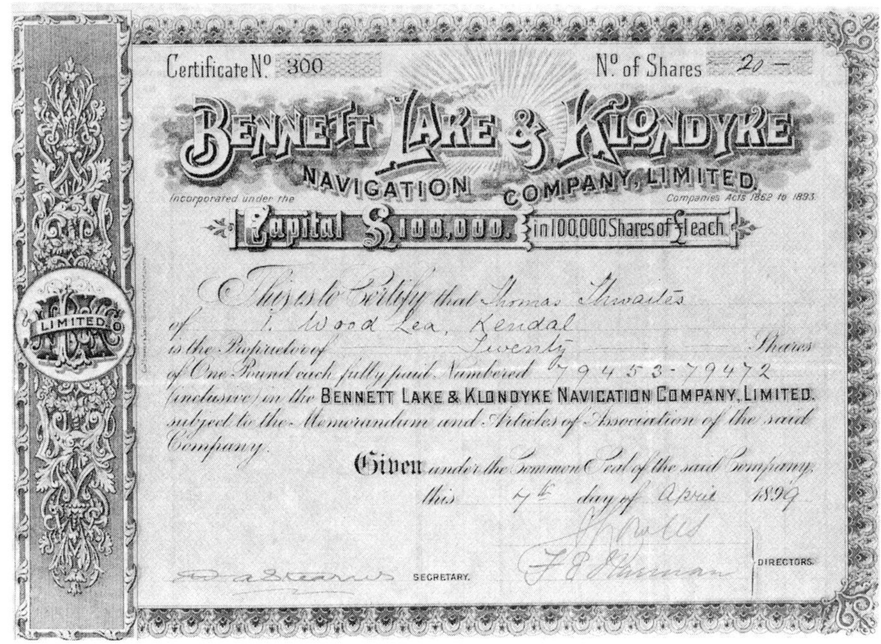
Certificate No. 300 No. of Shares 20 —

BENNETT LAKE & KLONDYKE NAVIGATION COMPANY, LIMITED.

Incorporated under the Companies Acts 1862 to 1893.

Capital £100,000. in 100,000 Shares of £1 each.

This is to Certify that Thomas Thwaites of 1 Wood Lea, Kendal is the Proprietor of Twenty Shares of One Pound each fully paid Numbered 79453-79472 (inclusive) in the BENNETT LAKE & KLONDYKE NAVIGATION COMPANY, LIMITED. subject to the Memorandum and Articles of Association of the said Company.

Given under the Common Seal of the said Company this 4th day of April 1899.

SECRETARY. DIRECTORS.

Fig 119 A souvenir of the Gold Rush.

tinental line had been completed and, because it competed with the Great Northern Railway, small wooden sternwheelers were used to work branch lines down towards the American border. The first was *Aberdeen* on Okanagan Lake in 1893, which also saw the last, *Okanagan II*, in 1972. Others followed on the Arrow Lakes (part of the Columbia river), Kootenay Lake and Trout Lake. At first passengers were many, but from the 1930s they moved to the road and later the air, and such services as continued were mainly for freight.

Exemplifying both in extreme form was the Yukon river. Navigable for four months a year in a favourable season, sternwheelers had worked since 1869. As gold-rush fever grew, they struggled up some 1600m to Dawson City, and eventually on to Whitehorse, head of navigation, where boats took on board those who had survived the struggle on foot over the Chilkoot or White passes from the south coast of Alaska. More and more were built, until, as the century ended, so did the rush. In 1898, however, the building of the near-impossible 110m White Pass & Yukon Railway was begun, and reached navigable water in 1900. People left the area, a road came, and in 1953 the last sternwheeler was hauled ashore. Canada has one survivor of the old steamboat services, the restored *Segwun*, built in 1887, rebuilt in 1927, propeller-driven by two reciprocating engines. She works from Gravenhurst on Ontario's Muskoka lakes. And so from *Segwun* we move to modern times.

15
SINCE 1945: SOUTH AND WEST

The years since 1945 have seen astonishing developments in North America. A great ship channel, the St Lawrence Seaway, a major barge-route, the Gulf Intracoastal, and a new line which incorporates a considerable canal section, the Tennessee-Tombigbee, have all been completed. Among the many river works the re-canalisation of the Ohio is outstanding, while improvements of the same character, though on a smaller scale, have been made on the Monongahela, Green and Cumberland rivers, on the Black Warrior-Tombigbee system and on the older sections of the Gulf Intracoastal. Along with track improvements has gone great growth of traffic, considerable technological change, but, in Canada and the United States, the breaching of the practice of free transit. Change has indeed been so fast that the system now shows signs of strain.

It is difficult to grasp the size and importance of an integral waterway system, that of the Ohio river, Illinois river, Mississippi and Gulf Intracoastal between Mobile and Galveston, that carries almost 90 per cent of all the shallow-draught inland waterway ton-miles of cargo in the United States as it peaked in the late 1970s, fell back before recession, and in the mid-1980s seemed likely to advance again. A glance at the advertisements of the weekly *Waterways Journal*, published in St Louis, helps one to realise that, for instance, one Ohio river firm in 1980 launched 528 vessels. By then the barge fleet was some 28,000, together with some 400 towboats and tugs, and the inland waterway system was employing, afloat and ashore, nearly 80,000 people.

Let us look first at some examples of the expansion of the system itself, and then of the technological changes that have accompanied it.

The System Expands

(a) *The Gulf Intracoastal Waterway*

There had long been links to the Gulf from Houston, the lower Mississippi, and by the Mobile river before the Gulf Intracoastal Canal Association was founded in 1905 to promote a waterway along the coast that should link these and others. It would use existing lagoons, rivers, and sheltered waters, supplemented by artificial cuts, and be 9ft deep throughout, a figure later increased to 12ft.

In 1934 a Coyle Line tow of 650 tons loaded at Galveston, Texas, made the first voyage from what was then the western to what was then the eastern end of the Intracoastal at Pensacola, Florida. Subsequently it was lengthened at one end to the Mexican border at the Rio Grande beyond Brownsville and at the other to Apalachee Bay, Florida, between them serving Corpus Christi, Freeport, Houston, Galveston, Port Arthur, Beaumont, Lake Charles (La), Morgan City, New Orleans, Gulfport and Mobile. Thus completed in 1949, by 1971 its waters carried some 106M tons, and traffic has maintained itself at about that level since. The GIWW is costly to maintain, thanks to the Gulf's weather patterns, but its economic value is great. Nearly half its traffic is in oil and its products, with chemicals and coal also important, its main competitor being probably the pipeline rather than the railroad.

A 12ft channel had been dredged from the Gulf for 50m to Houston as early as 1876. In 1914, however, the Ship Channel was opened which cut across the GIWW. Since then the city has become one of the biggest ports of the US, with barge terminals and building yards sited along the Channel, and much interchange of ship and barge cargoes.

New Orleans is central to the Waterway, which there crosses the Mississippi. Indeed, the Waterway might well be thought of as two lateral branches of the central Mississippi stream. The earliest to the west was the Harvey Canal running from the city past Bayou Barataria to the Gulf, which dated back to the 1840s. After being widened and given new locks, a barge tow first used it in 1934 to take 1400 tons of steel to Houston. In 1958 the Harvey was supplemented by the new and large 9m Algiers Canal, which leaves the Mississippi through a separate entrance lock, but later joins the Harvey's route.

For tows coming down the Mississippi and bound west on to the Intracoastal, two cut-offs shorten the distance compared to the canal routes. The first, by way of a lock at Port Allen opposite Baton Rouge and then by a 12ft channel for 64m to the GIWW at Morgan City, carried 18.1M tons in 1981; the other, by turning off above Baton Rouge on to the Atchafalaya river, and then by a 12ft channel and a lock in Old river, again to the GIWW at Morgan City, carried 4.9M tons in the same year.

Eastwards, the 5½m Inner Harbor Navigation Canal joins the Mississippi at New Orleans to Lake Pontchartrain, from which the Rigolets channel links with the Intracoastal. A second line runs from

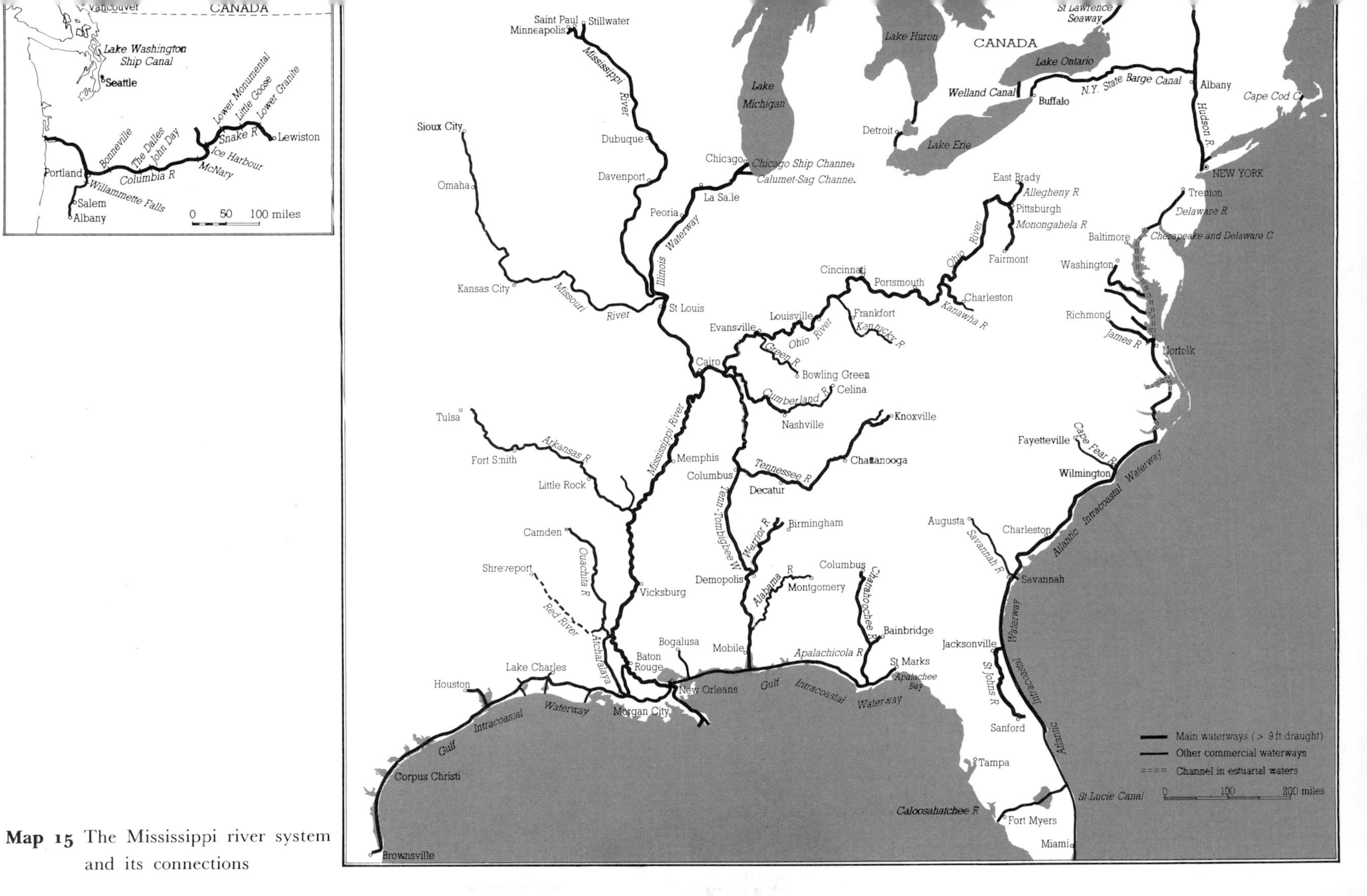

Map 15 The Mississippi river system and its connections

the Navigation Canal to join the first beyond the Rigolets. From this, the 36ft-deep Mississippi River-Gulf Outlet also turns off to run for 76m to the sea, against 120m by the river itself. Completed in 1965, terminals along it make it a New Orleans growth area.

Eastwards, the Gulf Intracoastal crosses Mobile Bay, from which Mobile can be reached by way of the Mobile river. Beyond is access to the Tombigbee river and so the Tennessee-Tombigbee Waterway, or by its tributaries the Black Warrior and Warrior rivers to Birmingham.

There has been a set-back. In February 1964 the Cross-Florida Barge Canal was finally begun, to connect the southern section of the Atlantic Intracoastal by a 9ft-deep, 107m-long cut with five locks to the Gulf coast of Florida at the mouth of the Withlacoochee river at Yankeetown. On the one hand it would have saved 600m of sea passage; but on the other a 100m gap of open sea crossing would have been left between the end of the new canal and the beginning of the GIWW at Apalachee Bay, which would probably have had to be worked by specially designed tug-hauled barges. After some $50M had been spent, the project was halted in 1971 because of its effect on the environment, especially the scenic Oklawaha valley.

(b) *Ohio river*

In 1955, the Ohio river carried 71.5M tons of cargo, yet 1000ft tows were having to be split to pass the many low-lift locks, often elderly, except when in flood-time they could pass over the wicket dams. Three years later the Engineers had agreed with river users that new structures should be fewer, with higher lifts to give longer navigation pools. Each would incorporate one 1200ft × 110ft lock (then the largest on any inland waterway), one 600ft × 110ft lock, and a fixed dam with tainter gates*.

Building began at Greenup locks in 1954, on a programme designed to replace 49 old structures with 19 new dams with lifts from 16ft to 35ft, plus providing a 1200ft lock at McAlpine at the site of the old Louisville & Portland Canal without altering the river level. In 1980 the Smithland facility near Paducah was opened with twin 1200ft × 110ft locks, the only such pair on the Ohio, by which year 14 new structures had been completed. In 1985, the lowest projected lock structure, Mound City, was still in abeyance pending a new study of the lower 61m of the Ohio in the light of what would be best done

*Named after J. B. Tainter, these are radial steel gates, the upstream face of which is in the form of an arc centred on the gate hinge. They are used to regulate the flow of water over a dam, and can be raised high enough to clear floods.

Fig 120 Back in 1963 a 19-barge coal tow enters the 1100 ft × 110 ft McAlpine lock on the Ohio. To the left is the site of the earlier Louisville & Portland Canal, built to by-pass the falls of the Ohio at this point.

near the mouths of the Cumberland and Tennessee rivers, both of which enter the Ohio in the projected Mound City pool, given expected changes in traffic patterns in consequence of the opening of the Tennessee-Tombigbee Waterway. Meanwhile, temporary 1200 ft locks have been built at old Dams 52 and 53. Much higher up, Gallipolis above Greenup remained in 1985 without a 1200 ft lock, a bottleneck to the system, where standard 15-barge tows have to be broken into two sections to pass. Studies for a single 1200 ft lock are, however, under way.

It is difficult to grasp the scale of such locks and dams as these. Here are a few details from Smithland. The overall length of the river lock wall, including the approach walls, is some ¾ m. Each lock-gate weighs 250 tons, and, along with the valve machinery, is hydraulically powered. Emptying a lock takes 9 minutes; filling, 8 minutes. The Smithland dam is some ¾ m long, consisting of a 2260 ft fixed weir and a section with eleven electrically operated tainter gates. A concrete service bridge runs above the gated section of the dam and also the upstream ends of the twin locks, and carries a rail-mounted crane which can lower emergency bulkheads (stop planks or logs).

By 1977 Ohio traffic had built up to 151 M, or 37,500 M ton-miles, such had been the speed of waterway growth. Pittsburgh was now the country's biggest inland port, in that year handling over 40 M tons. Ports on the Mississippi system handling more than 10 M tons were St Louis (upper Mississippi), Huntington (Ohio), Memphis (lower Mississippi), Cincinnati (Ohio) and Louisville (Ohio). In 1981, Ohio traffic was 158.7 M tons, a volume that would have been impossible without the new locks.

(c) *Arkansas river*

In steamboating days the Arkansas had played its part, but when a multi-purpose* programme for the river was initiated, political opponents called it 'the greatest boondoggle in history'. Nevertheless, construction began in 1957, and in 1971 the 9 ft navigation, with 17 locks of 600 ft and a total rise of 420 ft, was opened, and the first barge travelled the full 448 m from the Mississippi to the port of Catoosa near Tulsa, Oklahoma. A little over three years later, the investment of private and local government money in industrial plants and port facilities along the waterway had exceeded the total federal expenditure of $2.1 billion. By 1981 tonnage carried was running at some 9 M annually, and Catoosa had become a busy terminal.

Other new navigations have not been so lucky. A project authorised in 1968 for a part-canal, part-river route with 9 locks for 294 m from the Mississippi to Daingerfield, Texas, has not been begun, and though the first Red river lock was being built in 1981, its future was uncertain.

(d) *Tennessee-Tombigbee Waterway*

Authorised in 1946, begun in 1971, the 234 m Tenn-Tom is a major waterway of a new kind for its area. Part-canal, part-slackwater navigation, it provides a route for 8-barge tows from the Tennessee river and entire upper Mississippi basin, including the Ohio valley, to the Tombigbee river and so to the port of Mobile, Alabama, and the Gulf Intracoastal, alternative to that via New Orleans (where the Inner Harbor Navigation Canal is a bottleneck), and, for much of the

*The purposes included navigation, sedimentation control, hydroelectric power, flood control and water-based recreation, and involved a system of upstream dams and reservoirs on tributaries above the head of navigation.

Fig 121 The Tennessee-Tombigbee Waterway is opened on 18 January 1985 when a tow carrying 64,000 barrels of petroleum products enters the 600 ft × 110 ft Columbus lock.

Ohio and Mississippi valleys, one considerably shorter.

From the Tennessee it runs for 39 m by true canal through the high ground of the divide section, to an 84 ft falling lock. Thence it enters a lateral part-canal part-Tombigbee river section 46 m long with 5 falling locks, and then by 149 m of slackwater navigation with 4 locks to link at Demopolis with the existing Tombigbee-Warrior river system leading to Mobile. The Waterway has a 9 ft channel depth, with a planned 12 ft in the divide and lateral canal sections. Locks are 600 ft × 110 ft.

By July 1981 115 m from Demopolis to Columbus, Miss, were open to limited navigation; meanwhile, considerable anticipatory investments were being made at Mobile, Columbus and elsewhere. After a final scare lest economy would force stoppage of work, completion was agreed and the first commercial tow passed on 14 January 1985, after 307 M cu yd of earth had been shifted (as against 211 M on the Panama Canal). Traffic, much of it coal, is forecast to build up to some 28 M tons.

(e) *Willamette, Columbia and Snake rivers*

Far to the west, the Columbia river runs into the Pacific. At Vancouver, some 100m upstream, the Willamette comes in from the south; 224m higher, the Snake joins it from the east, at a point where the Columbia turns north through the state of Washington into British Columbia.

In steamboat days these rivers were navigated far inland, helped here and there by early portage canals and locks. The Willamette Falls locks are one example. Built privately in 1870–3, this 4-rise part-timber, part-masonry staircase plus upper guard-lock, is still open for navigation, the only such in the United States, current traffic being mainly in logs. On the Columbia steamboats quickly followed the discovery of gold in Idaho. Old canal systems, such as the two locks at the Cascades (1878–96), or the five on the Dalles-Celilo Canal (1905–15), have been inundated by modern lock and dam structures, in these cases Bonneville and The Dalles.

The present waterway runs from the sea past Portland, up the Columbia with four locks, and then up the Snake with four more, to Lewiston, Idaho, 465m away and 730ft higher. The lowest and oldest lock, Bonneville, opened in 1938, is 500ft × 76ft, the others 675ft × 86ft, the last on the Snake having been completed in 1975. Some are remarkable for very high lifts: John Day lock, at 105ft*, is closely followed by three others of 100ft and one of 98ft. The navigation can take seagoing craft as well as towboats and barges, traffic being mainly oil upbound, and grain downwards. Total all-commodity internal tonnage excluding oceangoing in 1982 was 11.9M.

Channel Depths are Standardised

A 12ft channel now extends the whole length of the Gulf Intracoastal Waterway, upon the lower Mississippi up to Cairo, and the links between the two by Port Allen or the Atchafalaya river. From Cairo the 9ft channel extends along the upper Mississippi and the Illinois Waterway to Chicago and Lake Michigan, up the Ohio to Pittsburgh, and from the Ohio 127m up the Monongahela, 72m up the Allegheny and 90m up the Kanawha, as also up the Tennessee to Knoxville. Similar depths obtain on the lockless Missouri to Sioux City and the Arkansas to Catoosa (Tulsa). The same depth applies through the Tenn-Tom, down to Mobile, and up to Birmingham. On other rivers, lesser depths are maintained.

*Maximum 113ft if river levels require it. John Day lock is unusual in America in having vertically moving gates at both ends, the upper one falling vertically, the lower rising between towers as commonly in Europe.

Technology Moves On

Now to glance at some technological changes that followed the war:

(a) *Towboats*
After the war diesel power continued to replace steam, till what had been two-thirds in 1947 had become four-fifths in 1950, and propellers to supersede sternwheels. In the 1950s, too, towboats became bigger and more powerful, able to handle heavier tows. This was partly the result of using the high- and medium-speed engines that had been developed during the war, which yielded much more horsepower for a given size and weight, to which supercharging and turbocharging could be added. Average horsepower per towboat on the Mississippi river network and the Gulf Intracoastal increased by almost two-fifths between 1950 and 1958. Engines also became more efficient in fuel terms. Parallel with this development came improvements in propeller, power transmission and steering technology, the introduction of radar and the use of electronic depth-sounders.

These trends continued after the 1950s. New towboats became more powerful, and old ones were re-engined. At the same time many towboat functions were automated, to improve safety and reduce labour costs. Between 1964 and 1969 towboat numbers increased by a quarter and total horsepower by half.

Tugs are still used, in harbour, on the Intracoastals, on the Great Lakes, and at sea, but most barge movements depend on towboats small and large. Small ones may have a single screw, large ones four propellers each with its own diesel, and power-operation of separate sets of steering and flanking rudders. Propellers are usually enclosed in tunnel sterns to concentrate the thrust. Horsepower may range from 100 to over 10,000, the largest being able to handle a tow carrying 40,000 to 50,000 tons. Big tows may also be given bow thrusters, able to provide lateral thrust, a practice that can be dated back to nineteenth-century powered bowboats, lashed sideways to the head of a lumber-raft to help it round bends. These are valuable not only at curves (and bridges are often found at curves, for there the river channel tends to stabilise) but also when the tow is manoeuvring at slow speed. The steeply rising cost of oil fuel helped to bring about other changes in design, all aimed at greater efficiency at less cost, especially since the size of towboats is now not far off the physical capacity of the rivers to take them.

We have seen push-towing at work in Europe. Let me summarise its advantages under American large-scale conditions. By lashing the barges together with cables, with the towboat at the back, so as to form a single rigid unit, the tow is operated as though it were a single vessel. It can quickly come to a stop by reversing the propellers and then remain stationary in the current, and move round sharp bends by flanking (see p. 341). The modern towboat is very manoeuvrable. With two or three screws, each operating independently, and each having two rudders ahead and one astern, a towboat can turn in its own length, or even move sideways, as to remove a barge from the centre of a multi-barge tow. Such manoeuvrability is essential when working large tows in narrow channels and congested fleeting areas.

Aids to navigation are the universal walkie-talkie radios, radiophone contact between towboats and the shore and between one towboat and another, so that pilots can arrange how to pass each other, depth-sounders, radar for night or fog navigation, a swing-indicator to show when the tow starts to deviate from course, an automatic pilot, and searchlights to pick out the reflecting navigation markers on the banks.

Modern highly automated towboats carry a crew of perhaps 10–13, working two six-hour spells a day in two alternate ships for some weeks, and then being given long leave, so that working hours over the year average 40–42 hours per week. Towboats may well not stop, except in locks, for two weeks or more. Running continues through the night, bunkering, watering and replenishing supplies being done from store boats while under way, and a speedboat sometimes being used to run ahead of the tow to do shopping and give crew members a short time ashore. Where tows may have to be split to pass locks, haulage units (powered capstans) may be fitted to the lock wall and upstream and downstream guide walls, as at the Pickwick lock on the Tennessee.

(b) *Barges*

After the war, the number and capacity of barges increased, as did the number of terminals with modern facilities for handling their loads. Barge shape was improved to increase capacity and speed, new special-purpose craft were designed, and labour- and time-saving devices introduced for assembling and breaking the tow.

Today the range is wide. Open hopper barges, virtually open rectangular boxes used for bulk cargoes, come in three usual sizes, 1000-, 1500- and 3000-tons capacity. Dry-cargo barges have watertight covers over the hold. Tank barges are of three main types; single-skin; double-skin for poisonous or hazardous liquids; and those with independent cylindrical tanks to carry or discharge liquids under

Fig 122 Container barges and towboat enter Bonneville lock on the West Coast's Columbia river.

pressure; and again in 1000-, 1500- and 3000-ton sizes. The variety is astonishing: barges carrying liquefied methane at −258°F; others molten sulphur at 300 – 50°F; others anhydrous ammonia at 250lb pressure; others again with collapsible plastic containers holding carbon black, or carrying cement mixed with air so that it can be pumped in and out. Decked barges ranging from 350 tons to 1500 tons or more, carry cargoes such as machinery, vehicles or heavy equipment, while miscellaneous barges include railway-wagon carriers, dump barges with bottom-doors, self-unloaders for cement and grain, derrick and crane barges. Hydrodynamically efficient integrated tows date back at least to their use by the Belgians on the Congo (see p. 267). Similar rakes of bigger barges, usually three in line ahead, achieving a 20 per cent reduction in resistance, are now increasingly seen.

What is now a successful container-on-barge (cob) service was begun on the Columbia/Snake rivers to and from Portland in 1975, and by 1984 was moving well over 2000 teus*, a month, carrying

* See note on p. 218.

mainly agricultural products, animal hides and paper. A tug container-barge service on the Atlantic Intracoastal Waterway followed. But on the Mississippi, port authorities, labour organisations and towing companies took a conservative view of its introduction; indeed, container-carrying lighters off Seabee vessels (see p. 390), run to St Louis in 1973–4, had to be taken off, partly because towing companies were reluctant to include singleton lighters in their tows.

However, a regular container service was started in 1982, working from a terminal on the Harvey Canal to others up the Mississippi, the Chicago Ship Channel at Joliet, and on the Arkansas, which had big enough cranes. Tricon, who initiated the service, was an agency for several shipping lines and had contracts with others, a link essential to success since transhipment at New Orleans could often be directly overside from ship and not via the terminal.

Combining now towboats and barges, the length of haul increased twentyfold in the years between 1934 and 1974, while normal tow size increased from 5000 tons to 50,000. By 1980, typical towing times were:

Pittsburgh – New Orleans, 1852m: upstream, 14 days 2 hours; downstream, 8 days 18 hours.
Pittsburgh – Houston via Atchafalaya river, 2186m: upstream, 16 days 6 hours; downstream, 10 days 14 hours.
Chicago – New Orleans, 1418m: upstream, 11 days 8 hours, downstream, 6 days 7 hours.

(c) *Ship-barge Transfers*

As we have seen throughout this book, inland waterways and the sea have always been closely linked. Seagoing ships have sailed up rivers, tugs have hauled or pushed barges on sea routes, and where the two meet, there also is a transhipment port for the transfer of cargoes. New Orleans, Baton Rouge and Mobile are examples of intermodal ports equipped to transfer cargoes between seagoing ships and barges by direct overside loading and unloading, as well as to or from rail or road. Nearly half the import/export traffic of New Orleans ends or begins in a barge movement: export grain, soybeans, farm machinery, road-building equipment, soda ash and salt are so exported; steel and sugar are among commodities imported.

In addition to New Orleans' bulk and grain terminals, both involved in barge traffic, the International Marine Terminals' coal transfer terminal was recently built about 40m below the port, to transfer coal from river- to oceangoing barges or to ships. The plant includes a ship-loader that can load direct from river barges at 4000 tons per hour. Mobile, too, offers a high proportion of ship-barge

Fig 123 Midstream coal loading at Baton Rouge on the lower Mississippi. The vessel is held on four mooring buoys as well as her own anchors, while four floating cranes load some 2000 tons per hour. As the barges are emptied, rubber-tyred articulated end-loaders are lowered by crane into each barge (as in that at top, left), to clean it up.

transfers; ore moving inland, grain and coal outwards, and a coal barge unloader with a capacity of 3000 tons per hour.

(d) *Barge-carrying Vessels (BCVs)*
At New Orleans in 1962, with the designs of the naval architect Jerome L. Goldman, a new technology was born, the barge-carrying vessel. Seven years later, in September 1969, the *Acadia Forest* was delivered to her owner, Central Gulf Lines of New Orleans. The world's first LASH vessel had been created.

The LASH (Lighter Aboard Ship) system enables large numbers of uniform lighters* (83 in the case of the *Acadia Forest*) to be lifted from the water by a gantry crane at the ship's stern, and positioned in a number of holds. Each lighter is 18.75 metres long by 9.5 metres wide, and has a loaded draught of 2.73 metres and a capacity of about 400 metric tons. These dimensions were chosen so that LASH lighters in multiple would have the same dimensions as standard Europa barges, and would be able to use European Class IV waterways. So the international barge/lighter appeared, one that might begin its journey on the Mississippi, be loaded on to a carrier at New Orleans to be off-loaded at, say Rotterdam, to finish its journey high up the Rhine.

*Usually 'lighter' in Britain, 'barges' in the USA. To be consistent with descriptions earlier in the book, I have used 'lighter' throughout.

A considerable number of further LASH vessels were built, for it was appreciated that the system was suited not only to sea/inland waterway interchange, but also to countries with insufficient port accommodation: LASH lighters could be left behind to be unloaded at any accessible wharf, and picked up on the next voyage.

Unfortunately, LASH developed alongside the rapid growth of container ships. Containers were therefore also carried on LASH vessels, partly in container cells and partly by carrying them inside lighters, though the vessels' structure did not altogether suit them, nor were they intended to come alongside at container ports, but to discharge lighters in open water.

Lykes Lines, also of New Orleans, then developed their Seabee system, which differed in several ways from the LASH design. Lighters were much larger (29.72 metres × 10.67 metres, with a service draught of 2.59 metres and a maximum of 3.25 metres), and could carry some 847 metric tons. They were raised out of the water by a semi-submersible stern elevator and not by crane, and were positioned not in holds but on three continuous decks. In this case lighter/barge size compatibility was based on American standards: two Seabee lighters are the same size as one 1500-ton jumbo. Recently flat-deck lighters for heavy lifts, and refrigerated lighters, have been added. The structure of the three Seabee vessels, which offers greater headroom, has after some modifications, more successfully enabled container and lighter carrying to be combined. Containers can be better stowed in lighters as well as on frames, the result being that a Seabee vessel on a normal voyage carries some 600 or more teus in addition to loaded lighters*.

In 1983 experiments were made with a Dutch seagoing heavy-lift submersible barge able to carry nine Mississippi barges. Fully laden with 16,000 tons of grain, these were enabled to cross the Gulf of Mexico from New Orleans to Progreso in Mexico's Yucatan peninsula without transhipment.

(e) *Tug-barge Systems*

Since about 1960, and considered here for completeness' sake though not specially relevant to the south and west, another new technology has been developed, that of the seagoing barge. Though barges have often made short sea crossings—for instance, when tug-worked and used to carry rail cars across San Francisco bay or in the Seattle-Vancouver area—they were long thought to be unsuitable for

* For later BCV systems, see pp. 223–4.

Fig 124 The ITB (integrated tug-barge) *Presque Isle* at work on the Lakes.

regular ocean work. Such barges are currently very large, and may either be tugged, one or two at a time, or be built in the form of an integrated tuge-barge (ITB). One large barge is joined to a tug, either non-rigidly because a projection from the letter's bow works into an equivalent notch in the barge's stern, or by a rigid from of linkage which firmly locks together tug and barge. Put simply, an ITB is a ship in which engineroom and hold have been separated.

Because tug-barge outfits are cheaper to build and run, and have smaller crews, they are appearing on the Great Lakes, and in the coastal and oceangoing North American trades. Perhaps the first was the *Carport* in 1950, a 4500 dwt rigid tug-barge system on the Great Lakes. Then came big hawser-towed barges, leading to the 30,000 dwt tank barge *New York* in 1969. In 1971 came the first ITB unit using the Breit system of interlocking, the oceangoing 11,250 hp tug *Martha R. Ingram* pushing a 30,000 dwt petroleum barge. In 1973 the ITB *Presque Isle* appeared on the Great Lakes, a 14,840 hp towboat pushing a coal-carrying 50,000 dwt barge. Still more recent, a catamaran-type towboat with rigid interlocking, the *Seabulk Challenger,* forms a unit with a 45,000 dwt barge. By 1981, the USA had 44 barges of 14,000 dwt or more, with others on order up to 55,000 tons. Destinations include the Caribbean or, on the West, Hawaii and Alaska. More recently still, four very large barges, 580 ft × 105 ft × 11 ft, with more than one deck, have been put into service to carry container-trailers between Jacksonville (Fla), Lake Charles (La) and Caribbean ports, and a 147.5 metres × 19.5 metres barge with 8,000 hp tug for carrying railroad wagons across Lake Michigan. Triple-decker trailer barges have been ordered for the Seattle-Seward (Alaska) run; it is planned to add a trailer deck to each of the six existing tugged rail-car barges working between Seattle and Whittier (Alaska), and in 1984 it was reported that orders had been placed to lengthen three barges from 400 ft to 730 ft, the longest so far to operate in the US.

So we move to the Seaway and the Great Lakes.

16

SINCE 1945: SEAWAY AND ON

The St Lawrence Seaway

We left the project to deepen the St Lawrence to 27ft blocked by opposition in the United States. It so remained till 1946, when the suggestion was made that tolls should be charged. This was a first step forward; the second was in 1949 when a railway was begun to carry ore from great new Quebec-Labrador iron deposits down to the St Lawrence. A third was the urgent need, especially in Ontario, for more electric power.

In 1951, therefore, Canada passed a St Lawrence Seaway Authority Act, and also made it clear to the United States that if necessary Canada would go it alone. Then, in 1952, she proposed to the International Joint Commission joint development of the power possibilities of the international section of the St Lawrence. This was agreed, and led to a change of mind of enough senators to ensure, in 1954, the passage of the Wiley-Dondero Act* to set up a counterpart of the (Canadian) St Lawrence Seaway Authority, the Saint Lawrence Seaway Development Corporation. On 10 August excavation began, and on 25 April 1959 the Seaway began to operate, though power had been generated a year earlier.

Built in five years, the Seaway was an enormous engineering work. Railway lines, roads, towns and villages had to be moved (as on the Danube for the Iron Gates development) to make way for new canals that would replace the old 14ft channels and for the altered water-levels of the remaining river stretches. Then the new waterway had to be built, with seven locks of Welland width and depth, but slightly shorter (766ft [730ft usable] × 80ft × 30ft over sills), together with their accompanying power plants and dams.

At the Seaway's upper end, Iroquois lock (Canadian) leads via the artificial Lake St Lawrence and the Wiley-Dondero approach channel to the two United States locks, Eisenhower and Snell. These three by-pass the once-existent Galop, Rapide Plat and Cornwall Canals, and lead the Seaway out into the natural Lake St Francis. Beyond lies the Beauharnois Canal (originally built in the early 1930s for

*Named after the senator and representative who led the way to success.

navigation and power generation) and the two Canadian Beauharnois locks. These lead to Lake St Louis and then to the South Shore Canal with its last two locks, Côte Ste Catherine and St Lambert, at the approaches to Montreal, a route that replaces the former Lachine Canal. Together, the Seaway and Welland Canal Locks lift ocean-going ships some 600ft. As Lionel Chevrier graphically put it: 'The seaway has given Canada a south coast stretching half way across the nation.'(1)

Iroquois lock on the Seaway, which controls the level of Lake Ontario as Lock 7 of the Welland, backing up the Port Colborne guard-lock, does of Lake Erie, are among those which have sector (see p. 79) upper lock-gates. In the event of a serious accident to the gates, sector gates have inherent stability as structures, so that, even though hit so severely that their hinge supports are damaged, they will not fall over—as would mitre-gates—but would still rest on their roller supports, each sector gate supporting the other. So the crisis would be contained until stop-logs could be lowered to hold back the water. Arrester cables carried on vertically rising booms are also features of the Seaway, Welland and Soo locks. They can stop a ship travelling at 3mph, and are lowered ahead of a moving craft and behind one that has stopped. Their presence now governs the usable length of locks.

The Seaway's opening had many repercussions. Transhipment at Prescott ended, and traffic patterns changed at such ports as Cleveland and Chicago. In recent years main traffics on the Lake Ontario-Montreal section have been Canadian grain (1984: 15.85M tonnes), downbound US grain carried in brisk competition with the Mississippi-Gulf route (1984: 7.65M tonnes) and iron ore (1984: 11.42M tonnes), with proportions on the Welland not very different. Though significant numbers of European ships had been moving into the Great Lakes through the 14ft canals ever since WWII, their numbers now increased, bringing with them complications arising from the unfamiliarity of their crews with Lakes navigation practices. On the other hand, a decline in US domestic traffic on the Lakes, mainly iron ore and coal, was mostly due to the higher iron content of Lake Superior ores, which required smaller quantities per ton of iron produced, and some overall decline in the production of the US steel industry. As for coal, northbound coal movements from Chicago and Lake Erie also declined, principally because more western coal become available to upper Lake communities. The replacement in 1968 of the old Poe lock at the Soo by another, 1200ft × 110ft × 32ft, was to permit the movement of new self-unloaders (too large to transit

Map 16 The St Lawrence Seaway

the Welland Canal) in the ore trade.

In the Seaway's early days the old 14-footers continued to trade, but were slowly replaced by '730s', ships of maximum permissible lock length and only 4ft under lock width. Later, new types of craft began to appear on the Lakes, such as self-unloading bulk carriers and integrated tug-barge (ITB) units. Therefore the number of annual transits through the Montreal-Lake Ontario section and the Welland continues to fall as ships become larger. On the Seaway, freighter transits fell from some 7000 to an average of 4475 for the recession years of 1980–3 and 3759 for 1984. On the Welland, transits, one-third of them oceangoing, fell from some 8000 to an average of 5612 for 1980–3 and 4750 in 1984. At the Soo, though long accustomed to big cargo-carriers working within the Lakes, the same tendency showed itself.

Cargo tonnages at the Soo have long remained constant at some 90–100M annually. Since opening, those on the Montreal-Lake Ontario section have been:

Dates	*Million Metric Tons average*	*Dates*	*Million Metric Tons average*
1959–61	19.4*	1971–3	49.6
1962–4	29.0	1974–6	44.3
1965–7	41.3	1977–9	56.6
1968–70	42.4	1980–3	47.0**
		1984	47.5

The Welland's pattern has been somewhat different:

Dates	*Million Metric Tons average*	*Dates*	*Million Metric Tons average*
1959–61	26.7	1971–3	58.7
1962–4	38.7	1974–6	53.4
1965–7	50.1	1977–9	65.7
1968–70	52.8	1980–3	54.4**
		1984	53.9

Because it had already been enlarged, it started from a higher base-line of traffic, and has maintained its lead. So busy, indeed, did the canal become that in 1967 a by-pass was begun to cut out the curve through the city of Welland, with its vertical lift-bridges and, worse, railway swingbridge with a concrete pier in mid-canal. The by-pass, 8.3m long, was opened in 1973, and reduced the canal's length to 26m. Instead of bridges, two tunnels run beneath it, and a

*As compared with 10.1 metric tons for 1953 through the old St Lawrence canals

** These mask the disastrous tonnage results of 1981–2 on both the Montreal-Lake Ontario and Welland sections.

4-tube syphon culvert carries the Welland river under the canal near Port Robinson. In this area, too, the canal was in 1981–2 widened from 183ft to 244ft over 4½m.

Operationally, the Welland presents a bigger problem than the Seaway, if ships are to be kept moving into and out of its approaches, and through its seven locks (especially those that are singletons), at night, in fog, and at the beginning and end of the winter. This is done from a central control room which is constantly in touch with ships before and during their transits.

Seaway and canal are closed by ice during the winter, the former roughly from 15 December to 1 April, the latter from 30 December to 31 March. Complications arise on the one hand from seasonal variations, and on the other from the bunching of craft at the beginning and end of the season, the latter often leading to the danger of a ship being iced in for the winter. At the Soo, problems are less because five parallel locks (four American, one Canadian) are available, but greater in that the ice-free season is approximately seven months, and that some 90 per cent of loaded ships move downwards from Lake Superior.

When the Seaway was opened, it was hoped that tolls would not only pay operating costs, but interest on capital. That did not happen, so that the Authority had to borrow from government in order to pay interest, thereby facing a mounting debt. In 1977 the financial structure was altered. A large debt was converted into government-held equity, and interest ceased to accrue on the remainder. Moreover, Canada and the United States agreed on a revised toll structure (tolls had remained unchanged since 1959), to be phased in over three years from 1978. As a result, the financial year 1979 produced the first surplus in the Seaway's history. However, recession then hit finances. A joint Canadian-US review in 1981 led to increased tolls, and a decision also to reintroduce lockage fees on the Welland. Traffic improved in 1983 and again in 1984. Meanwhile, an optimistic lobby seeks the abolition of Seaway tolls and extension of the navigation season.

Rivers to the Arctic

From railhead at Waterways (near McMurray), 300m north of Edmonton, Alberta, the Athabasca river runs to Lake Athabasca, whence the Great Slave river crosses the border into the Northern Territories, and so past the one obstruction, the Pelican rapids at Fort Smith, to the Great Slave Lake. Out of it, running north-east, the

Mackenzie flows to its delta in the Beaufort Sea, past the new town of Inuvik.

In the 1880s this 1600m route north was developed by the Hudson's Bay Company, based upon the needs of the fur trade. A sternwheeler ran from Waterways for 400m to Fitzgerald, whence a rough road round the rapids led to Fort Smith. Thence another service led to the Beaufort Sea. First steamer on this was the wood-burning sternwheeler *Grahame,* assembled at Fort Smith after her engines and boiler had been hauled north from railhead. The Arctic service, running for 16 weeks a year, was kept going with such sternwheelers—the *Grahame* having been succeeded by *Distributor* and then by *Distributor II,* sometimes pushing three or four barges—until after WWII, with the help of one or two diesel tugs and some barges also.

Gold-mining started near Lake Athabasca in the 1930s, followed by a uranium mine on Great Bear Lake, and later those for other minerals. Air services began, and the mining concerns initiated their own tug and barge line, the Northern Transportation Co, in 1931. The war came, and while the Canadian government took over the NTC, that of the United States built a new fleet of barges to get supplies to Norman Wells on the Mackenzie for the Canol pipeline from the oilfield there. After the war the NTC, still Canadian government owned, continued to run a transport service. Then there followed drilling for gas and oil in the Beaufort Sea and the Mackenzie delta.

In 1965 the railway had been extended northwards to Hay river on Great Slave Lake, and the southern section of the old waterway route was no longer needed. However, 1200m of river remain, indispensable for heavy traffic (though a road now runs to Inuvik) to the Beaufort Sea. Fortunately 90 per cent of loaded traffic is downstream, and oil fuel is available at Norman Wells. Because of the very short navigation season, from early June to late September, some 125 barges are needed to carry the traffic and return to the south before the freeze up. Traffic is some 350,000 tons pa, naturally at very high freight rates.

Shallow-draught (4ft or less) towboats of up to 4500hp are used, together with barges up to 1500 tons, with maybe six pushed in one train. Some craft can also be used for coastal navigation on the Beaufort Sea, though the town of Tuktoyaktuk has grown up where Mackenzie and oceangoing ships meet.

To the west, towboats work on the Yukon river where once the steamboats ran.

Canada's Canals for Pleasure

These, now called heritage canals, flourish and are developed under Parks Canada's control. On the Trent-Severn waterway, with its increasing summer pleasure traffic, the marine railway at Swift Rapids was replaced in 1965 by a 47ft-rise lock. The other, at Big Chute, was kept as a barrier to sea-lampreys from Lake Huron entering the interior lakes and damaging fishing. Because the marine railway's capacity (boats 50ft × 13ft6in × 4ft) was smaller than that of the waterway's other locks and lifts, a much bigger railway was brought into operation in 1978, its cradle working on a 20 per cent grade, yet carrying boats level, unlike the old one. Up to 20 boats of average (60ft) length can be carried together. Such is cruising

Fig 125 Big Chute marine railway on the Trent-Severn Waterway.

demand on the Trent-Severn that several locks have been mechanised and swingbridges replaced by high-level fixed spans.

The Rideau, kept in beautiful condition, is thronged with summer cruisers. Most of its locks are still manually operated by lock-masters, but two of the busiest, at Black Rapids and Smiths Falls, have been mechanised, the latter being a new deep lock opened in 1973 to replace the original flight of three. The Ottawa river carries far less traffic. In 1960 work began on the great Carillon hydroelectric dam, which was to submerge the old canals, and bring the upriver pool level with the top lock of the old Grenville Canal, still in place but gateless. The new lock, 200 ft × 45 ft, with 14 ft over the sill and a lift of 60 ft, was opened in April 1963. Unusually in North America, it has a vertically rising gate at the lower end, and two sector gates at the upper.

Elsewhere, Parks Canada control the Canadian lock at the Soo, and the Murray, Chambly and St Peter's Canals. In 1975 the Federal Minister of Indian and Northern Affairs (responsible for Parks Canada) and Ontario's Secretary for Resources Development signed an agreement to preserve the 425 m recreational corridor represented by the Trent-Severn and Rideau waterways, while optimising its development and use.

The Atlantic Intracoastal Waterway

Currently, the southern section of the Atlantic Intracoastal Waterway—often called The Inland Waterway—from Florida to Chesapeake Bay is lightly used by freight traffic, but heavily by motor cruisers, short-distance and those undertaking the long spring run from the south to north of the United States or on to Canada, and in the fall the reverse.

World War II, however, which brought intensive submarine attacks to the American coast, forced traffic on to the Intracoastal. Most passed through the Albemarle & Chesapeake Canal, but quite an amount by the Dismal Swamp, then in good condition. After the war, use of the Dismal Swamp fell away. In 1959 the Army Corps of Engineers proposed closure on the grounds that its operational costs were no longer justified, given that most of the traffic passed through the Albemarle & Chesapeake. The move was, however, defeated by the efforts of a local committee. In 1970 the Corps again proposed closure, and were again defeated.

Further north, WWII saw the Chesapeake & Delaware section of the Waterway used to utmost capacity, especially to bring oil from the

Gulf, much of it in quickly converted or newly built wooden tank barges. The tonnage figure for 1942, indeed, was 10.8M. This figure fell to 3.7M in 1945 and then rose to 10.7M in 1956. In 1962 further enlargement was begun to relieve congestion from the 20,000 vessels using the canal, and also to open it to larger freighters, especially container ships, who had to use the sea passage round Virginia Cape to pass between Baltimore and Philadelphia, American north-east coast ports or those of northern Europe. The 46m canal was now to be 35ft deep and 450ft wide at bottom, with reduced curvatures, high-level bridges instead of movable ones, and dredged approach channels. The Corps of Engineers finished the job in 1981, and in 1984 an 820ft container ship, the largest yet, successfully transitted the canal.

The Cape Cod Canal today handles about 30,000 transits a year from a ship control centre equipped with radar and closed-circuit television. Apart from pleasure craft, about 13M tons a year are carried in freighters, tankers and integrated tug-barge (ITB) (see pp. 127, 391) sets.

The New York State Barge Canal

The toll-free New York State Barge Canal system carried some 4 to 5M tons pa in the 1950s. By 1971, however, 55 cargo-vessels making 1120 loaded trips carried 2.5M tons, most of it oil from New York. Half this tonnage passed on the Champlain Canal, mostly to Lake Champlain ports. By 1982 traffic was down to 777,292 tons. On the other hand, there had been a steady increase in pleasure-craft use, in 1972 some 84,000 local passages by such boats being recorded. Currently, the position has not greatly changed.

Because the Seaway is icebound in winter, and offers only a circuitous course from the Great Lakes to the Atlantic seaboard, the Corps of Engineers in 1980 began investigations into alternative all-year large-barge routes, and hope to decide by 1985 whether one of four is practicable:

(a) From Lake Erie via Buffalo and Albany by a heavily modified New York State Barge Canal;
(b) Oswego via Syracuse to New York City;
(c) Lake Ontario via Lake Cayuga to Binghamton and by the Delaware river to Delaware Bay;
(d) Lake Erie via Silver Creek to Elmira to the Susquehanna river and Delaware Bay

The West Coast

We have already glanced at the developed barge traffic on the Columbia/Snake system. Also on the US west coast the less important 43m tidal Sacramento Deep Water Ship Channel runs from the sea at San Francisco via San Pablo Bay, Suisun Bay and the Sacramento river to the city of Sacramento, which has California's only important navigation lock, built in 1963. The Stockton Deep Water Channel leads from Suisun Bay via the San Joaquin river to smaller centre of Stockton. The delta of these two rivers provides a network of pleasure-cruising waterways.

The 8m Lake Washington Ship Canal at Seattle links Puget Sound and the city's freshwater harbour, Lake Washington, by way of Salmon Bay and Lake Union, all of which have been brought to the same water-level. The canal has two parallel locks built 1911 – 16, one 825ft × 80ft (it has a third set of centre gates), the other 150ft × 28ft, the rise being 6ft to 26ft depending on the tide. Inflow of salt water during lockage is ingeniously prevented.

Conservation and Recording

The last thirty years or so have seen growing interest in North America's waterway past. Canal and slackwater navigation history is being widely studied and written about, folklore recorded and photographs treasured. More, important structures and lengths of old canal are being preserved and in some cases restored by volunteers, local, state or federal bodies. Unlike in Britain, this movement is not aimed at restoring the waterways for navigation, except in short demonstration sections. Most are too far gone for that, and America has too many other cruising grounds. The aim is rather to preserve their memory as part of America's past. Recent years have, however, seen a great increase in the building of new cruise and trip-boats operating voyages of a few hours to a week and more—craft epitomised by the *Delta Queen* and the *Belle of Louisville*.

The movement perhaps began with the federal government's acquisition in 1938 of the whole 185m of the Chesapeake & Ohio Canal, fourteen years after its closure. The lower 22m, the Georgetown division near Washington, were rehabilitated soon afterwards. Then in 1971 the government set a precedent by creating the Chesapeake & Ohio Canal National Historical Park, the first of its kind, over the canal's whole length. The park is administered by the National Park Service, the towpath is open throughout as a footpath, a mule-drawn barge works on a section near Washington, structures are preserved, public information programmes presented for historians, naturalists and students, and camping facilities provided. Soon afterwards, in 1940 part of the Delaware Division Canal was transferred to Pennsylvania to become part of the Roosevelt State Park.

Elsewhere, other efforts have been made. The Illinois & Michigan Canal, for instance, became in 1984 the first US National Heritage Corridor, 100m long from Chicago to La Salle/Peru on the Illinois river. Again, the Delaware & Raritan is maintained intact and watered, though now used only for supply. At De Witt, the Erie Canal park, 30m eastwards to New London, shows many features of the old Erie, while at Canal Fulton, on the Ohio & Erie, the horse-drawn *Helena II* runs trips on a restored length. In Canada, work began in 1983 on turning a section of the Shubenacadie Canal between Lakes Charles and Micmac into a historical and recreational park. Elsewhere in North America, canal structures and lengths are recognised as National Historical Sites, included in the National Register of Historic Places, or maybe, as on the Middlesex Canal, designated as National Historic Civil Engineering Landmarks by the American Society of Civil Engineers. Canal museums have also been founded, such as those for the Erie Canal in the Weighlock building at Syracuse and at Rome, the Delaware & Hudson at High Falls, or the Allegheny Portage Railroad at Lemon House on the summit.

The American Canal Society was founded in 1972, and among its activities has been the listing of all known North American navigation works. It is supported by a number of state and local canal societies, which enable members to interest themselves in the history and preservation of their local waterway. There is now a Canadian Canal Society also.

The environmental movement of the last thirty years or so has, however, hindered the development of some commercial waterway improvements: the Cross-Florida stopped, the Tennessee-Tombigbee more than once threatened, and other improvements made more difficult, often in alliance with others whose motives are very different. On 29 August 1981, concerning an effort to stop a new Lock 26 on the Mississippi, *The Waterways Journal* wrote:

> What has taken place, however, is that a wave of environmental concern has swept the nation since the advent of the Environmental Protection Act and this wave, it is fair to say, ran amuck for some time. Like most movements, it has accomplished a great deal that is good. Unfortunately there has been a great amount of over-reaction along with it, and some

> more ardent members of the movement have made unreasonable demands.
>
> Nevertheless, recent years have seen a growing uneasiness over most projects, a growing distrust of big government and any agency that represents it, and growing efforts to slow down almost any and all big projects in the nation.

Maybe this book will make a small contribution towards giving historical perspective to such immediate controversies. Meanwhile, time drags out, transport costs to the consumer increase, and final expenditure rises.

Present and Future

The recession of the early 1980s had a serious effect upon waterways, and led to the closing by mid-decade of several old-established barge operators and boatbuilders. Its effects were then still apparent.

Controversy arose in 1983 and after over the proposed merger between a large railroad-holding conglomerate, CSX Corporation, and a big barge operator, American Commercial Barge Line Co, which brought to the fore a clause in the Panama Canal Act that prohibited railroad ownership of barge lines if competition were thereby reduced. It went to the Interstate Commerce Commission, which in August 1984 ruled in favour of the merger. As I write, an appeal has been lodged.

What, then, of the future of some 26,000 m of waterways improved and maintained by the Army Corps of Engineers? The contribution they made during WWII encouraged Congress to develop them further. Large appropriations were granted, private enterprise expanded fleets, local authorities spent money on terminals, until in the 30 years 1947–77 carryings by waterway grew from 700 million to 1.9 billion tons. This growth was largely the result of economies of size: the deeper and wider the channel, the bigger the locks and the more they operated in pairs, the larger the tows, the more sophisticated the navigation equipment, the more efficient the working, the cheaper the carriage and the greater the demand.

But just as in the years before the war, traffic expanded more slowly than infrastructure, so now infrastructure began to expand more slowly than traffic. Given locks' huge size and great cost, and the many years needed to design and build them, the shopping list lengthens. Takes two examples: the single aged 600 ft Gallipolis lock on the Ohio, the most critical of the five not yet rebuilt, causes tows to be divided to pass through, and so is a bottleneck in the whole river;

the Monongahela, which runs for 129 m from Fairmont, West Virginia, to the Ohio at Pittsburgh, works most of the coal brought from western Pennsylvania and West Virginia mines through 12 locks that are congested, small and time-expired.

Authorisation and funding for each improvement have become a legal and political conflict in which those interested in fishing, conservation and government economy form an ususual alliance with railroad interests in the prevention of waterway modernisation. That for the replacement of Lock 26 at Alton on the upper Mississippi, finally authorised in 1979 as one 1200 ft × 110 ft lock, enabled the idea of a fuel tax on waterway users to be accepted by Congress in 1978. The present tax, effective in 1980, of 8 cents per gallon, increasing to 10 cents in 1986, applies to commercial craft under way, other than deep-sea vessels, passenger-ships and LASH/Seabee barges. Even though the tax receipts are put towards the cost of repairing and improving the system, it ended a century and a half during which federally improved and maintained waterways had been free to all, and so had given low-cost service.

As an essential tool in deciding upon priorities, not just on a project-by-project basis, but throughout the nation, the Corps of Engineers in 1974 began a 3-year programme, Inland Navigation Systems Analysis (INSA) to simulate the operation of the entire inland waterway system on a computer by representing its component functions such as tow reconfiguration, approaches to locks, filling and emptying of locks, exiting, and lock-to-lock transit, in terms of time elements and their interdependencies. To build such a model and have it accurately represent the waterway system to be studied, accurate knowledge of the system was needed, including its physical description, component costs, method of operation, and time components.

While INSA was running, the Water Resources Development Act gave them another task, to define the need for waterways themselves nationally and not locally; to assess the capacity of what existed to meet current and projected needs; to make clear the relationship between the use of waterways for transport and for other worthwhile purposes; and then to prepare alternative plans, assess each, and make recommendations to Congress in the light of projected national needs.

In 1980 the Corps published their first results. These concluded that oil price rises would lead to heavy increases in coal tonnages, and also in agricultural exports to Gulf ports. (Already, in 1980, waterways carried 40 per cent of US grain exports.) Along the Gulf coast petrochemical and industrial development was likely. These all added up to increased use. Against this forecast, locks that arc too small or too old lay scattered over an ageing system*, lowering efficiency and raising maintenance costs, while too little dredging, too narrow channels, too little water and awkward bridges also hampered many navigations. In August 1981 the Corps followed this up with a preliminary summary of the Study which, in the light of traffic projections, set out alternative national strategies for action between 1982 and 2003, including construction work on 58 locks, three major channel deepenings, and much additional maintenance and safety work.

Forward planning, however well done, and economic forecasting, however comprehensively computerised, can never be made proof against recession and recovery, or against political changes at home and abroad. But they help.

*One-third of all locks and dams are over 50 years old, another one-third over 25.

17
LATIN AMERICAN CODA

Latin America is a land of rivers, upon which ships and barges travel long distances, often helped by channel dredging and buoying. Developing lock and dam systems are found mainly in Brazil. It has, too, in Lake Titicaca the highest large body of navigable water in the world; and it has the Panama Canal, its extra-territorial zone ended on 1 October 1979 under the Panama Canal treaty, its administration under a US-Panamanian Commission until the year 2000.

As I write the Canal has entered its eighth decade. Its track improved, its traffic-handling modernised in every way that sophisticated and ingenious men can devise to keep it at work every day of the year, it is still the canal that Bunau-Varilla saw from *Ancon* in 1914. So far, talk of a replacement canal, maybe at sea-level or on a different route, has come to nothing. We have seen on the Welland that ships have come to be built to fit the locks, not locks enlarged to suit the ships. So it has turned out at Panama. Ship size has quickly increased over the last twenty years, and transits have fallen. But size is now stabilising at dimensions the Canal can take. One exception did, however, show itself in 1984 when Alaska North Slope oil began to cross the Panamanian isthmus by pipeline instead of waterway.

We find only one true canal system, created by the Dutch. The eighteenth-century prosperity of Guyana was due to sugar, produced in the coastal districts with the help of dykes and small canals. Many of the latter still wind across the estates, carrying strings of small barges or punts laden with cane.

Britain's and North America's canal ages had some repercussions, as when in 1823 the government of Buenos Aires (Argentina) sought to recruit 200 Irish navvies for a canal to the city of Buenos Aires from Ensenada lower down the River Plate, or when about 1836 John Henry Freese, merchant of London and Rio, put out a 70-odd-page prospectus, complete with folding map and aquatint frontispiece, of the Imperial Anglo-Brazilian Canal, Road, Bridge and Land Improvement Company to build public works in Rio de Janeiro province, including about 20–25 miles of canal. His purpose was to improve transport for sugar and coffee, and also drainage. Naturally the Erie Canal was quoted. We shall find later one example of action, upon the Digue linking Cartagena with the Magdalena river in Colombia.

The Rio de la Plata or Plate river is in fact a great estuary lying between Argentina and Uruguay that serves the Rivers Paraná and Uruguay, which with their tributaries, form the fourth largest river system in the world. Yet it is so shallow that a deepened ship channel has to be maintained up to Buenos Aires. Above, navigation of the rivers of the plain is hindered by great seasonal variation in depth, and requires constant dredging.

The Uruguay is navigable for bigger seagoing ships to Concepción del Uruguay, for smaller to Salto. On the Paraná, 6000-ton ships can reach Rosario, 500 m above Buenos Aires. Beyond, smaller ships along with barge tows can navigate to Paraná, Santa Fé (where the Salado enters), and Corrientes. Here the Paraguay river branches off to enable craft to reach Asunción 250 m away, Corumbá in Brazil, and with enough water, Puerto Suárezover the Bolivian border. Most barges and passenger-boat traffic works with the limits of Asunción, Corrientes and Salto, but also some way up many of their tributaries. The Paraná itself runs up between Argentina and Paraguay and into Brazil. At Itaipú Brazil and Paraguay have jointly completed a hydroelectric dam; there navigation ends, though three locks have been planned for future construction.

Fig 126 A towboat and two 400 dwt barges carrying sugar-cane reach their destination at an alcohol-from-sugar plant on Brazil's Tietê river.

Above, in Brazil, three dams, each with one lock 689ft × 56ft, with 66ft rise, are built or building below the tributary Tietê that rises near São Paulo, and one (without a lock) above—in ascending order Ilha Grande, Porto Primavera, Jupiá and Ilha Solteira, above which navigation will continue to São Simão on the tributary Paranaiba, a distance of some 620m. Currently, some 311m are navigable between Jupiá and Ilha Grande, using a temporary lock at Porto Primavera, with completion of the Paraná works planned for 1992. Traffic now at some 250,000 tons pa, mainly in grain, cattle and oil by-products, is lower than it might be, due to problems with rail links. Towboats with 6-barge tows are used.

The navigable section of the Tietê is planned to end at Conchas, 354m from the Paraná. Thence dams and locks (all 446ft × 39ft) are being built downwards, with the first two, at Barra Bonita and Bariri (both 82ft rise)already completed, and two more, Ibitinga and Promissão (each with one lock of 82ft rise) due to be opened in 1986, giving a navigable length of 242m. On the existing stretch, traffic is some ½M tons a year of sugar cane (for distillation into industrial

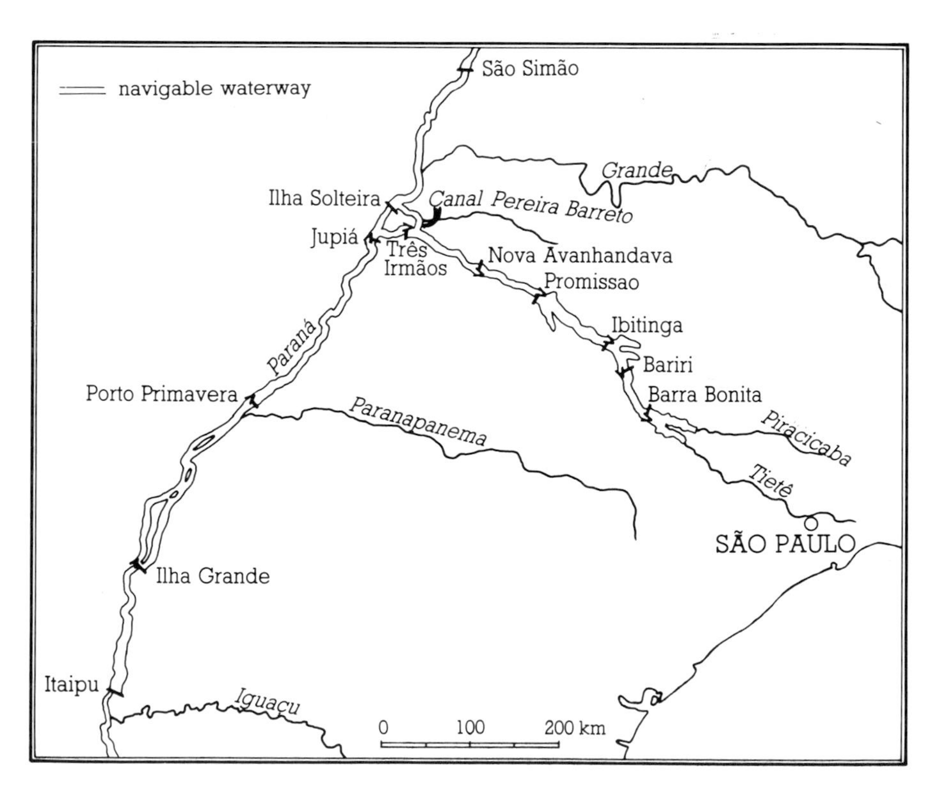

Map 17 Brazil's Paraná-Tietê lock and navigation system

Fig 127 Barra Bonita lock on the Tietê, 466 ft × 39 ft with a rise of 82 ft, and dam for its accompanying 123 Mw power plant.

alcohol mainly for fuel), sand and gravel, carried mainly in 2-barge push-tows. There are also passenger-craft. Two further dam and lock systems will complete the navigation, at Nova Avanhandava (two locks each of 59 ft rise) and Três Irmãos (two of 79 ft rise). Thence the Canal Pereira Barreto (planned for completion in 1991) will provide a link with the upper Paraná past the Ilha Soltcira dam, which in later years will also be given a lock. When completed, hopefully on the planned date of 1992, some thousand miles of interconnected navigable water will be available, for which traffic of some 20 M tons is expected.

Brazil has some 27,000 m of rivers in some sort navigable: from south to north notably also the Jacúi, São Francisco, Amazon and Paráiba rivers and their tributaries.

In the far south the Jacúi runs down past Porto Alegre into the Lagoa dos Patos, whence there is an opening to the sea. These two carry a considerable tonnage, partly because of coal carried to Porto Alegre. The São Francisco is navigable for self-propelled craft not drawing too much from Pirapora* north of Belo Horizonte for 852 m to Juazeiro in Bahia, below which are the 265 ft Paulo Afonso falls. Some 25 m above Juazeiro stands the enormous Sobradinho lock, 394 ft × 56 ft, its rise of 110 ft making it and the USA's John Day the highest of the New World. The final 148 m of the river are

*The *São Francisco*, a wood-burning sternwheeler built at Pittsburgh in 1913 which subsequently worked on the Mississippi and the Amazon, now operates from Pirapora as a trip-boat.

also navigable. Northwards, in the state of Piaúi, the Rio Paráiba is navigable for hundreds of miles for barges as it flows to the sea. Two locks, each 164ft × 39ft and with a 74ft rise, are at present being built at Boa Esperanca. Northwards again, two large locks, part of a hydroelectric system, were in 1985 being built at Tucurui on the Tocantins river running down to Belém.

The Amazon is greatest of South American rivers. Rising in the Andes, it flows first through Peru past Iquitos, and then across Brawil past Leticia, São de Olivença, Manaus and Santarém to the port of Belém on the tributary Pará river. Seagoing ships can reach Iquitos, 2300m from the sea, and those of lesser draught another 486m to Achual Point. Immense variations in depth occur according to the season, as also in width. When the floodwater comes down, large areas on each side of the main river can be flooded. With its tributaries, most notable of which are the Madeira, Purús, Juruá and Negro, there are about 7500m of navigable channels in the Amazon basin and others in Peru, notably the Ucayali, joining the Amazon at Iquitos and leading to the busy barge port of Pucallpa. From the Negro the Casiquiare canal runs for over 200m to the Orinoco; once thought to be artificial, it is now known to be a natural link, with a slope towards the Negro.

Organised steam navigation began in 1853 with three small steamers operated under government contract by the Amazonas company set up the year before. Other companies followed, to work on tributaries as well as main stream. Rubber production from wild trees brought growing trade from 1872 to 1910, and with it high demand for river transport. Then competition from plantation rubber in Malaysia and Indonesia caused prices to fall heavily; trade declined and whole towns were abandoned. Since then, however, other traffics, among them timber, have grown up.

The Orinoco lies entirely within Venezuela; 1281m long, and navigable for large craft to the Atures rapids some 700m from the sea, it varies greatly in width and flow. Of its many tributaries, navigationally the most important is the Caroni, which comes in to the lower river at Ciudad Guayana. A dredged channel has been made from the sea up the Orinoco to the Caroni's mouth, to which iron ore is brought for transport by ship and barge.

In 1852 the Orinoco Steam Navigation Co was formed in New York, and granted rights by Venezuela. After many difficulties regular steamboat services began in the 1860s.

Until the 1960s the Magdalena river, running from the north coast south in the general direction of the capital, Bogotá, had been central to Colombia's development. The head of navigation was at Neiva, whence boats could use the river for some 200 m past Girardot, where it improves, to the Honda rapids. Lower down, at La Dorada, it is navigable for another 600 m to the coast at Barranquilla.

Carriage originally was by pack animals on the land sections, keelboats of the Ohio/Mississippi type on the river. From 1822 steamboats began to operate, though they flourished only after the tobacco boom of the 1850s onwards. Then the land sections were filled in by railways, one from Bogotá to Girardot, the other from Beltran above the Honda rapids to La Dorada below.

Steamboats were often bought second-hand from the Mississippi, and in spite of shallows and snags, regular services were worked, especially after the Colombia Railways & Navigation Co had opened a railway in 1894 from Cartagena to Calamar on the river, in 1913 acquired the Colombia Navigation Co and in 1914 the Magdalena River Steamboat Co. However, the opening in July 1961 of the Atlantic Railway from Bogotá to Santa Marta on the coast, which partly parallels the river, and has branches to Barranquilla and Cartagena, ended passenger services. Cargo is still, however, carried on the river, and motor boats work a kind of taxi service.

Three cities near the Magdalena's mouth competed for its trade, Cartagena, Barranquilla and Santa Marta. Santa Marta, the oldest, had a good harbour, but a water connection with the river only though shallow and swampy channels. Barranquilla, the newest, was on a river mouth blocked by a sandbar. Cartagena had a good bay, and a tantalising water connection with the Magdalena, the Digue, formerly one of its channels. From 1650 to our own times, efforts had been made to canalise it, and by doing so to make Cartagena Colombia's principal port, in spite of silting and blockage. From the American canal age, and linked with the coming of steamboats, such efforts became more ambitious. J. B. Elbers, who from 1823 put steamboats on the river, tried. Then came George Totten, later builder of the Panama Railroad, who by 1848 had cut some new channel and built two locks. In 1847 a steamboat company was formed at Cartagena to use it and the river, a year after another company to navigate the river had been formed at Santa Marta. But within a few years both Cartagena company and canal had failed. Then in late 1879 an American engineer, James J. Moore, reopened it for river craft drawing up to 5 ft. For ten years or so Cartagena's trade flourished, until again nature began to damage the canal. Barranquilla having in the 1890s linked itself to the sea by a railway, Cartagena built one to Calamar on the Magdalena. However, in the 1920s and 1930s more efforts were made to cut a sizeable Digue to avoid transhipment from rail at Calamar, and by 1935 an 11½ft-deep channel existed which in 1946–51 carried an average 58,000 tonnage. Once more the Digue deteriorated, and once more in 1952 it was rebuilt—and so served Cartagena until the Atlantic Railway was opened. Today some cargoes from Cartagena still pass along it to be shipped at Calamar.

Among Latin American lakes that have been used for passenger and goods transport are Managua (38m long, 16m wide) and Nicaragua (c100m long, 5m wide) in Nicaragua, and Titicaca, shared by Peru and Bolivia. Lake Nicaragua and sometimes Managua also became known as part of a potential transport line across the Central American isthmus. Napoleon planned a route using both; indeed, in 1851 the American Accessory Transit Co got a line opened, using the San Juan river from the Caribbean and then across Lake Nicaragua, with two small steamers, and finishing by road coach. It flourished for some years. Later, in 1882, regular steamship services on the lake began.

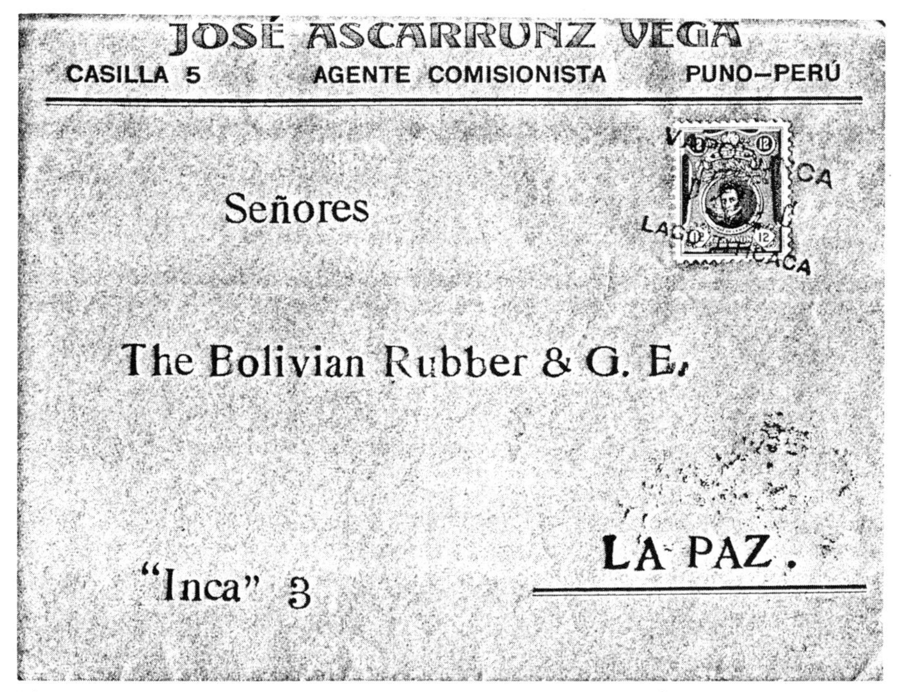

Fig 128 A letter carried by the steamer *Inca* on Lake Titicaca in 1911. The Peruvian stamp is cancelled VAPOR INCA LAGO TITICACA.

Fig 129 The wharves at Guaqui, Bolivia, on Lake Titicaca, in 1974, from the deck of *Ollanta*. *Inca* lies beyond the cranes.

Lake Titicaca is extraordinary: 110m long, divided into two parts by a strait, it lies 12,497ft above sea-level. Long used for passenger- and freight-carrying, services are worked between a Peruvian railhead at Puno and a Bolivian at Guaqui. Its first ship, the 170-ton steamer *Yavari*, built in 1861, had to be carried in pieces up from the coast on muleback, and assembled at Puno. Though later diesel-engined, it was in use not long ago. Later the 650-ton British-built *Inca* of 1905 was brought by rail also to be assembled lakeside. Currently, the 850-ton passenger- and freight-carrying *Ollanta* of 1929 is at work together with the *Manco Kapac*. Tourist craft also carry passengers between the two countries, but only on overnight visits.

So, at the world's highest navigable spread of water, we end a story that began on the Nile. To reach the one from the other, we have travelled the world.

ACKNOWLEDGEMENTS

I cannot hope to thank here a tithe of those very many who have helped me over so long, by enabling me to visit new waterways, lending me books, sending me articles and press cuttings, making pictures available, or introducing me to helpful friends.

I have been for many years encouraged and helped by Mr David Edwards-May, who has put his wide knowledge of European waterways at my disposal, Dr Mark Baldwin who has found me books and answered engineering questions, and Mr & Mrs Ron Oakley, whose organised visits abroad under Inland Waterways Association auspices enabled me to visit waterways I should not otherwise have seen. On one occasion, indeed, they gave me a purpose-made time of canal exploration in North Germany and the Low Countries. Three colleagues of the IWA's Inland Shipping Group, Dr David Hilling, Mr Derric Webster and Mr John Fricker, have long supplied me with press cuttings, while other ISG members have also helped me.

Many specialists have been generous of their knowledge and time: notably Dr Robert Legget (Canada), Dr Thomas Hahn, Mr William Shank and Dr William Trout (USA), Dr Arnaldo Giraldo (Brazil), Mr Yiliang Zheng (China), Dr Eiichi Aoki (Japan), Mr J. Diandas (Sri Lanka), Mr S. K. Ghosal (India), Mr Brian Seymour (Australia), Mr Norman Halfar (Berlin), and Mr E. Switek (Poland). I have leaned heavily upon M. R. Thélu's articles on early pound-locks and inclined planes, Dr N. A. F. Smith's work on Roman canals, Dr Joseph Needham's on Chinese canals, and Professor A. W. Skempton's in Vol 3 of *A History of Technology*.

Portions of the manuscript have been read by Dr Michael Lewis, Professor A. W. Skempton, Mr David Edwards-May, Dr Robert Legget, Mr William Shank, Mr William E. Gerber, Professor Marvin J. Barloon, Dr Roger Squires, and Dr Paul K. Walker. I have greatly benefited from their comments, but responsibility for the final text is of course entirely mine.

Three research assistants have helped me: Miss Anthea Beamish (now Mrs Webb), who got me started back in 1971; Mr Stanley Holland, who searched for and had copied dozens of articles from *The Engineer*, *Engineering* and the *Transactions* of the Institution of Civil Engineers; and more recently, Mrs Penny Beck, who has sought early pictures held in Continental collections. Three secretaries, whose work it would be difficult to praise enough, have typed seemingly endless drafts and redrafts, and devotedly produced photocopies: in London Mrs Dawn Bijl, in South Cerney Mrs Jean Cowdall and Mrs Susan Cadwallader.

Finally, I would like to acknowledge my debt to two old friends, Mr David St J. Thomas of David & Charles, who long ago agreed to publish this book, and has given unwavering support ever since, and the book's editor, Mrs Pam Griffiths, upon whose skills and perception I have so often relied.

Mr David Edwards-May has drawn the maps. I do not think I could have provided them without his technical ability and wide knowledge of waterways. Mr Hugh McKnight of Hugh McKnight Photography, Shepperton, and Mr A. J. Griffiths of Photography 2000, Newton Abbot, have greatly helped me. I should here like to note the sources of pictures used in the plates and text, and thank those who have provided them. Fig 1, Professor R. E. H. Mellor; 3, Schiffleutzechbuch, Stadtarchiv, Passau; 4 and 5, Comune di Milano, Ripartizione Cultura e Spettacole; 6, Phot. Bibl. nat. Paris; 7, by Reinier Nooms (1623–64) from the series *Verscheyde Schepen en Gesichten van Amstelredam*, No 7: Nederlandsch Historisch Scheepvaart Museum, Amsterdam; 9, Ch Florange, *Etude sur les Messageries et des Postes* 1925, fig 624; 10, Hugh McKnight Photography; 11, Gösta Hallin and Allhems Förlag; 12, *Picturesque Europe*, III, ed Bayard Taylor, New York, (1879), Hugh McKnight Photography; 15 (a), (b) and (c), Hugh McKnight Photography; 16, DDSG, Vienna; 17, Alexander Marx, *Pittoreske Ansichten des Ludwig-Donau-Main-Kanals*, 1845; 18, Hugh McKnight Photography; 19, *Illustrated London News*, 3 April 1869; 20, Mr and Mrs W. Kalaugher; 21, *Engineering*, 26 January 1894, Hugh McKnight Photography; 22, Hugh McKnight Photography; 23, Focke-Museum, Bremen; 24, *The Engineer*, 15 February 1867; 25, Deutsches Museum, Munich; 26, Hugh McKnight Photography; 27, National Reference Library of Science and Invention; 28, Hugh McKnight Photography; 29, Roger Pilkington; 30, Hugh McKnight Photography; 31, *The Engineer*, 15 February 1889; 32, Hugh McKnight Photography; 34, *Engineering*, 12 October 1888; 35, A. M. C. Switek; 37, Hugh McKnight Photography; 38, *Illustrated London News*, 20 August 1881, Hugh McKnight Photography; 39, (a), (b) and (d) and 40, Hugh McKnight Photography; 41, *Engineering*, 18 October 1878; 42, *The Engineer*, 11 August 1893; 43, *The Engineer*, 17 April 1896; 45 (a), Museum of Transport, Budapest; 45 (b) and (c), *Geographical Magazine*; 47, *The Engineer*, 8 October 1886; 48, *Illustrated London News*, 6 June 1885; 51, Inchcape & Co Ltd and Sir Percival Griffiths; 55, *Geographical Magazine*; 56, Murray River Developments Ltd; 62, J. H. Boyes; 63, H. M. Doughty, *Our Wherry in Wendish Lands*, *c.* 1892; 65, *Geographical Magazine*; 66 and 69, Hugh McKnight Photography; 70, Mr Robert Cowley; 71 (b), Alan Wrigley, Köln; 72, Luftbild Bischof & Broel, Nürnberg; 74, Cave Wood Overseas Transport Ltd; 76, Harold Sumption; 77, Mark Baldwin; 78, Bryn Weightman; 79, Harold Sumption, 80, Frank Holroyd; 81, Hugh McKnight Photography; 83, J. and G. Carlisle; 85, Hugh McKnight Photography; 86, Mr David Edwards-May; 87, Photo Hugues, brevets J. Aubert; 91 and 92, David Hilling; 93, Netherlands National Tourist Office; 96, Ironbridge Gorge Museum Trust; 97, Hugh McKnight Photography, 100, *Engineering*, 31 July 1868; 103 and 104, Hugh McKnight Photography; 105, *Illustrated London News*, 6 October 1860, Hugh McKnight Photography; 106, *Illustrated London News*, 4 May 1861, Hugh McKnight Photography; 109, Public Archives Canada; 110, National Film Board of Canada, courtesy of St Lawrence Seaway Authority; 113, *Illustrated London News*, 16 June 1888; 114 and 115, Hugh McKnight Photography; 118, Robert F. Legget; 120, *The Waterways Journal*, St Louis; 121, Winston Thompson,

Commercial Dispatch, Columbus, Mississippi, and W. Shank, American Canal Society; 122, Port of Portland, Oregon (Jim Douglas); 123, Ryan-Walsh Stevedoring Co. Inc., Baton Rouge; 124, Litton Great Lakes Corporation; 125, Roger Squires; 126 and 127, A. Giraldo, Companhia Energética do São Paulo; 129, John Marshall. All others are from my own collection.

Map 3 owes much to those in Jan de Vries' article in *A.A.G. Bijdragen 21*, 1978, Map 5 to those in Sir Percival Griffiths' *A History of the Joint Steamer Companies,* Map 12 to that in G. R. Taylor's *The Transportation Revolution, 1815–1860*, 1964, and Map 13 to that in William E. Shank's *The Amazing Pennsylvania Canals*. I am indebted to Dr Robert Legget's *The Canals of Canada* and *The Seaway* for help with Maps 12, 14 and 16, and to Dr Arnaldo Giraldo for the material upon which Map 17 is based.

SOURCES OF QUOTATIONS

Chapter 1

1. Robert Payne, *The Canal Builders*, 1959, p. 12–13, quoting J. H. Breasted, *Ancient Records of Egypt*, 1906, I, 324.
2. *Op. cit.*, 16–17, based on E. A. Wallis Budge, *The Book of the Dead*, 1913.
3. Catalogue of Papyri, BM, iv. 19. Information from Dr M. J. T. Lewis.
4. Quoted by Dr N. A. F. Smith, 'Roman Canals', *Trans. Newcomen Soc*, 49 (1977–8), 82.
5. Herodotus, *History*, 7, 24. I am grateful to Dr M. J. T. Lewis for the reference.
6. Quoted by Dr Joseph Needham from the Sung Shih, in 'China and the Invention of the Pound-Lock'. *Trans. Newcomen Soc.* XXXVI (1963–4), 91.
7. Translated from the Swedish of J. H. Kellgren (1751–95) in 'Dumboms Inferne' (A fool's life), quoted *Transport Technology and Social Change*, Stockholm, 1980, 63.
8. *John Smeaton's Diary of his Journey to the Low Countries*, 1755, reprinted Newcomen Society, 41.
9. R. Thélu, 'Les écluses avant le 17e siècle-Recherches sur les origines des écluses à sas', *Navigation, Ports et Industries*, 25 October, 10 November 1978, 631 (my translation).
10. *Op. cit.*, 633–4.
11. *Op. cit.*, 633 (my translation).

Chapter 2

1. *Cambridge Economic History of Europe*, V, 217–8.
2. Charles Wilson, 'Transport as a Factor in the History of Economic Development', *Journal of European Economic History*, 1973, 328–9.
3. Jan de Vries, 'Barges and Capitalism. Passenger transportation in the Dutch economy, 1632–1839', *A. A. G. Bijdragen* 21, 1978, 54.
4. *Ibid.*, 356.
5. *Ibid.*, 79.
6. *John Smeaton's Diary of his Journey to the Low Countries*, 1755, reprinted Newcomen Society, 33, quoted by de Vries, 'Barges and Capitalism', *op. cit.*, 88.
7. The MS is in the National Library of Ireland, Dublin (MS 2737). It was reprinted in *Transport History*, March and July 1972.

Chapter 3

1. *Le Canal de Nantes à Brest dans l'Histoire de Châteaulin* (pamphlet), 1977 (my translation).
2. Arthur Young, *Travels in France during the Years 1787, 1788 and 1789*. C.U.P., 1950, 41.
3. *An Embassy to China, Lord Macartney's Journal, 1793–94*, ed. J. L. Cranmer-Byng, 1962, 266–7.
4. Sir George Staunton, *An Authentic Account of an Embassy... to... China, taken chiefly from the papers of ...The Earl of Macartney*, 1797, 382–3.
5. *Ibid.*, 383.
6. *Ibid.*, 392.
7. *Ibid.*, 426–7.
8. *Ibid.*, 451.
9. A. P. Herbert, *Why Waterloo?* 1952, 127.
10. G. B. W. Jackson, 'Description of the Great North Holland Canal', *Procs. Inst. C.E.*, 9 February 1847, 94–5.

Chapter 4

1. Arthur Young, *Travels in France during the Years 1787, 1788 and 1789*, C.U.P., 1950, 6.
2. Robert Southey, *Journal of a Tour in the Netherlands in the Autumn of 1815*, ed. 1903. Saturday 23 September.
3. Peter Stevenson, *Sketch of the Civil Engineering of North America*, 2nd ed., 1859.
4. A. B. van Meerten Schilpevoort, *Reis door het Koningrijk der Nederlanden*, 2vv, 2nd ed., 1829, 1, 162. Quoted de Vries, *op. cit.*, 104.
5. *The Diary of a Cotswold Parson*, ed. David Verey, 1978, 158 (4 May 1839).
6. Laurence Oliphant, *The Russian Shores of the Black Sea*, 1853, 48.
7. *Ibid.*, 23.
8. Basil Greenhill, *Boats and Boatmen of Pakistan*, 1971, 128.
9. M. W. Manès, *Notice sur la Navigation à la Vapeur de la Saône et du Rhône* (pamphlet), 1843, 18 (my translation).
10. *Ibid.*, 14.
11. Cecil Torr, *Small Talk at Wreyland*, rpt 1979, 73.
12. Laurence Oliphant, *The Russian Shores of the Black Sea*, 1853, 353–4.

Chapter 5

1. W. J. C. Moens, *Through France and Belgium by River and Canal in the Steam Yacht 'Ytene'*, 1876.
2. Quoted Philip Ziegler, *King William IV*, 1971, 234.
3. Negley Farson, *Sailing Across Europe*, ND [1926], 42–3.
4. J. Macgregor, *A Thousand Miles in a Rob Roy Canoe*... [1866], 18.
5. L. F. Vernon-Harcourt, 'Some Canal, River and other Works in France, Belgium, and Germany' *Procs ICE*, 12 February 1889.
6. Anon, *Six Weeks on the Loire*, 1833, 274.
7. Laurence Oliphant, *The Russian Shores of the Black Sea*, 1853, 47–8.

8. *Ibid.*, 93–4.
9. J. Macgregor, *A Thousand Miles in a Rob Roy Canoe...*, 222.
10. *Ibid.*, 88
11. H. M. Doughty, *Friesland Meres and Through the Netherlands*, 1889, 14.
12. H. M. Doughty, *Our Wherry in Wendish Lands*, ND [c1892], 162.
13. Donald Maxwell, *A Cruise across Europe*, 1907, 192.
14. E. P. Warren & C. F. M. Cleverly, *The Wanderings of the Beetle*, 1885, 31–2.
15. 'The Captain', *Our Cruise in the Undine*, 1854, 27.
16. H. M. Doughty, *Our Wherry in Wendish Lands*, 19.
17. Donald Maxwell, *The Log of the Griffin*, 1905, 188.
18. Donald Maxwell, *A Cruise across Europe*, 74.
19. Robert Louis Stevenson, *An Inland Voyage*, 1919, 10.
20. *Ibid.*, 74.

Chapter 6

1. *Proceedings of the Institution of Civil Engineers*, 1898–9, Pt II, 294.
2. Charles Tower, *The Moselle*, 1913, 1.
3. E. F. Knight, *The 'Falcon' in the Baltic*, 2nd ed. 1892, 177.
4. Archibald Williams, *Victories of the Engineer*, ND [c1908].

Chapter 7

1. *Picturesque Europe*, ed. Bayard Taylor, III, 414, New York, 1879.
2. F. S. V. Donnison, *Burma*, 1970, 25.
3. *Report on a route from the mouth of the Pakchan to Kraw, and then across the isthmus of Kraw to the Gulf of Siam, by Capt. A. Fraser and Capt. J. G. Furlong*, 1863.
4. Christopher Hibbert, *The Dragon Wakes*, 1970, 328.
5. Lt Viscount Kelburn, RN, 'Mother Yangtse—I', *Geographical Magazine*, 1937.
6. *Ibid.*
7. Personal communication to author.
8. Professor Middleton Smith, 'The Waterways of China', *Canals and Waterways Journal*, August 1919, 73.
9. Lord Charles Beresford, *The Break-up of China*, 1899, 130.
10. Professor Middleton Smith, *op. cit.*
11. Norman Smith, *A History of Dams*, 1971, 221.

Chapter 8

1. Negley Farson, *Sailing through Europe*, 1926, 29.
2. C. S. Forester, *The Annie Marble in Germany*, 1930, 70.
3. Negley Farson, *op. cit.*, 126.

Chapter 9

1. *Seatrade*, February 1980, 16.
2. Lord Kinross, *Between Two Seas*, 1968, 281.

Chapter 11

1. J. C. Kendall, 'The Construction and Maintenance of Côteau du Lac: The First Lock Canal in North America', *Journal of Transport History,* February 1971.
2. Quoted from Ronald E. Shaw, *Erie Water West,* 1966, 12.
3. Quoted from Ralph D. Gray, *The National Waterway: a History of the Chesapeake and Delaware Canal, 1769–1965,* 1967, 3.
4. Quoted in Mary Stetson Clark, *The Old Middlesex Canal,* 1974, 32.
5. Mary Stetson Clark, *op. cit.,* 10.
6. Ed. Charles G. Haines, *Public Documents Relating to the New York Canals,* 1821, 87–8, quoted Hugh G. J. Aitken, *The Welland Canal Company,* 1954, 19.
7. Karl Bernhard, Duke of Saxe-Weimar Eisenach, *Travels Through North America During the Years 1825 and 1826,* 1828, I, 61–2.
8. Quoted from Ronald E. Shaw, *Erie Water West,* 1966, 90.
9. National Library of Wales, Aberystwyth, Edward Griffith Collection.

Chapter 12

1. Paul Fatout, *Indiana Canals,* 1972, 153.
2. Ralph D. Gray, 'The Canal Era in Indiana', in *Transportation and the Early Nation,* 1982, 129, quoted Ronald E. Shaw, 'Canals in the Early Republic: a Review of Recent Literature', in *Journal of the Early Republic,* Summer 1984, Vol. 4, No. 2.
3. Lt-Col George Phillpotts, *Report on the Canal Navigation of the Canadas,* 1842, 158.
4. Peter Stevenson, *Sketch of the Civil Engineering in North America,* 2nd ed, 1859, 25.
5. Robert Legget, *The Ottawa River Canals* (MS).
6. Peter Stevenson, *op. cit.,* 118.
7. Charles Dickens, *American Notes,* X.
8. N. E. Whitford, *History of the Canal System of the State of New York,* 1906, I, 616–25.
9. *Harrisburg Reporter,* 2 June 1832, quoted in William H. Shank, *The Amazing Pennsylvania Canals,* edition of 1981, 78–9.
10. Peter Stevenson, *op. cit.*
11. Charles Dickens, *American Notes, X.*

Chapter 13

1. Charles Dickens, *American Notes,* 1867 ed, 108.
2. Peter Stevenson, *Sketch of the Civil Engineering in North America,* 2nd ed, 1859, 40–1.
3. Charles Dickens, *op. cit.,* 118–19.
4. G. R. Taylor, *The Transportation Revolution, 1815–1860,* 1964, 79.
5. *Pittsburgh Daily Union,* 5 December 1856.
6. 'American Waterways', *The Annals of the American Academy of Political and Social Science,* Vol XXXI, No. 1, January 1908, 120.
7. Jerome K. Laurent, 'Trade, transport and technology: the American

Great Lakes, 1866–1910', *The Journal of Transport History*, March 1983, 10.
8. William R. Willoughby, *The St Lawrence Seaway: a Study in Politics and Diplomacy*, 1961, 72–3.

Chapter 14

1. 'American Waterways', *The Annals of the American Academy of Political and Social Science*, Vol XXXI, No. 1, January 1908, 188.
2. *Ibid.*, 1–3.
3. *Ibid.*, 50.
4. *Ibid.*, 139.
5. *Ibid.*, 55.
6. Michael C. Robinson, 'The Federal Barge Fleet: an Analysis of the Inland Waterways Corporation, 1924–1939, *National Waterway Roundtable Papers*, 1980, 156.
7. *Owen Sound Advertiser*, 24 April 1884, quoted by George Musk, *Canadian Pacific*, 1981.

Chapter 16

1. Lionel Chevrier, *The St Lawrence Seaway*, 1959, 6.

INDEX

Pressure on space prevents me indexing countries; please use the subheaded Contents List on pp 5–7.